Handbook of Geology

Handbook of Geology

Editor: Joe Carry

R CALLISTO REFERENCE

www.callistoreference.com

Callisto Reference,
118-35 Queens Blvd., Suite 400,
Forest Hills, NY 11375, USA

Visit us on the World Wide Web at:
www.callistoreference.com

ISBN: 978-1-63239-970-0 (Hardback)

Cataloging-in-Publication Data

Handbook of geology / edited by Joe Carry.
    p. cm.
Includes bibliographical references and index.
ISBN 978-1-63239-970-0
1. Geology. I. Carry, Joe.
QE26.3 .H36 2018
550--dc23

Table of Contents

Preface

Geology is a primary branch of earth science. It studies in detail the structure of Earth. It provides prominent theories and evidence concerning the evolution and history of our planet as well as life on the planet. Geology can be applied to diverse areas such as petroleum exploration, hydrology, mining, etc. The aim of this book is to present researches that have transformed this discipline and aided its advancement. Scientists, geologists and students actively engaged in this field will find this book full of crucial and unexplored concepts.

Significant researches are present in this book. Intensive efforts have been employed by authors to make this book an outstanding discourse. This book contains the enlightening chapters which have been written on the basis of significant researches done by the experts.

Finally, I would also like to thank all the members involved in this book for being a team and meeting all the deadlines for the submission of their respective works. I would also like to thank my friends and family for being supportive in my efforts.

Editor

Evaluation of Combating Desertification Alternatives Using Promethee Model

Mohammad Hassan Sadeghravesh[1], Hassan Khosravi[2], Azam Abolhasani[3] & Sahar Shekoohizadeghan[2]

[1] Department of Environment, College of Agriculture, Takestan Branch, Islamic Azad University, Takestan, Iran

[2] Faculty of Natural Resources, University of Tehran, Tehran, Iran

[3] Natural Resources Engineering - Living with the Desert-University of Tehran, Tehran, Iran

Correspondence: Hassan Khosravi, Department of Arid and Mountainous Reclamation Region, Faculty of Natural Resources, University of Tehran, Tehran, Iran. E-mail: hakhosravi@ut.ac.ir

The research is financed by (Sponsoring information)

Abstract

According to the extent of damage, various effects and complexity of desertification process, selecting appropriate alternatives considering all effective desertification criterions is one of the main concerns of Iran in the field of natural resources. This can be effective in controlling, reclamation of disturbed lands and avoiding destruction areas at desertification risk. This paper tries to provide a systematic and optimal alternatives in a group decision-making model. For this aim, PROMETHEE II method was used for ranking desertification alternatives. At the first in the framework of Multiple Attribute Decision-making (MADM), normalized decision matrix was provided by Delphi model. Then, to ease and accuracy in estimating the criteria preference and alternatives priority, the normalized decision matrix data were entered in Visual PROMETHEE software. Based on the results, the alternatives of prevention of unsuitable land use changes (A18), vegetation cover development and reclamation (A23) and modification of ground water harvesting (A31) with pure out ranking progress of phi =0.3660, 0.1909 and -0.0887 were selected as the main combating desertification altarnative in the study area, respectively. Therefore, it is suggested that the obtained results and ranking should be considered in projects of controlling and reducing the effects of desertification and rehabilitatyion of degraded lands plans.

Keywords: preference, rank, matrix, criteria, model, pirewise comparison, Iran

1. Introduction

Management of desert ecosystem is a collection of various management actions which is done for optimal control of desertification phenomena and decline of economic, social and environmental damages. Decision making issues of desert area management are complex issues because of various decision making criteria. There are several ways in order to achieve a specific goal that each of them provides different preferences for different economic, social, environmental, political and institutional issues. These requirements cause the use of Multiple Attribute Decision Making method which its goal is selection of the best answer among the different solutions. So the goal of this study, due to the limitation of inputs, is assessment of de- desertification alternatives in order to achieve the optimal alternatives in the context of sustainable management of desert area. In order to achieve this goal in the context of Multiple Attribute Decision Making models, Preference Ranking Organization Method for Enrichment Evaluation (PROMETHEE) that is one of the most important (Chou et al., 2004) and understandable (Pomerol and Romero, 2000) methods of Multiple Attribute Decision Making for ranking of de-desertification alternatives, was considered. Facility, clarity and stability of results are some of advantages of PROMETHEE method. This method can evaluate a set of alternatives either as a partial ranking or as a complete ranking. Clear impact of each criteria and its weight value on answers, high-performance of algorithms despite the simplicity in this method and its basis that is on the importance of the performance difference between the two answers, are some of Characteristics of this method (Mohaghar and Mostafavi, 2007).

The history of the use of decision making models for optimal alternatives presentation, in the context of desert area management is limited to Grau et al. and Sadeghiravesh et al. researches. Grau used Elimination Et Choice

Translating Reality (ELECTRE), Analytical Hierarchy Process (AHP) and PROMETHEE models in his research in order to select optimal alternatives for providing an integrated plan for erosion and desertification control (Grau et al, 2010). His study results indicated high- performance of these models in providing optimal alternatives for de- desertification, and the results were the same despite the use of complex methods in each model.

Sadeghiravesh also used AHP (Sadeghiravesh et al, 2010) technique for order preference by similarity to ideal solution (TOPSIS) (Sadeghiravesh et al, 2012) ، Electera (Sadeghiravesh et al, 2014) and fuzzy analytical hierarchy process (Sadeghi ravesh et al, 2015) and prioritized de- desertification alternatives in Khezer Abad region. The results of these studies were identical and also similar to the results of the present study.

Relating to the use of PROMETHEE model, Researches have been done for decision making and ranking in different science. Selection of an appropriate transportation system in an underground mine, (Elevli and Demirci, 2004), Selection of suitable alternatives for optimal utilization of energy (Shirsikar and Patil, 2005), Assessment of famous alternatives in production of energy (Diakoulaki and Karangelis, 2007), The choice of appropriate alternatives for profitable investment in stock market (Albadavi, et al., 2007), Assessment of suitable location for a recycling factory (Queiruga, et al., 2008), Evaluation of potential of alternatives for a construction company (Ginevicius et al., 2010), Choosing a suitable system to dry material (Prvulovic et al, 2011), Selection of appropriate alternatives for profitable investment in stock market (Chen et al., 2011), Selection of suitable alternatives in designing rescuer robots (Taillandier and Stinckwich, 2011), Selection of raw materials supplier (Safari et al, 2012), Selection of a suitable way for machining (Karandea and Chakraborty, 2012) and Choosing the best alternatives for determining the location of storage (Fontana and Cavalcante, 2014) are samples of these researches.

In recent years, also, some researches have been considered in this context in Iran. Selection of project team members (Omidi et al, 2011), Evaluation of innovation condition in southwest Asia countries (Bakhshi et al, 2011) and Assessment of ecological potential of agriculture (Nasiri et al., 2012) are samples of these researches. In all of these studies, researchers used Entropy method, Linmap, Weighted average method, the least squares method and Eigenvector method and assessed weight value of criteria and then prioritized alternatives by the use of PROMETHEE technique (Asgharpour, 1999). Extensive investigations in this study show that PROMETHEE model has not been used either in Iran or other countries for desert area management issues. This paper tries to provide systematic and optimal de-desertification alternatives in a group decision-making model for controlling, reclamation of disturbed lands and avoiding destruction areas at desertification risk

2. Method

PROMETHEE model is one of the new out ranking methods of Multiple Attribute Decision Making systems that is used for ranking of alternatives among inconsistent criteria (Behzadian and Pirdashti, 2009). For the first time, this method was introduced by Barns in 1982 at a conference in Laval University in Quebec City (Barns, 1982; Halouani, 2009) and then was developed by Barns and Vincke (Barns and Vincke, 1985; Barns, et al, 1986) and Barns and Mareschal (Brans and Mareschal, 1994; Brans and Mareschal, 2005). The development of this method in different situations created copies that are named as PROMETHEE family and are included PROMETHEE I, II, III, IV, V, VI.

Because the goal of this study is selection of de- desertification optimal alternatives on the basis of a set of effective criteria, PROMETHEE II, with the ability of separated alternatives complete ranking, has been used (Chou et al., 2004). This method is effective and compatible in where numerous alternatives should be assessed based on quantitatively, qualitatively and often contradictory criteria (Albadavi et al, 2007).

This method is simpler than other methods conceptually and practically (Albadavi et al, 2007; Pomerol and Romero, 2000). Also, Software features that are provided for this method as a support, ready appropriate analysis for determiner (Keyser and Peeters, 1996). There is limitation in the use of PROMETHEE technique for compensation of a criteria disadvantage or strong point, so an ideal alternative should gain minorities from all criteria. This means that an alternative can only gain the best rating when it has a comprehensive attitude. So PROMETHEE method is more effective than other methods which use algebraic sum. In addition, PROMETHEE method can use criteria with different scales without needing to match the criteria scales and it defines the distinct six functions commensurate by criteria scale and information, so it is a strong point for determinant in Multiple Attribute **Decision Making that usually criteria are measured with different scales. Also when assessing alternatives are not identical, for example one alternative has better function for a special criteria and the other alternative for another criteria, like this study, the use of** PROMETHEE is useful like other Multi- criteria methods such as ELECTERA (Soltanmohammad et al, 2008). PROMETHEE disadvantage is that there is no approach for weighting criteria and it is assigned to the determinant (Macharis et al, 2007). So at first, weight of criteria should be assessed with Entropy method, weighted average, Linamp, The least squares and Eigenvector method and then alternatives should be prioritized by the use of PROMETHEE technique (Asgharpour, 1999).

In this method for assessing several alternatives, based on some criteria, type of criteria, preference **function**, preference threshold and indifference threshold, Should be specified. For increasing the PROMETHEE performance, Graphical Analysis for Interactive Assistance technique or GAIA is used. This technique helps determinant about criteria differences and weight of criteria on final results and for this purpose, Visual PROMETHEE software should be used.

Briefly, steps of this method are as follow:

2.1 Selection of Effective Alternatives and Criteria

Choosing criteria and alternatives from a lot of proposed criteria and alternatives in de- desertification process could be done either, according to the experience of the expert, information sources and field studies or by the use of Delphi technique and preparation of questionnaire then asking experts familiar with the study area to express effective alternatives and criteria and also score them form 0 to 9. Finally by gaining the average of all criteria or alternatives any alternative with the mean value less than 7 ($\overline{X} < 7$), would be removed but alternatives with the mean value more or equal to 7 ($\overline{X} \geq 7$) would be used (Azar and Rajabzadeh, 2002; Azar and Memariani, 2003).

2.2 Calculate Local Priority of Criteria and Alternatives and Establish Group Pirewise Comparisons Matrix

In continuation, to achieve local priority, a second questionnaire entitled "pirewise comparisons questionnaire" was designed and Experts were asked to conduct pirewise comparison on obtained results of first questionnaire regarding the nine-point Saaty's scale (Table1) based on importance to goal and priority to each criteria respectively. After forming pairwise comparisons matrix of each expert about criteria importance and alternatives priority (Table2), by the use of geometric mean and assumption of uniform expert's opinion, pirewise comparisons of each expert were composed according to Eq. 1; and pirewise comparisons were formed regarding to group (Azar and Rajabzadeh, 2002, Ghodsipour, 2002).

$$\overline{a}_{ij} = \left(\pi_{k=1}^{N} a_{ij}^{\ k} \right)^{\frac{1}{N}} \tag{1}$$

Table1. Importance and priority degree of nine-point Satty's scale (Azar & Rajabzadeh, 2002)

Score	Importance Degree	Priority Degree in Pirewise Comparison
1	Non-importance	Equal
2	Very low	Equal-Moderately
3	Low	Moderately
4	Relatively low	Moderately - Strongly
5	Medium	Strongly
6	Relatively high	Strongly-Very strongly
7	High	Very strongly
8	Very high	Very strongly-Extremely
9	Excellent	Extremely
1/2, 1/3,1/4, …., 1/9	Mutual Values	

Table2. pirewise comparaisons matrix (Azar & Rajabzadeh, 2002)

$$A = \begin{vmatrix} a_{11} & a_{12} & \cdots & a_{1n} \\ a_{21} & a_{22} & \cdots & a_{2n} \\ \vdots & \vdots & \vdots & \vdots \\ a_{m1} & a_{m2} & \cdots & a_{mn} \end{vmatrix} \qquad \begin{array}{l} A = [a_{ij}^{\dagger}] ,\ i = 1,2,.....m \\ \qquad\quad j = 1,2,.....n \end{array}$$

$a_{ij} =$ preference of i criteria to j criteria

2.3 Compute the Priorities Based on Group Pirewise of Comparisons Tables

At this stage, the numbers of group pirewise comparisons matrix (values of criteria importance and alternatives priority to each criterion) were calculated based on weighted average or average of each level of normalized matrix (Fig. 3 and 4) after normalization by using Eq. 2.

$$\bar{r}_{ij} = \frac{\bar{a}_{ij}}{\sum_{i=1} \bar{a}_{ij}} \tag{2}$$

2.4 Formation of Normalized Decision Matrix (NDM)

The Weighs of criteria importance (W_j) and alternatives priority (P_{ij}) were entered according to decision matrix (Table3).

Table3. Normalized Decision Matrix (NDM)

Alt	Criterion				
	C_1	C_2	C_3	----------	C_n
	W_1	W_2	W_3	----------	W_n
A_1	P_{11}	P_{12}	P_{13}	--------	P_{1n}
A_2	P_{21}	P_{22}	P_{23}	----------	P_{2n}
:	:	:	:	:	:
A_m	P_{m1}	P_{m2}	P_{m3}	----------	P_{mn}

Note. In this matrix m= the number of choices or alternatives, N= number of criteria, C= title of criteria, P_{ij} = weight value that each alternative gains in relation to related criteria, $W_{j=}$ Weight value that each criteria gains in relation to the goal.

2.5 Calculating the Difference Between Pirewise Sizes of Alternatives in Each Criterion (Dj) and Forming Table of Sizes Difference

The difference between pirewise sizes of alternatives (i) in each criterion (j) was calculated according to Eq. 3 (Table 4).

$$d_j(a,b) = P_{aj} - P_{bj} \tag{3}$$

$d_j(a,b)$ = The difference between sizes of alternatives a and b in different criteria

P_{aj} = Alternative a priority based on each criterion

P_{bj} = Alternative b priority based on each criterion

Table 4. The difference between pirewise sizes of alternatives in each criterion

		Criterion				
		C_1	C_2	C_3	----------	C_n
The difference of A_1 with other alternatives	A_2	$d_{1(1,2)}$	$d_{2(1,2)}$	$d_{3(1,2)}$	--------	$d_{n(1,2)}$
	A_3	$d_{1(1,3)}$	$d_{2(1,3)}$	$d_{3(1,3)}$	----------	$d_{n(1,3)}$
	:	:	:	:	:	:
	A_m	$d_{1(1,m)}$	$d_{2(1,m)}$	$d_{3(1,m)}$	----------	$d_{n(1,m)}$
The difference of A_2 with other alternatives	A_1	$d_{1(2,2)}$	$d_{2(2,2)}$	$d_{3(2,2)}$	--------	$d_{n(2,2)}$
	A_3	$d_{1(2,3)}$	$d_{2(2,3)}$	$d_{3(2,3)}$	----------	$d_{n(2,3)}$
	:	:	:	:	:	:
	A_m	$d_{1(2,m)}$	$d_{2(2,m)}$	$d_{3(2,m)}$	----------	$d_{n(2,m)}$

Note. m= the number of alternatives , n= the number of criteria, C= title of criteria, $d_{ij=}$ the difference between pirewise sizes of alternatives in each criterion

2.6 Calculating Alternatives Preference Function (P_j) and Forming Table of Preference Function

Preference functions exchange the difference between two alternatives, for a specific criterion, to a degree of preference that varies from 0 to 1. Determiner should choose a preference function for each criterion. So many preference functions can be defined in this case but usually these six preference functions consist of line function, Gaussian function, u-shaped function, v-shaped function, flat function and normal function are used (Barns et al, 1986).

Preference functions have been shown in figure1. In each of these functions two characteristics of following characteristics should be setup.

These characteristics are as follow:

1. Indifference threshold q: the largest negligible deviation that is considered in criterion.

2. Preference threshold p: the smallest deviation that is considered for determining full preference by decision makers. Indifference threshold and preference threshold are respectively small and large amounts according to measurement scale.

3. Gaussian threshold s: the turning point of preference threshold is Gaussian. Amount of Gaussian threshold usually is considered between p and q.

Preference function or alternatives preference (i) in each criterion (j) was calculated according to Eq.4 and in regard to preference functions.

$$P_j(a,b) = P\big[d_j(a,b)\big] \qquad (4)$$

$P_j(a,b)$ = Preference amount of alternatives a and b in different criteria

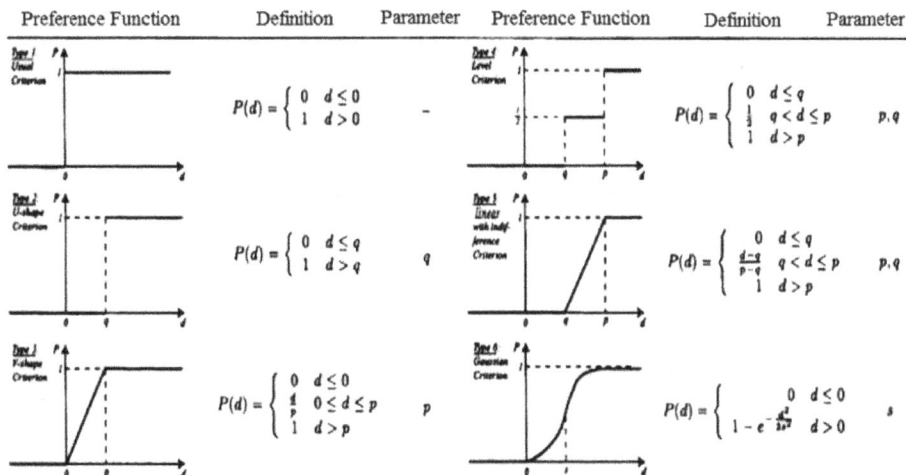

Figure1. The six preference functions (Brans et al., 1986)

Table5. Alternatives pirewise preference in each criterion

		C₁	C₂	C₃	----------	Cₙ
	A_2	$P_{1(1,2)}$	$P_{2(1,2)}$	$P_{3(1,2)}$	--------	$P_{n(1,2)}$
Preference amount of A_1 with	A_3	$P_{1(1,3)}$	$P_{2(1,3)}$	$P_{3(1,3)}$	-----------	$P_{n(1,3)}$
other alternatives	:	:	:	:	:	:
	A_m	$P_{1(1,m)}$	$P_{2(1,m)}$	$P_{3(1,m)}$	-----------	$P_{n(1,m)}$
	A_1	$P_{1(2,2)}$	$P_{2(2,2)}$	$P_{3(2,2)}$	--------	$P_{n(2,2)}$
Preference amount of A_2 with	A_3	$P_{1(2,3)}$	$P_{2(2,3)}$	$P_{3(2,3)}$	-----------	$P_{n(2,3)}$
other alternatives	:	:	:	:	:	:
	A_m	$P_{1(2,m)}$	$P_{2(2,m)}$	$P_{3(2,m)}$	-----------	$P_{n(2,m)}$

Note. m= number of alternatives, n= number of criteria, C= title of criteria, P_{ij}= preference of alternatives pirewise   sizes in each criterion

2.7 Estimating Weighted Sum of Alternatives Priority Relative to Each Other and Forming Table of Alternatives Priority

Weighted sum of alternatives priority relative to each other was calculated based on table 5 and according to Eq.5.

$$\begin{cases} \pi(a,b) = \sum_{j=1}^{k} P_j(a,b) W_j \\ \pi(b,a) = \sum_{j=1}^{k} P_j(b,a) W_j \end{cases} \tag{5}$$

$\pi(a,b)$ = Weighted sum of alternative a priority relative to b

$\pi(b,a)$ = Weighted sum of alternative b priority relative to a

W_j = Weights of criteria

Table6. Alternatives priority based on total criteria

		Criterion						*Priority of an alternative relative to total alternatives*
		C_1 W_1	C_2 W_2	C_3 W_3	-----	C_n W_n		
Weighted priority A1 relative to other alternatives	A_2	$P_{1(1,2)}.W_1$	$P_{2(1,2)}.W_2$	$P_{3(1,2)}.W_3$	------	$P_{n(1,2)}.W_n$	$\pi_{(1,2)}$	
	A_3	$P_{1(1,3)}.W_1$	$P_{2(1,3)}.W_2$	$P_{3(1,3)}.W_3$	-----	$P_{n(1,3)}.W_n$	$\pi_{(1,3)}$	$\pi_{(1,X)}$
	$\vdots$	$\vdots$	$\vdots$	$\vdots$	$\vdots$	$\vdots$		
	A_m	$P_{1(1,m)}.W_1$	$P_{2(1,m)}.W_2$	$P_{3(1,m)}.W_3$	-----	$P_{n(1,m)}.W_n$	$\pi_{(1,m)}$	
Weighted priority A2 relative to other alternatives	A_1	$P_{1(2,2)}.W_1$	$P_{2(2,2)}.W_2$	$P_{3(2,2)}.W_3$	------	$P_{n(2,2)}.W_n$	$\pi_{(2,2)}$	
	A_3	$P_{1(2,3)}.W_1$	$P_{2(2,3)}.W_2$	$P_{3(2,3)}.W_3$	-----	$P_{n(2,3)}.W_n$	$\pi_{(2,3)}$	$\pi_{(2,X)}$
	$\vdots$	$\vdots$	$\vdots$	$\vdots$	$\vdots$	$\vdots$		
	A_m	$P_{1(2,m)}.W_1$	$P_{2(2,m)}.W_2$	$P_{3(2,m)}.W_3$	-----	$P_{n(2,m)}.W_n$	$\pi_{(2,m)}$	

Note. m= number of alternatives, n= number of criteria, C= title of criteria, w= weights of criteria, P_{ij}=preference of alternatives pirewise sizes in each criterion

Finally priority of each alternative (alternative a) relative to total alternatives was calculated according to Eq. 6 (Table6)

$$\pi(a,x) = \sum_{i=2}^{m} \pi_i(a,i) \tag{6}$$

2.8 Calculating Positive out Ranking Current and Negative out Ranking Current of Each Alternative Relative to Other Alternatives

After calculating priority of each alternative relative to total alternatives, positive out ranking current and negative out ranking current of each alternative was calculated according to Eq.7 and Eq.8. Positive out ranking current shows an alternative preference like a relative to other alternatives. The higher value of it shows that this alternative would be better.

$$\varphi^+(a) = \frac{1}{n-1} \sum_{x \in A} \pi(a,x) \tag{7}$$

n= number of alternatives

Negative out ranking current shows other alternatives preference relative to alternative a. The lower value of it shows that this alternative would be better.

$$\varphi^-(a) = \frac{1}{n-1} \sum_{x \in A} \pi(a,x) \tag{8}$$

n= number of alternatives

2.9 Alternatives ranking

In order to rank alternatives completely, PROMETHEE II was used. For determining final priority of alternatives, net out ranking current of each alternative that was obtained from Eq.9, was used.

$$\phi(a) = \phi^+(a) - \phi^-(a) \tag{9}$$

$\varphi(a) = $ Net out ranking current of alternative a preference

So, the better net out ranking current of an alternative, the better that alternative (Eq.10, 11).

$$a \; P^{II} \; b \; iff \; \varphi(a) \rangle \varphi(b) \tag{10}$$

P^{II} = Sign of preference

$a \; P^{II} \; b$ = Alternative a is preferred relative to alternative b because of alternative b limitation

$$a \; I^{II} \; b \; iff \; \varphi(a) = \varphi(b) \tag{11}$$

I^{II} = Sign of indifference

$a \; I^{II} \; b$ = Alternative a and alternative b are equally preferred

In this method all alternatives are comparable.

For calculating output and input current and also preference net current in order to estimate alternatives preference, Visual PROMETHEE software was used.

3. Results

3.1 Selection of Important Criteria and Preferred Alternative According to Group and Design Hierarchical Decision Structure

In order to achieve important criteria and preferred alternative among several criteria and alternatives, a questionnaire was prepared in two parts according to the Delphi method. Finally, by gaining the average of given scores to each criteria and alternative, any of them with the mean value less than 7 were removed and others were used for designing hierarchical decision structure (Figure 2).

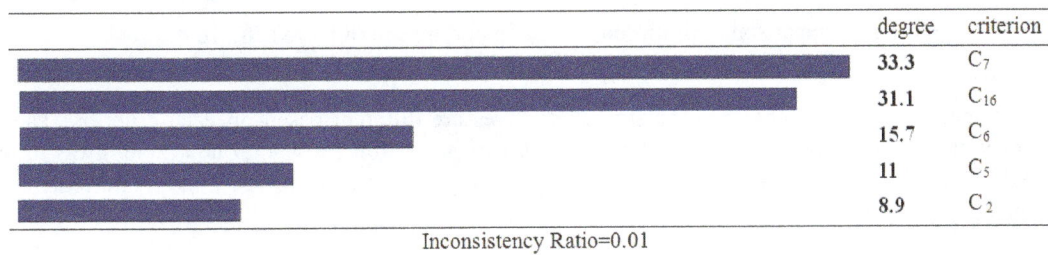

degree	criterion
33.3	C_7
31.1	C_{16}
15.7	C_6
11	C_5
8.9	C_2

Inconsistency Ratio=0.01

Figure 2. Hierarchical decision structure to select optimal de-desertification alternatives in Kezr Abad region

3.2 Calculate Relative Weight of Criteria and Alternatives and Format Group Decision Matrix (DM)

In order to calculate relative weight of criteria and alternatives or their priority, pirewise comparisons questionnaire was prepared and distributed among experts. Then by combining experts' opinions and by the use of geometric mean, group pirewise comparisons matrix of criteria importance relative to the goal and alternatives priority relative to all criteria was formed. Here, only criteria importance matrix is presented (Table7).

Table7. Group pirewise comparisons matrix of criteria importance relative to the goal of "offering optimal de-desertification alternatives in Kherz Abad region"

criterion	C_{16}	C_6	C_5	C_2
C_7	1/2	2/5	2/5	3/4
C_{16}		2/3	3/1	3/1
C_6			1/7	2
C_5				1/3

In continuation, matrix values of criteria importance and alternatives priorities were entered EC software based on each criterion, and importance and priority of de-desertification criteria and alternatives were obtained according to group in the study area as bar graphs Based on percentage using normalization and weighted mean (Figure 3).

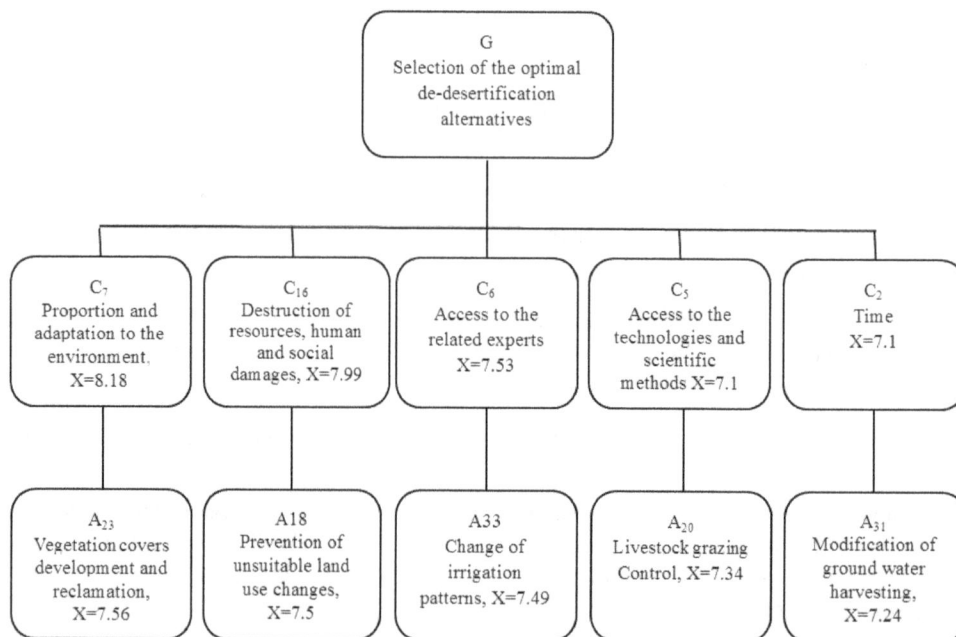

Figure 3. Comparison of proposed criteria importance in order to achieve the goal

Considering these graphs it is observed that the alternatives are different based on each criterion. Therefore, decision making matrix of optimal de-desertification alternatives according to the group (Table8) was formed to select final alternatives and classification of their priorities, in general framework of multiple attribute decision making methods (Table3).

Table 8. Decision matrix of optimal de-desertification alternatives according to group

Criteria importance (C) ▶ Alternatives priority (A) ▼	C_2	C_5	C_6	C_{16}	C_7
	0.0892	0.1095	0.1576	0.3074	0.3365
A_{23}	0.2509	0.2387	0.2488	0.1805	0.2257
A_{18}	0.1960	0.1635	0.1983	0.2383	0.2643
A_{33}	0.1620	0.2565	0.2093	0.1510	0.1599
A_{20}	0.2229	0.1762	0.1608	0.2209	0.1582
A_{31}	0.1682	0.1633	0.1826	0.2092	0.1918

3.3 Determination of Positive, Negative and Net Currents of Alternatives and Final Prioritizing of Criteria

In order to achieve final ranking of alternatives, after determination of criteria preference and alternatives priority in the framework of normalized decision making matrix (table.8) data of this table were entered Visual PROMETHEE software (Fig4). In terms of preference function, Barns and Vincke (1985) emphasized to use Gaussian function for practical application especially continuous data. So, in this study, for criteria of time, Gaussian preference function was used and s threshold was calculated according to Eq.12.

$$S = \left(\overline{X} - X_{\min} \right) + 2\delta \tag{12}$$

$\overline{X} =$ Average of total alternatives in each criterion

$X_{\min} =$ The minimum amount of total alternatives in each criterion

$\delta =$ Standard deviation of total alternatives in each criterion

Also V-shaped function was used for other criteria. Amount of P threshold for this preference function was calculated based on the difference between maximum and minimum amount of alternatives. Then data of table9 were entered Visual PROMETHEE software and PROMETHEE Table (9) key was selected and input, output and net currents of each alternative were calculated relative to other alternatives (Fig5).

Table9. Necessary characteristics for criteria

criteria	Effect type	Preference function	S	P	Weight of criteria (%)
C_7	Maximize	V-shape	-	0.1061	0.3365
C_{16}	Maximize	V-shape	-	0.0873	0.3074
C_6	Maximize	V-shape	-	0.088	0.1576
C_5	Maximize	V-shape	-	0.0932	0.1095
C_2	Maximize	Gaussian	0.154955	-	0.0892

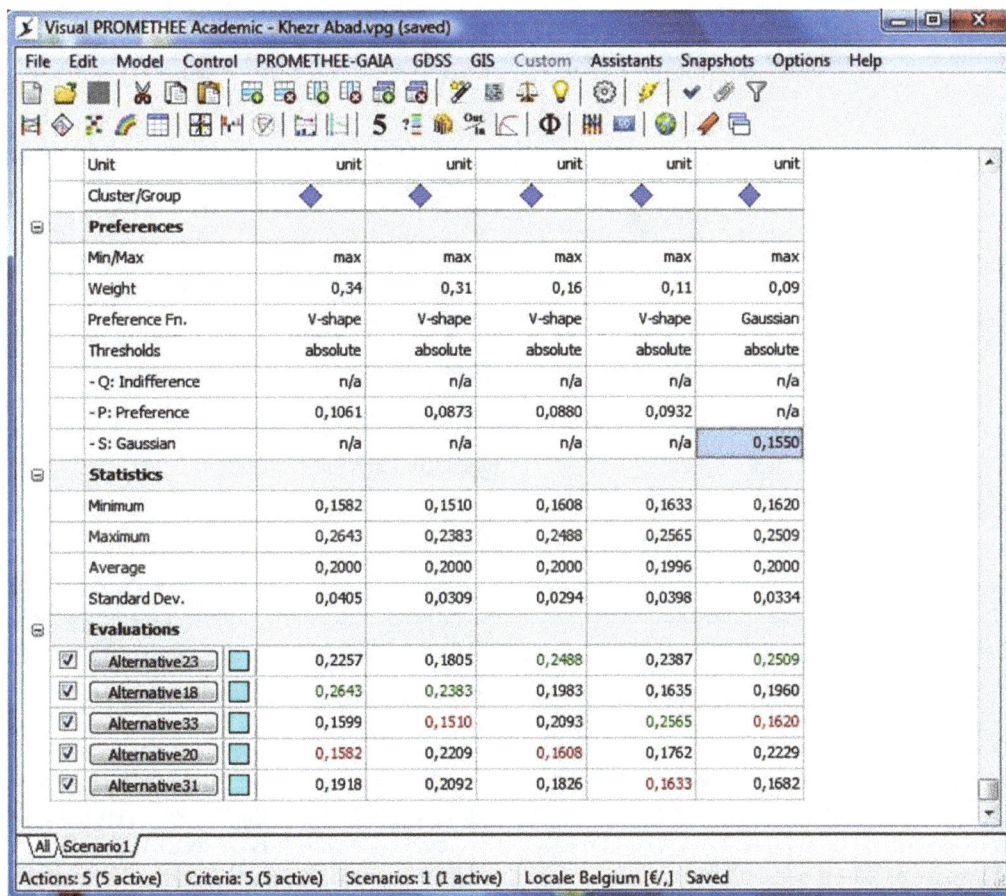

Figure 4. Decision matrix data in Visual PROMETHEE

PROMETHEE Flow Table

Rank	action	Phi	Phi+	Phi-
1	Alternative18	0,3660	0,4483	0,0823
2	Alternative23	0,1909	0,3385	0,1475
3	Alternative31	-0,0887	0,1382	0,2269
4	Alternative20	-0,1926	0,1182	0,3109
5	Alternative33	-0,2756	0,1234	0,3990

Figure 5. postitive, negative and net out ranking currents of de-desertification alternatives in study area

Finally in order to analyze the results better, GAIA graph was used in Visual PROMETHEE software (Fig.6). This graph is a powerful tool for analyzing multivariate issues that indicates criteria preference, alternatives priority, quality of each alternative and opposite and positive aspects of criteria clearly.

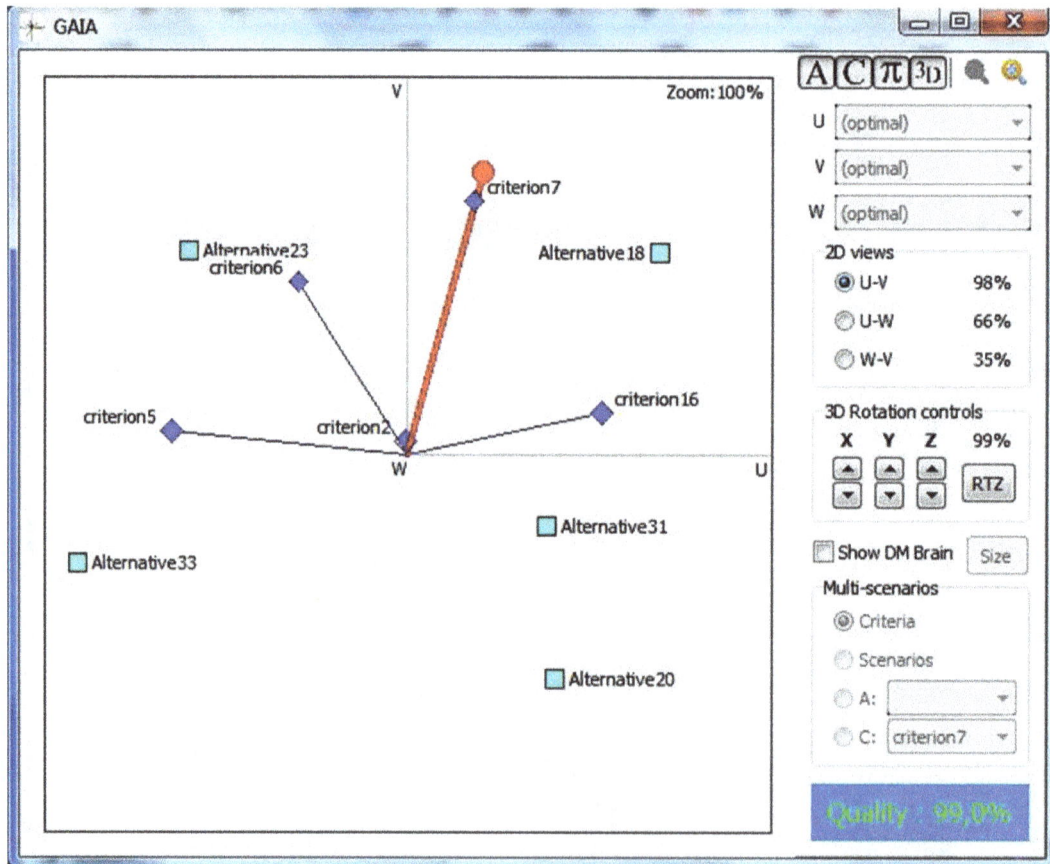

Figure 6. GAIA graph for ranking of de-desertification criteria and alternatives

Criteria C_7, C_{16}, C_6, C_5, C_2 were respectively preferred according to figure and suggested alternatives were different according to each criterion. For example in relation to criteria C_7, C_{16} alternative A_{18} was preferred and in relation to criteria C_5 alternative A_{33} was preferred. So Promethee decision axis (**π** axis) was obtained in order to select alternatives based on total criteria. The Promethee decision axis indicated priority of alternative A_{18} in order to achieve the goal. Alternatives A_{23}, A_{31}, A_{20} and A_{33} were the next order respectively (Fig6, 7). According to figure 7, obtained results are reliable with 99% accuracy.

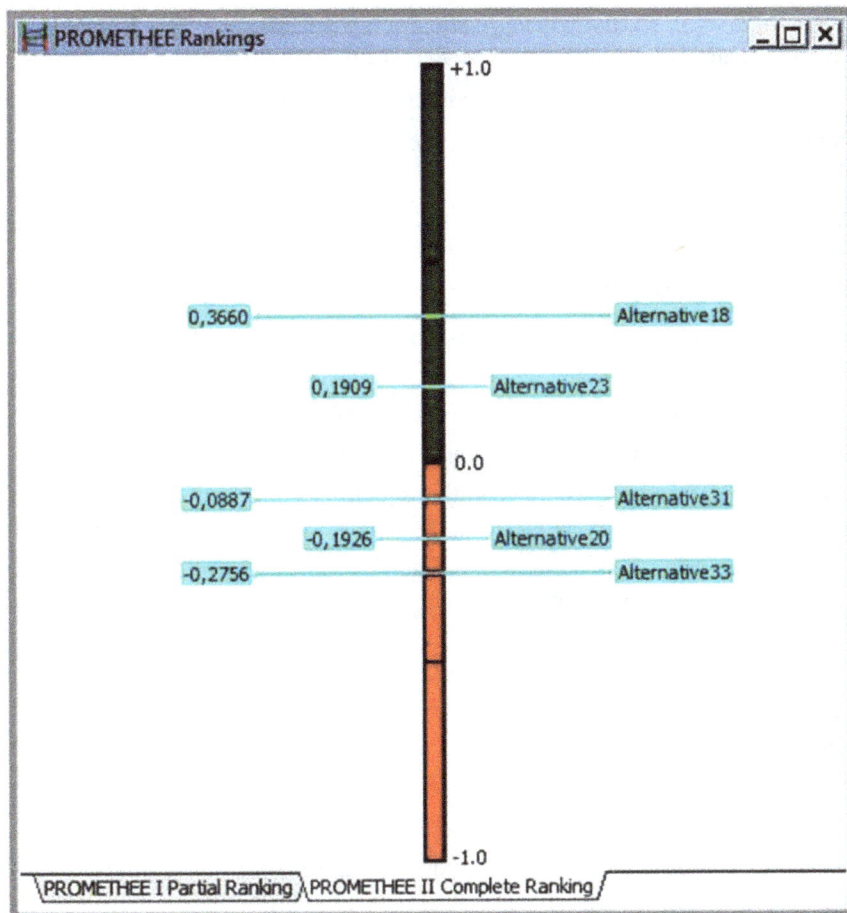

Figure 7. complete ranking of alternatives based on PROMETHEE II

4. Discussion

In this study a new method was represented in order to rank alternatives priority for combating desertification process. Obtained results of PROMETHEE method for final prioritizing of alternatives emphasize on obtained results of AHP ،TOPSISs ،ELECTERE and WSM for prioritizing alternatives and alternatives A18, A23, A31 were placed in first, second and third order respectively.

PROMETHEE method, like above mentioned method, has the limitation of ignoring determinants fuzzy judgment. Also some criteria have unknown structure or qualitative structure and cannot be measured accurately. In this situation, in order to achieve evaluation matrix, we can use fuzzy numbers. So the mentioned prioritizing method can be developed by the use of fuzzy numbers. Also software features that are provided as a support for this method, prepare appropriate analysis for determinant. The results of this method are closer to reality because there is a comprehensive development attitude in it.

Based on the results, the alternatives of prevention of unsuitable land use changes (A18), vegetation cover development and reclamation (A23) and modification of ground water harvesting (A31) with pure out ranking progress of phi =0.3660, 0.1909 and -0.0887were selected as the main de-desertification altarnative in the study area, respectively. So by the use of these alternatives in de-desertification projects, we can prevent desertification phenomenon in this area and do something for reclamation of degraded land.

In study area, land use changes are developing strongly because of population grow, unemployment, industrial and urbanization grow. Usually land use changes has occurred as conversion of pasture land to farm and garden because of deep and semi-deep wells development, conversion of garden to agriculture land on the effect of successive droughts and conversion of pasture land to urban and industrial land because of industrial and urbanization grow in recent years.

Pasture types density is 6 to 15 percent that is influenced by human actions strongly, and 40 to 50 percent of vegetation cover is destroyed because of cutting to compile livestock, fuel and building materials.

Irrigation for agriculture land is usually flood irrigation and basin irrigation, so over 50 percent of water is wasted and irrigation efficiency is calculated less than 40 percent in the farm.

So following executive suggestions are recommended in the framework of discussed macro alternatives:

Land use planning and estimating the ecological potential in national, regional and local levels and land use conformity with land potential.

Preventing inappropriate conversion of poor pasture land to garden with low efficiency and high potential for degradation and erosion.

Preventing development of industrial infrastructure in sensitive lands of desert areas and marginal areas.

The use of resistant and pasture species and modern irrigation systems for vegetation cover development and reclamation.

Prevention of Tamarix habitat destruction and trying for their reclamation

Observing the balance between number of livestock and pasture capacity

Conformity between livestock type and pasture condition should be considered and the number of goats should be reduced in poor pastures, because these animals increase pastures destruction, potentially.

It is recommended that de-desertification projects in the study area be focused on these alternatives to get better and suitable results, avoid investment wasting and increase control, reclamation and reconstruction project efficiency.

References

Albadavi, A., Chaharsooghi, S. K., & Esfahanipour, A. (2007). Decision making in stock trading: An application of PROMETHEE. *European Journal of Operational Research, 177*, 673–683. http://dx.doi.10.1016/j.ejor.2005.11.022

Asgharpour, M. J. (1999). *Multiple criteria decision making*, University of Tehran, Tehran, p. 397.

Azar, A., & Memariani, A. (2003), AHP new technique for group decision making, *Journal of Knowledge Management, 28*, 22-32.

Azar, A., & Rajabzadeh, A. (2002). *Applied decision making* (1st ed., p. 185), Tehran: Negahe Danesh.

Bakhshi, M., Panahi, R., Mollaei, Z., Kazemi, H., & Mohammadi, D. (2011). Evaluation of Innovation in South West Asia Countries and Determine of Iran'S Position: Application of PROMETHEE, *Journal of Science and Technology Policy, 3*, 19-33.

Barns, J. P. (1982). Lingenierie de la decision. *Elaboration dinstruments daide a la decision. Method PROMETHEE.* In: Nadeau, R., Landry, M. (Eds.), Laide a la Decision: Nature, Instrument set Perspectives Davenir. Presses de Universite Laval, Quebec, Canada, 183–214.

Behzadian, M., & Pirdashti, M. (2009). Selection of the Best Module Design for Ultrafiltration (UF)Membrane in Dairy Industry: An Application of AHP and PROMETHEE. *International Journal of Engineering, 3*, 126-142.

Brans, J. P., & Mareschal, B. (1994). The PROMCALC and GAIA Decision Support System for Multicriteria Decision Aid, *Decision Support Systems, 12*, 297–310. http://dx.doi.10.1016/0167-9236(94)90048-5

Brans, J. P., & Mareschal, B. (2005). *Multiple Criteria Decision Analysis: State of the Art Surveys*, Chapter 5: ROMETHEE METHODS, Springer, New York, 195.

Brans, J. P., & Vincke, Ph. (1985). A Preference ranking Organization Method, The PROMETHEE Method for Multiple Criteria Decision-Making, *Management Science, 31*, 647–656.

Brans, J. P., Vincke, Ph., & Mareschal, B. (1986). How to select and how to rank projects: The PROMETHEE method, *European Journal Operational Research*, 4, 138–228.http://dx.doi.10.1016/0377-2217(86)90044-5

Chen, C. T., Hung, W. Z., Cheng, H. L. (2011). Applying linguistic PROMETHEE method in investment portfolio decision making, *International Journal of Electronic Business Management, 9*(2), 139-148.

Chou, T. Y., Lin, W. T., Lin, C. Y., Chou, W. C., & Huang, P. H. (2004). Application of the PROMETHEE technique to determine depression outlet location and flow direction in DEM, *Journal of Hydrology, 287*, 49–61.

Diakoulaki, D., & Karangelis, F. (2007). Multi-criteria decision analysis and cost-benefit analysis ofalternative scenarios for the power generation sector in Greece. *Renewable and Sustainable Energy Reviews, 11*, 716–727. http://dx.doi.10.1016/j.rser.2005.06.007

Elevli, B., & Demirci, A. (2004). Multicriteria choice of ore transport system for an underground mine: Application of PROMETHEE methods. *Journal of the South African Institute of Mining and Metallurgy, 104*, 251–256.

Fontana, M. E., & Cavalcante, C. A. V. (2014). Use of PROMETHEE method to determine the best alternative for warehouse storage location assignment, *The International Journal of Advanced Manufacturing Technology, 70*, 1615-1624. http://dx.doi.10.1007/s00170-013-5405-z

Ghodsipour, S. A. H. (2002). *Analytical Hierarchy Process (AHP)*, Amirkabir University, Tehran, 220 p.

Ginevicius, R., Podvezko, V., & Novotny, M. (2010). The use of PROMETHEE method for evaluating the strategic potential of construction enterprises. *The 10th International Conference of modern building materials, structures and techniques*, Vilnius, Lithuania, 19- 21 May, 407- 413.

Grau, J. B., Anton, J. M., Tarquis, A. M., Colombo, F., Rios, L., & Cisneros, J. M. (2010). Mathematical model to select the optimal alternative for an integral plan to desertification and erosion control for the Chaco Area in Salta Province (Argentine), *Journal of Biogeosciences Discuss, 7*, 2601–2630. http://dx.doi.10.5194/bg-7-3421-2010

Halouani, N., Chabchoub, H., & Martenez, J.M. (2009). PROMETHEE-MD-2T method for project selection. *European Journal of Operational Research, 195*, 841–849. http://dx.doi.10.1016/j.ejor.2007.11.016

Karandea, P & Chakraborty, S. (2012). Application of PROMETHEE-GAIA method for non-traditional machining processes selection, *Management Science Letters, 2*, 2049–2060. http://dx.doi.10.5267/j.msl.2012.06.015

Keyser, W. D., & Peeters, P. (1996). Theory and Methodology a note on the use of PROMETHEE multicriteria methods, *European Journal of Operational Research, 89*, 457-461.

Macharis, C., Springael, J., Brucker, K. D., & Verbeke, A. (2007). PROMETHEE and AHP: The design of operational synergies in multicriteria analysis, Strengthening PROMETHEE with ideas of AHP. European *Journal of Operational Research, 153*, 307–317.

Mohaghar, A., & Mostafavi, A. (2007). Designing a Model for Selecting Project Team Based on Fuzzy Approach, *Management Research in Iran*, 11, 207-232. http://dx.doi.10.1016/S0377-2217(03)00153-X

Nasiri, H., Alavipanah, S. K., Matinfar, A., Azizi, H. R., & Hamzeh, M. (2012). Implementation of Agricultural Ecological Capability Model Using Integrated Approach of PROMETHEE II and Fuzzy-AHP in GIS Environment (Case Study: Marvdasht county), *Journal of Environmental Studies, 38*, 109-122.

Omidi, M., Razavi, H., & Meh Paikar, M. (2011). Selection of Project Team Members based on Effective Criteria using PROMETHEE Method, *Journal of Industrial Management Perspective, 1*, 113-134.

Pomerol, J. C., & Romero, S. B. (2000). *Multi-Criterion Decision in Management: Principles and practice*, (International Series in Operations Research & Management Science) Netherlands: Kluwer Academic, Dordrecht. Kluwer, Massachusetts, USA.

Prvulovic, S., Tolmac, D., & Radovanovic, L. (2011). Application of PROMETHEE-Gaia Methodology in the Choice of Systems for Drying Paltry-Seeds and Powder Materials, *Journal of Mechanical Engineering, 57*, 778-784. http://dx.doi.10.5545/sv-jme.2008.068

Queiruga, D., Walther, G., Benito, J. G., & Spengler, T. (2008). Evaluation of sites for the location of WEEE recycling plants in Spain, *Waste Management, 28*, 181–190. http://dx.doi.10.1016/j.wasman.2006.11.001

Sadeghiravesh, M. H., Ahmadi, H., Zehtabian, G. H., & Tahmores, M. (2010). Application of analytical hierarchy process (AHP) in assessment of de-desertification alternatives, case study: Khezrabad region, Yazd province. *Marta & Biaban Journal, 17*(1),35-51.

Sadeghiravesh, M. H., Khosravi, H., & Ghasemian, S. (2015). Application of fuzzy analytical hierarchy process for assessment of combating-desertification alternatives in central Iran, *Natural Hazards, 75*, 653-667. http://dx.doi.10.1007/s11069-014-1345-7

Sadeghiravesh, M. H., Zehtabian, G. H., & Khosravi, H. (2014). Application of AHP and ELECTRE models for assessment of dedesertification alternatives, *Desert, 19*, 141-153.

Sadeghiravesh, M. H., Zehtabian, G., Ahmadi, H., & Khosravi, H. (2012). Using analytic hierarchy process method and ordering technique to assess de-desertification alternatives. Case study: Khezrabad, Yazd, IRAN, *Carpathian Journal of Earth and Environmental Sciences, 7*, 51 – 60.

Safari, H., Sadat Fagheyi, M., Sadat Ahangari, S., & Fathi, M. R. (2012). Applying PROMETHEE Method based on Entropy Weight for Supplier Selection. *Business Management and Strategy, 3*, 97- 106.

Shirsikar, S. G., & Patil, S. (2005). Optimization of energy charges using improved PROMETHEE method, *Journal of Electronics and Communication Engineering, 5*, 36-42.

Soltanmohammad, H., Osanloo, M., Rezaei, B., & Aghajani, B. A. (2008). Achieving to some outranking relationships between post mining land uses through mined land suitability analysis, *International Journal of Environmental Science and Technology, 5*, 535-546.

Taillandier, P., & Stinckwich, S. (2011). Using the PROMETHEE Multi-Criteria Decision Making Method to Define New Exploration Strategies for Rescue Robots. *IEEE International Symposium on Safety, Security, and Rescue Robotics (SSRR)*, Kyoto: Japan.

2

Geotechnical and Chemical Evaluation of Tropical Red Soils in a Deltaic Environment: Implications for Road Construction

Akaha C. Tse[1] & Adunola O. Ogunyemi[1]

[1] Department of Geology, University of Port Harcourt, Nigeria

Correspondence: Akaha C. Tse, Department of Geology, University of Port Harcourt, Nigeria. E-mail: akaha.tse@uniport.edu.ng

Abstract

Tropical red soils which occur in the dry flatlands and plains of the eastern Niger Delta Nigeria were evaluated using combined conventional engineering geological investigation with major oxide geochemistry to determine their properties and evaluate their engineering performance in road construction. Laboratory test results indicate that the brownish materials are uniformly graded, silty clayey sandy soils. The silica to sesquoxide ratio values of 3 to 4.37 indicate that they are non-lateritic tropically weathered soils. The average values of the specific gravity, liquid limit, plasticity index and shrinkage limits are 2.67, 37%, 10% and 7.6% respectively. They are soils of low to medium plasticity. The unsoaked and soaked CBR values range from 14-38% and 3-9% respectively whereas the average undrained shear strength is 172kN/m^2. Maximum dry density and optimum moisture content values fall between 1680 to 1880kN/m^2 and 13-16% respectively. Generally the soils classify as A-7-6 to A-2-4 subgroups of the AASHO classification. The overall implication of these composite engineering properties is that the non-lateritic soils rate as poor to fair subgrade materials.

Keywords: geotechnical, red soils, sesquioxide, lateritic, subgrade, Niger Delta

1. Introduction

The Niger Delta wetland area in Nigeria consists of three geomorphologic zones including the coastal or Lower delta zone, Transition or Mangrove zone and the Upper deltaic plain or freshwater zone. The freshwater zone consists of dry flatlands and plains (Akpokodje 1989 and Teme 2002) and its subsurface soil profile consists of a top lateritic clay layer underlain by silty clays and sands which are in turn succeeded by poorly graded sand and gravel. Akpokodje (1989) described the lateritic layer as a reddish brown/brown soft to firm clay which becomes stiffer with depth and is mottled with shades of grey or brown and also tends to contain reddish, lateritic gravelly concretions with a varying proportion of silt and clay. Although all of the Niger Delta area is characterized by tropical rainfall conditions, annual rainfall ranges from 2000mm in the freshwater zone to over 4000mm at the coast which also accounts for nearly 85% of the annual rainfall. Thus the coastal and mangrove zones contain nearly 70% of the numerous marshes and back swamps that occupy as much as 50% by area of the entire delta region. Also they are usually submerged during the wet season (April to October). However the relatively less "wet" conditions in the freshwater zone coincide with the occurrence of lateritic or tropical red solids. These deltaic lateritic soils differ markedly from the other lateritic soils because of some mode of formation related peculiarities (Omotosho, 2015). Alabo et al. (1984) studied these soils in Port Harcourt area and its environs which belong to the freshwater zone. They referred to the soils as 'deltaic red soils' on the basis of their colour, one of the several identifying criteria for laterites and lateritic soils. These tropical red soils occur exclusively in the freshwater zone and are widely used in highway and other earthen work constructions, and as back fill materials. The performance of the soils for engineering construction in the delta region is varied. Early failure of pavements constructed with these soils is common. Studies by Alabo et al. (1984) showed that locally in Port Harcourt area, the soils belong to the A-2-7 class of AASHO classification and this correlates well with their good performance as fill materials. Adeyemi (2002) has emphasized the imperative of a comprehensive evaluation of any lateritic soils prior to utilization for any engineering purpose. This is because of the unique set of physical, chemical and engineering properties the soils exhibit in response to the different climatic, geomorphologic and geological conditions of their origin. Due to the abundance of these soils and ready availability, they have been widely used in the construction of foundations, roads, airfields, low-cost housing, and compacted fill in earth embankments. This affects their variable performance as soil materials for road earthwork, and often require stabilisation (Omotosho and Eze-Uzomaka 2008, Ugbe 2011, Adeyemi et al. 2015) compared to other matured tropical lateritic soils (Elarabi et al. 2013, Elsharief et al. 2013, Carvalho et al 2015,) which achieve better success under mechanical compaction in engineering construction. Therefore, for engineering purposes it is important that the geological and engineering properties as predicted or derived from testing are reliable (Kamtchueng et al. 2015). This paper will attempt to determine the major oxide geochemistry and geotechnical properties of these red soils. The new data will help to classify the lateritic nature of the examined soils and also determine their quality as fill materials.

2. Geological Setting

The studied area is located in the freshwater geomorphologic zone of the eastern Niger which lies between latitudes 4° 30' and 5° 20' and longitudes 6° 15' and 7° 35' (Fig.1). The geology of the Niger Delta including its tectonic framework, stratigraphy and sedimentation pattern is well known from the published accounts of several authors, including Allen (1965) and Short and Stauble (1967), Doust and Omatsola (1990), Reijers et al 1997, Reijers (2011) and Nwajide (2013). The Niger Delta is a large arcuate delta of the destructive, wave dominated type. Its development is a function of the balance between the rate of sedimentation and the rate of subsidence. Tectonic evolution of the delta was controlled by Cretaceous fracture zones formed during the triple junction rifting and opening of the south Atlantic. The sedimentary fill of the southern Nigeria sedimentary basin was controlled by three major tectonic stages together with epirogenic movements which led to major episodes of transgressions and regressions. The cycles accounted for the sedimentary units in both the Cretaceous and Tertiary Southern Nigerian sedimentary basins, among which is the Niger Delta (Odigi, 2007). The delta represents the regressive phase of the third cycle of deposition which began during the Paleocene and has continued to the present day. The basin contains Cenozoic to Recent deposits emplaced in high energy constructive deltaic environments. The delta is underlain by three lithostratigraphic units comprising from bottom of Akata Formation deposited under marine conditions and consisting of shale. The formation has an approximate range of thickness from 0 – 6,000m and ranges from Paleocene to Holocene in age. It is the main source rock in the petroliferous Niger Delta. Shale diapirism due to loading of poorly compacted, over-pressured, prodeltaic and delta-slope clays resulted in the deposition of the Akata Formation. It is overlain by a paralic facies of shale and sand intercalation known as Agbada Formation deposited under mixed marine and continental environments. This sequence consists of an upper predominantly sandy unit with minor shale intercalations and a lower shale unit which is thicker than the upper sandy unit. The Agbada Formation is over 10,000ft thick and ranges in age from Eocene to Recent. All hydrocarbon accumulations in the Niger Delta are found in this formation. It is a pro-delta deposit characterized by gravity tectonics structures such as shale diapirs, roll-over anticlines, collapsed growth fault crests, and steeply dipping closely spaced flank faults. The top Benin Formation, Oligocene to recent age, is predominantly sandy although some shale intercalations are common. It was deposited under continental conditions. Various types of Quaternary deposits overly the Benin formation especially in the coastal and Mangrove zone and their nature have been discussed extensively by Akpokodje (1989) and Tse and Akpokodje (2013). Generally they consist of a top stratum of clay, silt or organic matter and sand, silt-clay mixture overlying a sandy/sand substratum which occurs at variable depths, ranging from 5 to 30m or more in the subsurface.

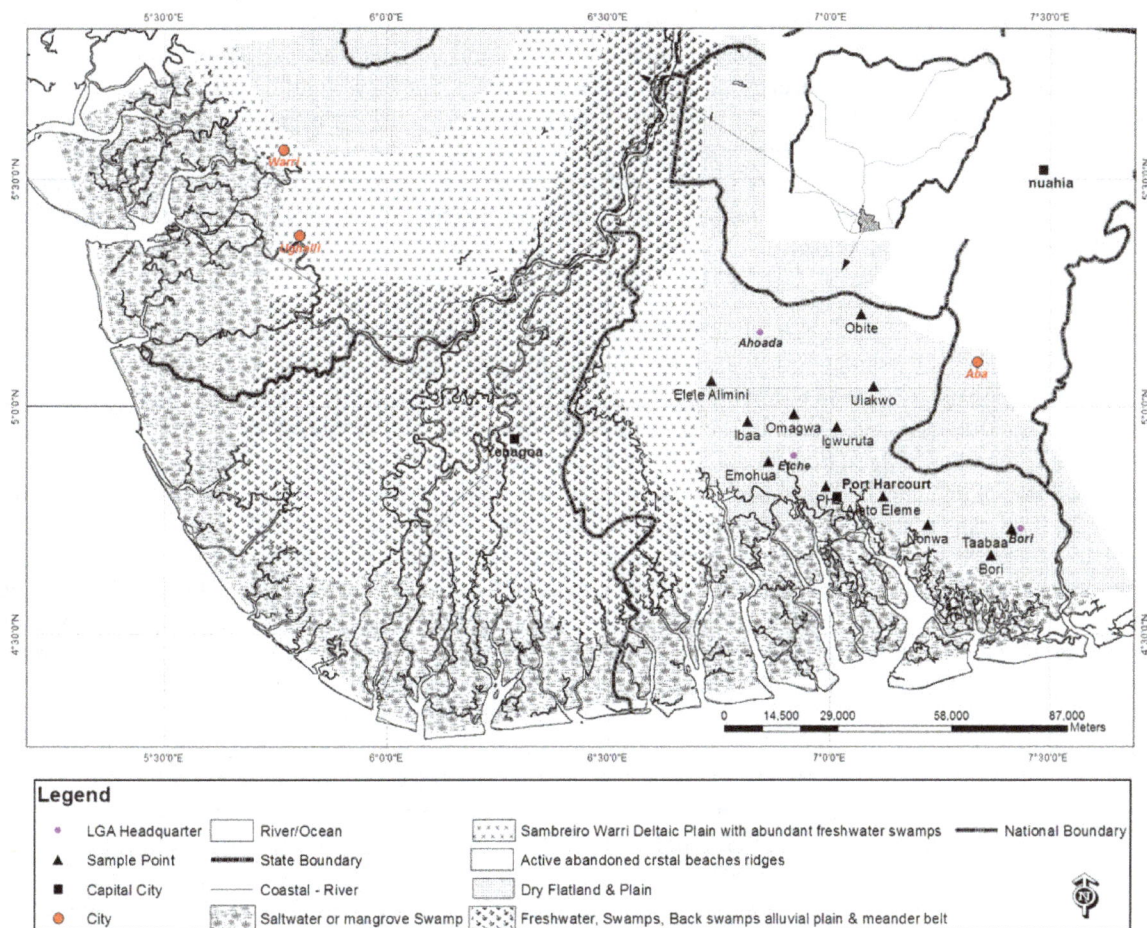

Figure 1. Map of Niger Delta showing geomorphologic zones and sampling points

In the freshwater zone, the Benin Formation is covered by appreciable thicknesses of red soils. Assez (1976) suggested they were formed as a result of the weathering of sedimentary deposits of sands and poorly cemented sandstones with some clay and their subsequent ferrugenisation. The dry flat land and plains is characterized by freshwater rivers, creeks and seasonal marshes. Soil profiles comprises of a top lateritic clay layer succeeded by poorly graded sand and gravel. The zone is comparatively the best drained of all the other geomorphic zones and is thus relatively drier. The permeability of soils is low to high and aquicludes, perched and normal aquifers are common. Overall, the soils have variable foundation potentials as discussed by Akpokodje (1979), Teme (2002), and Tse and Akpokodje (2013).

3. Materials And Methods.

Soil samples used for this study were obtained along major roads in areas which geomorphologically lie in the dry flatland and plains in the eastern Niger Delta along an east-west direction as shown in Fig. 1. The samples were collected from active and abandoned borrow pits used for mining of backfill and pavement materials by civil construction companies. The soils range from light yellow to greyish brown to reddish brown clayey-silty-sandy earth materials. At each location, three disturbed samples were collected at the bottom, middle and top in a vertical profile along the face of the pits which were first scraped to obtain fresh samples. The soils were described by visual inspection in the field and were subjected to laboratory tests according to methods and procedures specified by BS 1377 of 1990. Classification and strength tests carried out included particle size distribution, Atterberg limits, linear shrinkage, specific gravity by density bottle method, standard proctor compaction test, quick undrained unconsolidated triaxial test in standard triaxial test cells 76mm high and 38mm in diameter, and soaked and unsoaked California Bearing Ratio (CBR). Sieve analysis first involved washing of the soil samples through ASTM sieve number 200 (0.075mm mesh) to separate sand fraction from the fines (clay and silt) fraction. Thereafter, mechanical sieve analysis was performed to separate the sand into the various particle sizes. The moisture content – density relationships of the soils were determined by the standard Proctor test where 3kg of the soil was compacted in 3 equal layers in a cylindrical metal mould of volume $0.00956m^3$ using a 4.5kg rain falling through a height of 0.45m. Linear shrinkage tests were carried out to determine the water content at which no further decrease in the volume of the soil masses will not be experienced. The major oxides composition of the soils was determined by X-ray fluorescence analysis using Thermoscientific Advant 1200 model equipment. Results were used to determine the lateritic nature of the soils using the silica sesquioxide ratio presented by Rossiter (2004) as described in Alayaki (2015). Finally statistical regression analysis was performed on the test results and correlation coefficients tests were used to establish the relationship among the properties tested.

4. Results

A field examination and visual observation shows that the soils range from light yellow to greyish brown to reddish brown clayey-silty-sandy earth materials. The major oxides composition of the soils are shown in Table 1. The soils are made up of 62.66 to 69.94 wt.% of SiO_2, 0.11 to 0.32 wt.% of Fe_2O_3, and 27.20 to 34.12 wt.% of Al_2O_3. Thus quartz, iron and aluminium oxides are the dominant components of the soils. Results of the geotechnical properties of these tropical red soils which consist of variable proportions of gravel, sand, and fines fraction are shown in Table 2. Typical particle size distribution curves of the soils shown in Fig. 2 reveal that the amount of fines ranges from 22 to 76%. Soils in Ulakwo and Ogrike have the least amount of fines while Igwuruta and Emohua have the highest. A quantitative measure of the range of soil particles in a given sample is the uniformity coefficient. This varies from 2 to 9 with most of the samples having values of below 5 indicating that the soils are poorly graded. The moisture-density relationship results are summarized in Fig. 3. Maximum dry density of 1680 to 1880 kg/m^2 were obtained at optimum moisture content (OMC) of 13 to 16%. Soils in the northern and western northern parts of the study area attained the least MDD of approximately 1700kg/m^3 at OMC of 15%, while the ones in the south eastern axis from Igwuruta to Tabaa gave the highest MDD of approximately 1850kg/m^3 at an average OMC of 13%. Lateritic soils which give OMC between 8-10% and above 10% are rated by Philips (1952) as average to poor respectively for use under bituminous surfacing. Atterberg limits are important factors in the use of lateritic soils in pavement construction as sub-grade and sub-base materials. The average values of liquid limit of 26 to 48% classifies as clays of low to intermediate plasticity. Plasticity index range from 5 to 15%. These are soils of low swelling potentials according to the Ola (1982) rating. Thus, when the Skempton activity was calculated from the results of the plasticity index and percentage of clay obtained from the hydrometer tests, values of 1.01 to 5.56 with an average of 1.23 were obtained (Table 3). These classify as normal to active clays of low to medium sensitivity (Skempton, 1953). The California Bearing Ratio (CBR) test is used in the empirical estimation of the bearing capacity of sub-grade and sub-base materials under soaked and dry conditions. The soaked and unsoaked CBR values range between 3-9% and 14-38% respectively, and their relationship is graphically shown in Fig. 4.

Table 1. Major oxide geochemistry and silica:sesquioxide ratios

Location	% SiO₂	% Fe₂O₃	% Al₂O₃	% MgO	% CaO	% NaO	% K₂O	% MnO₂	% T₁O₂	% P₂O₅	% Total	SSR
Ogrike	69.94	0.11	27.20	0.08	0.19	0.08	0.10	0.08	2.02	0.020	99.82	4.37
Ahoada	68.43	0.13	28.60	0.05	0.25	0.10	0.13	0.10	2.02	0.023	99.83	4.06
Elele Alimini	65.51	0.15	31.55	0.04	0.21	0.08	0.11	0.10	2.12	0.026	99.90	3.52
Igwuruta	66.43	0.18	30.60	0.11	0.25	0.10	0.13	0.07	2.01	0.017	99.90	3.69
Ulakwo	67.08	0.16	29.75	0.10	0.30	0.12	0.10	0.08	2.15	0.022	99.85	3.82
Eleme	68.22	0.20	28.45	0.06	0.37	0.10	0.16	0.05	2.24	0.030	99.88	4.07
Nonwa	66.83	0.32	30.09	0.10	0.29	0.11	0.15	0.06	2.05	0.029	99.88	3.74
Bori	62.66	0.24	34.12	0.07	0.28	0.14	0.14	0.09	2.19	0.018	99.95	3.09
Tabaa	64.88	0.17	31.80	0.08	0.33	0.09	0.12	0.05	2.21	0.032	99.76	3.45

SSR = silica:sesquioxide ratio

Table 2. Geotechnical properties of the soils

S/No	Location	GS	LL (%)	PL (%)	PI (%)	LS (%)	% Fines	% Sand	CU	Cohesion (kN/m²)	FA (°)	OMC (%)	MDD (Kg/m³)	Soaked CBR	Unsoaked CBR
1	Ogrike	2.68	30	23	7	7.7	31	69	9	160	3	15	1780	3.61	18.79
2	Obite	2.66	35	24	11	7.5	50	51	7	20	23	16	1680	2.71	13.83
3	Abarikpo	2.67	26	21	5	7.6	53	47	2	35	17	16	1710	3.61	15.48
5	Elele Alimini	2.65	47	35	12	7.5	58	42	4	30	22	15	1700	5.11	14.13
6	Emohua	2.67	41	32	9	7.7	76	22	3	60	14	15	1730	6.61	18.49
7	Ibaa	2.68	44	33	11	7.5	51	51	5	135	5	15	1770	3.41	17.89
8	Omagwa	2.68	48	33	15	7.5	60	41	1	60	14	15	1770	4.36	19.39
9	Igwuruta	2.67	42	29	12	7.5	69	38	2	115	8	15	1820	5.56	24.80
10	Ulakwo	2.67	40	33	7	7.5	37	67	5	120	12	14	1850	9.17	25.54
11	PH	2.67	39	28	12	7.6	51	49	6	30	17	16	1690	3.76	13.68
12	Eleme	2.66	34	24	10	7.7	43	58	7	150	6	14	1820	5.86	23.00
13	Nonwa	2.67	28	24	10	7.7	51	49	6	40	25	14	1860	6.31	16.68
14	Bori	2.67	34	28	12	7.7	46	54	5	130	10	14	1880	7.21	29.16
15	Taabaa	2.66	40	29	11	7.8	62	39	2	100	7	13	1880	7.67	37.76

Average results for 3 samples at each location. GS = specific gravity, LL = liquid limit, PL = plastic limit, LS = linear shrinkage, CU = coefficient of uniformity, FA = frictional angle

Table 3. Classification and description of soil activity (after Skempton 1953)

S/No	Location	Clay amount %	Plasticity Index (%)	Activity value	Activity classification	Activity description
1	Ogrike	5.73	7.4	1.29	Active	Low sensitivity
2	Obite	4.30	11	2.56	Active	Medium sensitive
3	Abarikpo	4.86	4.9	1.01	Normal	Insensitive
5	Elele Alimini	5.08	12.8	2.51	Active	Medium sensitive
6	Emohua	4.03	9.2	2.28	Active	Medium sensitive
7	Ibaa	2.52	11	4.37	Active	sensitive
8	Omagwa	2.68	14.9	5.56	Active	sensitive
9	Igwuruta	5.08	12.4	2.47	Active	Medium sensitivity
10	Ulakwo	1.38	7.2	5.21	Active	sensitive
11	Port Harcourt	7.88	11.7	1.48	Active	Low sensitivity
12	Aleto Eleme	6.11	10.2	1.66	Active	Low sensitivity
13	Nonwa	4.28	8.00	1.87	Active	Low sensitivity
14	Bori	6.41	9.7	1.86	Active	Low sensitivity
15	Taabaa	10.08	10.60	1.05	Normal	Low sensitivity

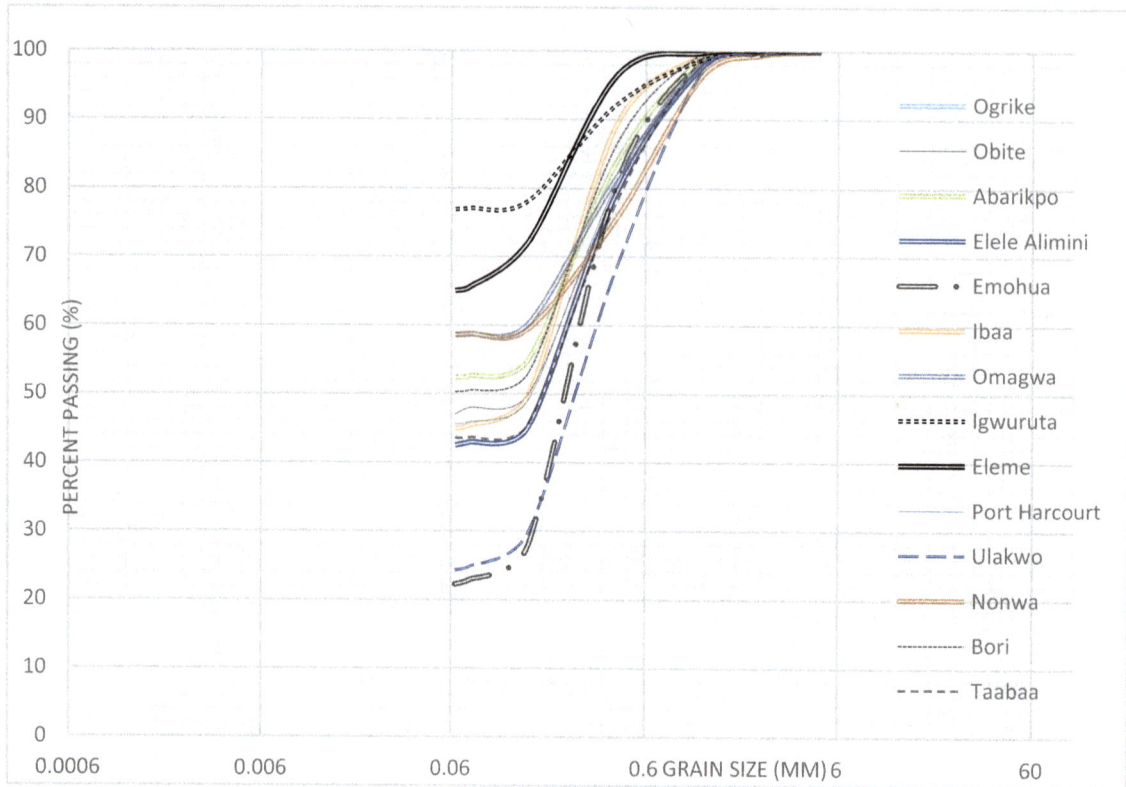

Figure 2. Particle size distribution of the soils

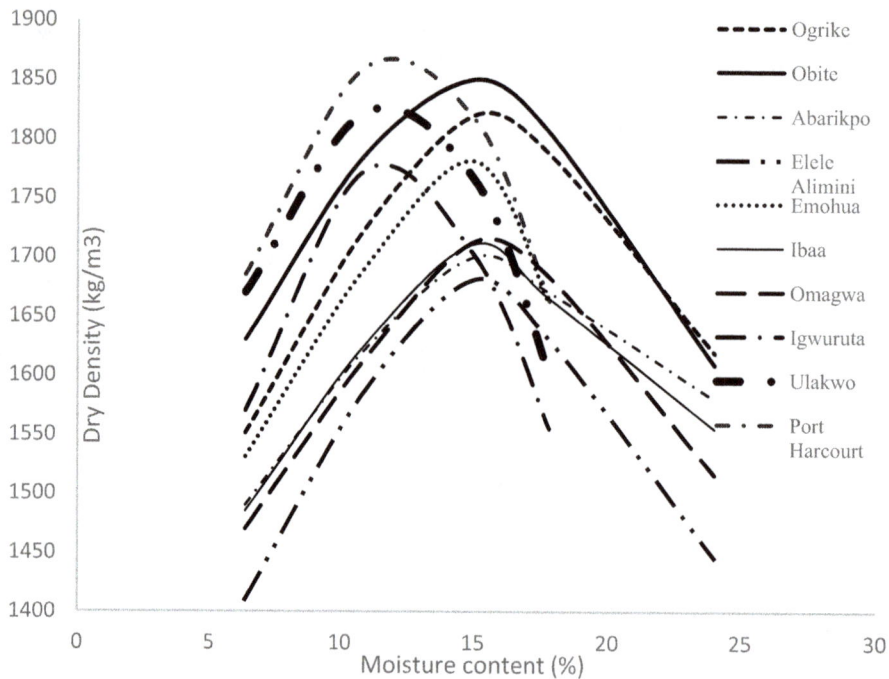

Figure 3. Density-moisture content relationships

5. Discussions

The studied soils are rich in oxides of silica, iron and aluminum, therefore the red color is derived from the presence of iron oxides. The relation between silica and oxides of iron and aluminium, is known as silica-sesquioxide ratio (Rossiter 2004). This ratio is extensively used as a criterion for the classification of soils as laterites. Soils are classified as true laterites if the ratio is less than 1.33 but ratios between 1.33 and 2.00 classify as lateritic soils whereas those with ratio above 2.00 are classified as non-lateritic tropically weathered soils. When the ratio was computed, the range of values fell between 3.09 and 4.37 indicating that the soils are non-lateritic. This agrees with the results of Akpokodje (1986) and Alayaki et al (2015) which show that the red deltaic soils are non-lateritic. The large amount of silica in proportion to the alumina indicates poor lateritisation (Rossiter 2004). Increase in sesquioxides ratios result in decrease in cation exchange capacity and moisture retentivity of soil. Sesquioxides accumulate in soil profiles as a result of leaching of silica. According to Gidigasu (1976), the high aluminum oxide content compared to iron oxides in soils is one of the features common with soils formed in a continuously wet climate characteristic of the Niger Delta. A higher degree of lateritisation is associated with higher specific gravity values which, for laterites, range from 2.50 to 3.60 (De Graf Johnson 1972). Also, laterite evolution increases with iron content and decreases in alumina content. These indicators imply that the soils in this study have not undergone any appreciable degree of lateritisation, given their relatively low specific gravity values of 2.65 to 2.68 and also the relative enrichment of aluminium oxide over iron oxides.

The more the fines proportion in a soil, the poorer the quality as a construction material. Fine grained soils in the silt and clay class may undergo volume changes on contact with water which results in weakening of the soil structure and consequent reduction in overall strength. The grading characteristics of soils are important factors for soils to attain high dry density when compacted. Well graded soils achieve very high density when compacted at the appropriate optimum moisture content. The larger the plasticity index of a soil, the greater the engineering problems associated with the use of the soil as an engineering material (e.g. foundation support, road sub-grades, etc.). Pavement waviness, a type of road failure which results from plastic flow of the underlying wet soil on application of axle load is common among the major paved roads in the study area. It occurs from the high fines content of the soils. The Nigerian Federal Ministry of Works and Housing specifications (1997) require that sub-base materials should have maximum values of 30% for liquid limit and plasticity index of 12%. However, studies of superficial soils used in pavement construction in the Niger Delta have identified high fines content in the soils as one of the leading factors that promote early pavement failure in the region. The classification scheme developed by AASHO is a very useful guide to determine the quality of soils as subgrade materials for pavement construction. It utilizes texture and plasticity to rate the quality of a soil as subgrade material, decreasing from A-2 to A-7. The soils fall within the silt-clay class of sub-groups A-2-4 to A-7-6 of the AASHO classification scheme, with majority classifying in the A-2-7 and are thus rated as poor sub-grade materials. This is because texturally they are A-2 soils but are A-7 in plasticity. It implies they can achieve low to moderate performance as sub-grade materials. The Nigerian Federal Ministry of Works (1997) specifies minimum values of soaked and unsoaked CBR at 10% and 15% respectively for sub-grade soils compacted using standard proctor method. This means soaking will cause reduction in the strength of the soils because the soaked CBR falls below the minimum requirements. Under unsoaked conditions, the soils will retain the field compactive strength. On the other hand, the CBR should be less than 30% after 24 hours of soaking at OMC indicating that the soils are not suitable for the sub-base. Under quick unconsolidated undrained conditions, triaxial test gave cohesion ranging from 20 to $180kN/m^2$ and frictional angles of 6 to 23°. To establish the relationship between the various geotechnical properties of the soils, a regression analysis was attempted. Results summarized in Table 4 indicate that there is high positive correlation between unsoaked CBR and OMC (r = 0.77), unsoaked CBR and MDD (r = 0.83), soaked CBR and MDD (r = 0.08) and shear strength and MDD (r = 0.71). This implies an increase in one variable is accompanied by an increase in the other and indicates a direct proportional relationship. However, a highly negative correlation exists between soaked CBR and OMC (r = -0.77), and between shear strength and OMC (r = -0.58). This means an increase in one variable is accompanied by a decrease in the other, indicating an inversely proportional relationship. The overall implications of these results is that although these non-lateritic soils possess fair to good engineering properties, considerable strength reduction will be experienced under wet conditions. Due to high precipitation prevalent in the Niger Delta, adequate drainage at the shoulders of the roads will be imperative during construction to forestall the loss of subgrade strength due to ingress of water into the pavement

Table 4. Bivariate relationship between some geotechnical properties

Parameter	Regression Equation	r
Soaked CBR vs OMC	$y = 27.77 + 9 (1.52) x$	-0.77
Unsoaked CBR vs OMC	$y = 18.70 + (-0.02) x$	0.77
Soaked CBR vs MDD	$y = -30.10 + (0.02) x$	0.08
Unsoaked CBR vs MDD	$y = -108.4 + (0.07) x$	0.83
Shear Strength vs OMC	$y = 650.81 + (-32.44) x$	-0.58
Shear Strength vs MDD	$y = -751.86 + (0.52) x$	0.71

The classification scheme developed by AASHO is a very useful guide to determine the quality of soils as subgrade materials for pavement construction. It utilizes texture and plasticity to rate the quality of a soil as subgrade material, decreasing from A-2 to A-7. The soils fall within the silt-clay class of sub-groups A-2-4 to A-7-6 of the AASHO classification scheme, with majority classifying in the A-2-7 and are thus rated as poor sub-grade materials. This is because texturally they are A-2 soils but are A-7 in plasticity. It implies they can achieve low to moderate performance as sub-grade materials. The California Bearing Ratio (CBR) test is used in the empirical estimation of the bearing capacity of sub-grade and sub-base materials under soaked and dry conditions. The soaked and unsoaked CBR values range between 3-9% and 14-38% respectively, and their relationship is graphically shown in Fig. 4.

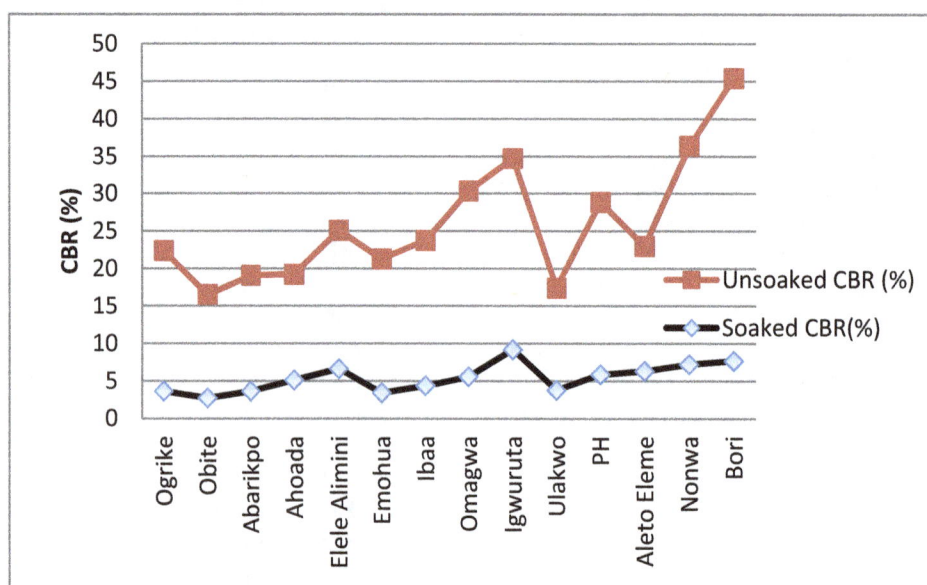

Figure 4. Relationship between unsoaked and soaked CBR

The Nigerian Federal Ministry of Works (1997) specifies minimum values of soaked and unsoaked CBR at 10% and 15% respectively for sub-grade soils compacted using standard proctor method. This means soaking will cause reduction in the strength of the soils because the soaked CBR falls below the minimum requirements. Under unsoaked conditions, the soils will retain the field compactive strength. On the other hand, the CBR should be less than 30% after 24 hours of soaking at OMC indicating that the soils are not suitable for the sub-base. Under quick unconsolidated undrained conditions, triaxial test gave cohesion ranging from 20 to 180kN/m^2 and frictional angles of 6 to 23°. To establish the relationship between the various geotechnical properties of the soils, a regression analysis was attempted. Results summarized in Table 4 indicate that there is high positive correlation between unsoaked CBR and OMC (r = 0.77), unsoaked CBR and MDD (r = 0.83), soaked CBR and MDD (r = 0.08) and shear strength and MDD (r = 0.71). This implies an increase in one variable is accompanied by an increase in the other and indicates a direct proportional relationship. However, a highly negative correlation exists between soaked CBR and OMC (r = -0.77), and between shear strength and OMC (r = -0.58). This means an increase in one variable is accompanied by a decrease in the other, indicating an inversely proportional relationship. The overall implications of these results is that although these non-lateritic soils possess fair to good engineering properties, considerable strength reduction will be experienced under wet conditions. Due to high precipitation prevalent in the Niger Delta, adequate drainage at the shoulders of the roads will be imperative during construction to forestall the loss of subgrade strength due to ingress of water into the pavement layers below the wearing surface.

6. Conclusions

The non-lateritic, tropical red soils in the dry flat lands and plains of the Niger Delta are composed of generally poorly graded soils consisting of clay, silt and sand. They yield moderate dry density when compacted at the optimum moisture content of the standard proctor method and are fair to good back fill material for pavement construction. The CBR values indicate significant strength reduction of the soils when soaked. Therefore adequate drainage must be provided during road construction to prevent the ingress of water into layers under the pavement. Generally, the poor soil properties may be improved by stabilization to increase their usefulness in pavement construction.

References

Adeyemi, G. O. (2002). Geotechnical properties of lateritic soils developed over quartz schist in Ishara area, Southwestern Nigeria. *Journal of Mining and Geology*, 38, 65-69

Adeyemi, G. O., Afolagboye, L.O., & Chukwuemeka, C. A. (2015). Geotechnical properties of non-crystalline coastal plain sand derived lateritic soils from Ogua, Niger Delta, Nigeria. *African Journal of Science, Technology, Innovation and Development,* 7(4), 230-235. http://dx.doi.org/10.1080/20421338.2015.1078105

Akpokodje, E. G. (1986): The Geotechnical Properties of Laterite and Non-Laterite Soils of South-Eastern Nigeria and their Evaluation for Road Construction. *Bull. Int. Assoc. Eng. Geol., 33*, 115 – 121.

Akpokodje, E. G. (1989). Preliminary Studies on the geotechnical characteristics of Niger Delta subsoils. *Engineering Geology, 26*, 247-257.

Alabo, E. H., FitzJohn, W. H., & Ogaree, F. A. (1984). Geotechnical Index properties of a Tropical Red Soil from parts of the Eastern Niger Delta, Nigeria. *Journal of Mining and Geology, 21*(1& 2), 35-39.

Alayaki, F., Al-Tabbaa, A., & Ayotamuno, J. (2015). Defining Niger Delta Soils: Are They Laterites? *Civil and Environmental Research, 7*(5), 21-26

Assez, L. O. (1976). *Review of Stratigraphy, Sedimentation and Structure of the Niger.* Kogbe (Ed.) Geology of Nigeria. Elizabethan pub. Co. Lagos. 436pp.

ASTM, American Society for Testing and Materials. (1992). *Standard Method for Classification of Soils for Engineering Purposes (Unified Soil Classification System):* D2487, American Society for Testing and Materials publication.

British Standards Institution-BS 1377. (1990). *Methods of testing for soils for civil engineering purposes.* BSI London

Carvalho J. C, de Rezende, L. R., Cardoso, F. B., Lucena, L. C., Guimarães, & Valencia, R. C. (2015). Tropical soils for highway construction: Peculiarities and considerations. *Transportation Geotechnics, 5,* 3–19

De Graft-Johnson, J. W. S. (1972). Lateritic gravel evaluation of road construction. *J. soil Mech. Div. Amst. Soc. Civil Eng.*, 98, 1245–1265.

Doust, H., & Omatsola, E. (1990) *Niger Delta:* in J. D. Edwards and P.A. Santogrossi, Eds. Divergent/passive margin basins: AAPG Memoir, 48, p. 239-248.

Elarabi, H., Taha, M., & Elkhawad, T. (2013). Some geological and geotechnical properties of lateritic soils from Muglad Basin located in the South-Western Part of Sudan. *Research Journal of Environmental and Earth Sciences, 5*(6), 291-294

Elsharief, A. M., Elhassan, A. A. M., & Mohamed, A. E. M. (2013). Lime Stabilization of Tropical Soils from Sudan for Road Construction. *Int. J. of GEOMATE,* 4 (2) (Sl. No. 8) 533-538

Federal Ministry of Works and Housing. (1997). *Nigerian General Specification for Roads and Bridges (Revised Editio*n), 2, 137-275.

Gidigasu, M. D. (1976). *Laterite soil Engineering.* Elsevier Scientific Publishing Company, Amsterdam Oxford, New York.

Gidigasu, M. D. (1983). Development of acceptance specifications for tropical gravel paving materials. *Eng. Geo.,* 19, 213–240.

Kamtchueng B. T., Onana, V. L., Fantong W. Y., Ueda A., Ntouala R. F. D., Wongolo, M. H. D., Ndongo, G. B., Ze, A. N., Kamgang, V. K. B., & Ondoa, J. M. (2015). Geotechnical, chemical and mineralogical evaluation of lateritic soils in humid tropical area (Mfou, Central-Cameroon): Implications for road construction. *International Journal of Geo-Engineering, 6*, 1 http://dx.doi.org/10.1186/s40703-014-0001-0

Nwajide, C. S. (2013). Geology of Nigeria's Sedimentary Basins. CSS Bookshops Limited. Lagos.

Odigi M. I. (2007). *Facies Architecture and Sequence Stratigraphy of Cretaceous Formations, Southeastern Benue Trough, Nigeria (*Unpublished PhD Thesis). University of Port Harcourt, Nigeria.

Ola, S. A. (1982). Mechanical properties of concretionary laterites from rainforest and savannah zones of Nigeria. *Bull. Of Int. Assoc. of Eng. Geol., 21*, 21-26

Omotosho, O. (2015). *Effects of Stabilisation on the performance of Deltaic Lateritic Soils as a Road Pavement Material.* University of Nigeria Institutional repository. Theses and Dissertations. Retrieved from http://repository.unn.edu.ng:8080/jspui/handle/123456789/1117

Omotosho, O., & Eze-Uzomaka, O. J. (2008). Optimal stabilization of deltaic laterite. *J. S. Afr. Inst. Civil Eng., 50*(2), 10-17.

Reijers, T. J. A. (2011). Stratigraphy and Sedimentology of the Niger Delta. *Geologos*, 17(3), 133-162.

Reijers, T. J. A., Petters, S. W., & Nwajide, C. S. (1997). *The Niger Delta Basin.* In R. C. Selley (Ed.): African basins. *Sedimentary Basins of the World (Elsevier, Amsterdam), 3*, 151-172

Rossiter, D. G. (2004). Digital soil resource inventories: status and prospects. *Soil Use & Management, 20*(3),296–301

Short, K. C., & Stauble, A. J. (1967). Outline of the geology of the Niger Delta, *Bull. Americ. Ass. Petrol. Geol., 51*, 761 – 779.

Skempton, A. W. (1953). *The colloidal activity of clays.* Proceedings of the third International Conference of soil Mechanics and Foundation Engineering. *Zurich, 1*, 57-60

Teme, S. C. (1999). Some geotechnical considerations for siting of land jetties (Ferry Terminals) in the Niger Delta region, Nigeria. 35th Annual Conference of Nig. Min. and Geosc. Soc. Programme and Book of abstracts. P. 29.

Teme, S. C. (2002). *Geotechnical considerations on foundations in the Niger Delta.* Paper presented at the special technical session, 39th Annual International Conference of the Nigeria Mining and geosciences Society, NMGS. Port Harcourt, Nigeria.

Tse, A. C., & Akpokodj, E. G. (2010). Subsurface Soil Profiles in Site Investigation for Foundation Purposes in Parts of the Mangrove Swamps of the Eastern Niger Delta. *Journal of Mining and Geology, 46*(1), 79-92

Ugbe, F. C. (2011). Basic Engineering Geological Properties of Lateritic Soils from Western Niger Delta. *Research Journal of Environmental and Earth Sciences, 3*(5), 571-577

3

Landscape Changes and Fragmentation Analysis in a Guinea Savannah Ecosystem: Case study of Talensi and Nabdam Districts of the Upper East region, Ghana

Steve Ampofo[1], Isaac Sackey[2] & Boateng Ampadu[1]

[1] Department of Earth and Environmental Sciences, University for Development Studies (UDS), Ghana

[2] Department of Applied Biology, University for Development Studies (UDS), Ghana

Correspondence: Steve Ampofo, Faculty of Applied Sciences (FAS), University for Development Studies (UDS), P. O. Box 24, Navrongo, Ghana. E-mail: sampofo@uds.edu.gh

Abstract

Landcover change is an observed natural change dynamics at both the local and regional levels. However, its scales are exacerbated by human interaction with its natural environment. The study examines these spatio-temporal changes in landcover and the level to which the change is accompanied by fragmentation of the identifiable cover types in the Talensi and Nabdam districts in Northern Ghana. The research uses digital classification of Landsat satellite imagery for 1999 and 2007 to produce the cover types which results in good accuracy levels of 66.39% and 63.03% respectively. Fragmentation analysis of the landscape was computed using FRAGSTATS® software for categorical maps obtained from the classified landcover maps for the two years. All cover types increased marginally. However, Bare areas decreased by as much as 17.17% and that of water decreased from 3% to 1%. The changing landscape involving conversions within and among various cover types is accompanied by fragmentation in all classes but more pronounced in the Bare class. The Bare class type which has more patches corresponds to the class with increased cover size and rather strangely decreases in the mean path size.

Keywords: Fragmentation, Landscape, Landcover/landuse change, Patches, Class

1. Introduction

The presence of humans on earth and their interaction with nature is constantly changing landscapes and is having a great effect on the natural environment. The changes in landscapes are in response to socio-economic development such as agriculture, mining, logging and construction. These are primary components of many current environmental concerns of land use and land cover changes (Hong, 1999). Land use/land cover change is gaining recognition as a key driver of environmental change. Concerns about land use/land cover change emerged in the research agenda on global environmental change several decades ago with the realisation that land surface processes influence climate (Southworth and Munroe, 2004). Changes in land cover can be attributed to natural factors such as change in climatic conditions and geomorphologic processes for instance earthquakes and volcanoes (Alves and Skole, 1996). In recent times, although natural factors contribute to land cover changes, the interaction of humans with the natural environment to improve livelihoods, have transformed land use and consequently land cover (Ampofo et al., 2015).

Land cover change and landscape fragmentation in the Talensi and Nabdam Districts in Ghana is occurring as a result of anthropogenic activities. Change in land cover of natural landscapes to satisfy human needs is one of the apparent modifications resulting from human activities (Bizimana et al., 2004). Talensi and Nabdam Districts are found to be in a fragile ecology prone to perennial bushfires and other high socio-economic extractions despite its enormous and varied natural resources endowments like vast large tracts of arable land for cropping and rangelands, rock and mineral deposits, forest reserves, and rivers. However, these rich resources are being adversely impacted upon by anthropogenic activities, particularly mining, logging (Frimpong, 2011), poor farming practices and expansion (Vogt et al., 2007; Bizimana, 2004), quarrying, road transport infrastructure development (Seiler et al., 2006) leading to a potentially irreversible land cover change and fragmentation of landscapes (Talensi-Nabdam District SEA, 2012).

Landscape fragmentation is a product of the interaction between ecological and socio-economic processes and cannot be ignored in the process of enumerating the effects of land use/land cover (LULC) change. Humans have actively managed and transformed the world's landscape for millennia (Ampofo, 2008; Bizimana, 2004). Population growth and migration to areas deemed favourable for agriculture is a major concern in tropical regions worldwide because of the resultant deforestation and landscape fragmentation (Berry et al., 1994).

The consequence of landscape fragmentation is a heterogeneous landscape, composed of smaller and isolated patches of habitats within a matrix less suitable for species that lived there originally (Fariq, 2003; Berry et al., 1996). A good understanding of the causes and effects of land use/land cover change and landscape fragmentation will help policy makers gain insight into the issues that contributes to the problem and decide which management practices should be instituted to improve on landscape ecology. According to Dwivedi & Sreenivas. (2005) and Honnay et al. (2003), fragmentation of land and such corresponding potential threats as global climatic change, land use dynamics, change in biogeochemical cycles, and loss of habitats and biodiversity have currently become a central issue. This paper therefore investigates the dynamics of landcover change using multi-temporal Landsat satellite imagery. Specifically, the paper evaluates landscape change in terms of the proportions and fragmentation using various indices, this is done to reveal the pattern and specific areas of change for land management purpose.

2. Materials and Methods

2.1 Study Area

Talensi and Nabdam Districts, with Tongo and Nangode as capitals, respectively, are newly created districts which were carved from the erstwhile Talensi-Nabdam District which was also carved from the Bolgatanga Municipality which has Bolgatanga as its capital and the regional seat of administration. The districts are bordered to the North by the Bolgatanga District, South by the West and East Mamprusi Districts, Kassena-Nankana District to the West and the Bawku West District to the East.

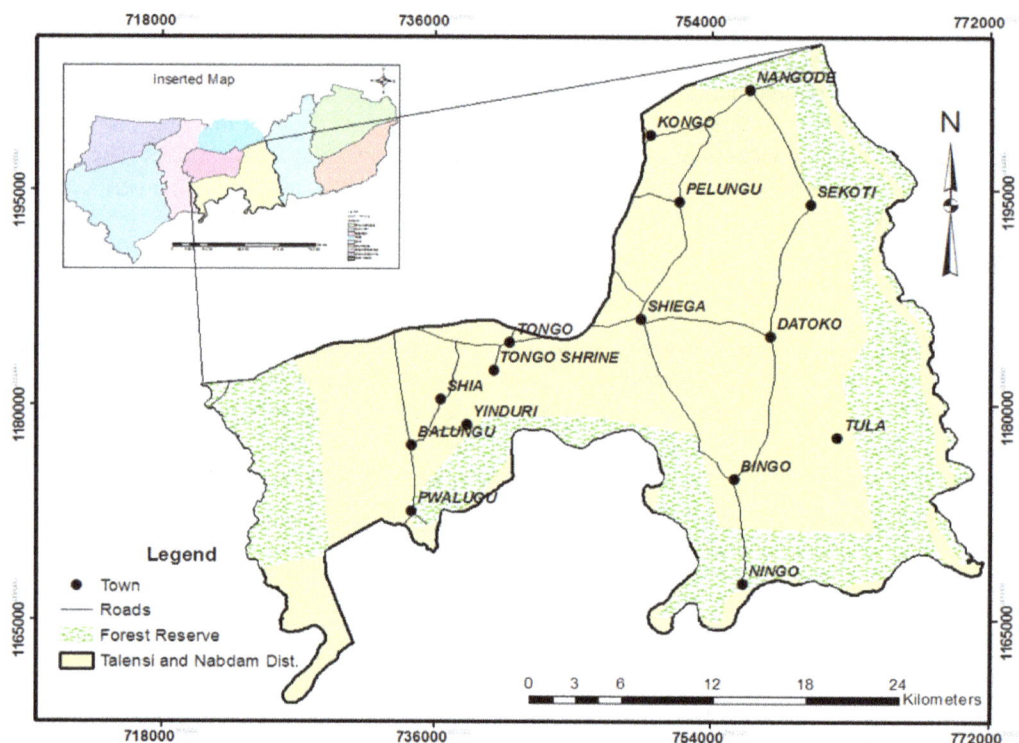

Figure 1. Map of Upper East Region of Ghana showing the location map of the study areas

2.1.1 Topography and Drainage

The districts have gentle slopes ranging from 1% to 5% gradient with some isolated quartzite and gneiss rock formations along some hilly terrain which have upland slopes of about 10% at the Tongo and Nangodi areas. The districts are drained mainly by the Red and White Volta and their tributaries. These physical characters have given rise to dry season farming activities along some parts of the Volta basin stretches of the districts. Similarly, the existence of mineral deposits has resulted in the proliferation of small scale artisanal mining activities in the districts and the rock outcrops have also resulted in the establishment of quarries.

2.1.2 Climate

The climate is classified as tropical, and has two distinct seasons, a wet rainy season, which is erratic, and runs from May to October, and a long dry season that stretches from October to April with hardly any rains. The mean monthly rainfall ranges between 88 mm and110 mm and a mean annual rainfall of 950 mm. The area experiences a maximum temperature of 45° C in March and April and a minimum of 12° C in December (Talensi-Nabdam District Profile, 2011).

2.1.3 Forest and Ecosystem

The vegetation is guinea savannah woodland consisting of short, widely spread deciduous trees and a ground flora of grass, which gets burnt by fire or the scorching sun during the long dry season. The extreme temperatures and prolonged dry season facilitate bush burning which affects tree recruitment and survival processes and promote land degradation (Talensi-Nabdam District Profile, 2011). The major savanna forest cover types in the districts are usually classified based on whether the savanna is closed or open, cultivated or riverine, grassland and the tree density per hectare. These results in the following discernible categories (Figure 2):

Closed cultivated savanna woodland (< 20 trees/ha)

Grass/herb/without scattered trees (0-5 trees/ha)

Grassland with/without scattered tree/shrub

Open cultivated savanna woodland (11-20 trees/ha)

Open savanna woodland (< 25 trees/ha)

Riverine savanna vegetation

Widely open cultivated savanna woodland (6-10 trees/ha)

The riverine savanna vegetation is mostly found along the channels of the White Volta and Red Volta rivers and other low lying areas such as dam sites and natural depressions which accumulate water. On the vast expanse of rocky hills that traverses the districts are grassland, shrub and very few isolated trees, particularly on the range of hills referred to as Tongo hills.

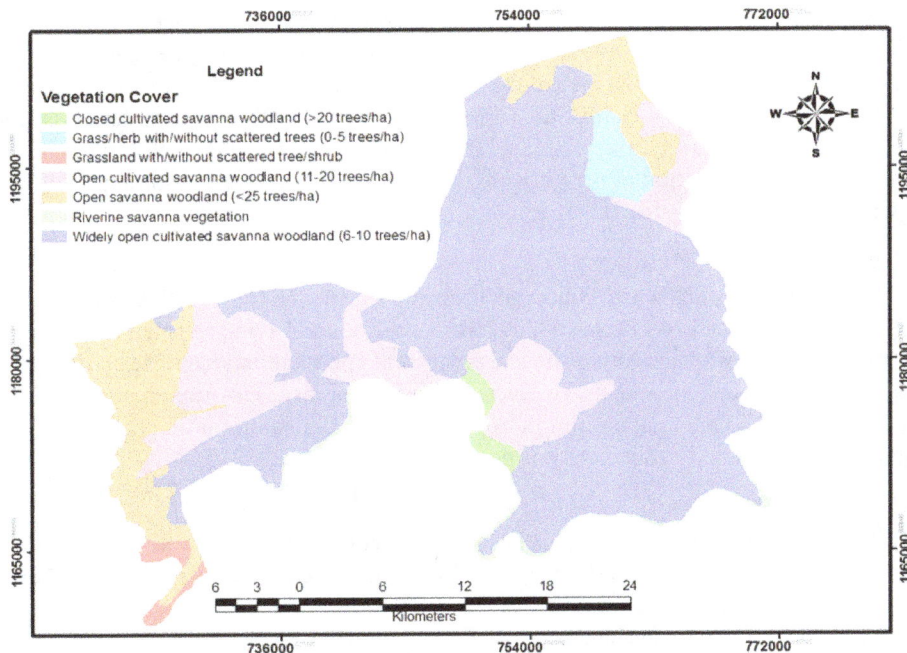

Figure 2. Vegetation map of the Talensi and Nabdam Districts

The natural environment is fairly degraded as it faces threat of severe drought with high temperatures and perennial outbreak of bush fires. It is evident that high population densities (especially in towns) with high demand for land for constructional activities, extensive cultivation, over-grazing, erratic rainfall and the extent of devastation do affect the natural environment thereby exposing it to desertification (Talensi-Nabdam District Profile, 2011).

2.2 Materials

The research is basically an analysis of spatial change; a pattern and trend analysis involving the use of multi-temporal Landsat satellite imagery. Landsat images of 1999 and 2007 with the same sensor and pixel resolution were used. The research materials used in this study and their characteristics are listed in Table 1.

Table 1. Satellite imagery and their acquisition dates

Satellite/sensors	Acquisition Date	Spatial Res.(M)	Remarks
Landsat TM	9th December, 1999	30	Dry Season
Landsat ETM	5th January, 2007	30	Dry Season

The processing of the satellite imagery was carried out using the ERDAS Imagine© software version 9.1. Some specific image processing operations were done using the ArcGIS® software version 10.1, while some fragmentation statistics of the land use/land cover types were computed using Fragstats® software version 3.3. However, Microsoft Office® Suite version 2010 and Microsoft Visio® version 2007 were used for data input and graphical representation.

2.3 Methods

2.3.1 Sampling Techniques

Stratified random sampling was employed to collect field data with an approximately equal sample size in each land cover class. In each stratum, a simple random sampling was used to collect training and accuracy assessment samples, which were used for the classification and the validation of the classification accuracy respectively. Global Positioning System (GPS) readings were used to record the training and validation samples representative of all identifiable cover types. All the data were stored as ArcMap shapefiles and used later in the data analysis process for the extraction of image spectra from the multi-spectral datasets (Ampofo et al., 2015; Frimpong, 2011).

During the data collection, the distributions of the Ground Control Points (GCPs) were taken into consideration and small sample points (GPS points) of a land use/land cover type were randomly selected in the field. This was done throughout the study area by moving around the feature with the GPS instrument and a total of 238 samples were collected and used for the study. Half of the GPS sample points (119) were used for classification and the remaining points were used for accuracy assessment. Topographic maps and a Global Positioning System (GPS) receiver were used to find the location of the GCPs in the field. The geographic co-ordinates of the observation sites from the GPS reading were recorded at an accuracy of ± 3 m.

2.3.1 Image Acquisition and Processing

Both Landsat Thematic Mapper (TM) and Enhanced Thematic Mapper (ETM) satellite images of the study area captured on 11th January 1999 and 24th February 2007, respectively, and having a spatial resolution of 30 m were acquired from the USGS global Visualisation site and processed. The satellite images selected and used were dry season images captured between the months of December and February. This was because of the difficulty in obtaining cloud-free images in the rainy season in tropical regions. It also meant that data on cover types in the cropping season were not captured as accurately as possible (Ampofo et al., 2015).

2.3.2 Image Geo-referencing

The Landsat ETM 1999 image was geo-referenced using GCPs collected in the dry season and this was used for an image-to-image registration of the other image using the same Area of Interest (AOI). This resulted in the same Root Mean Square Error (RMSE) of 0.015 a pixel. The following national topographic map projection properties were used for the geo-referencing:

Projection: Transverse Mercator

Datum: D_WGS_1984

Central Meridian: -3.000000

Scale factor at central meridian: 0.999600

False Easting: 500000.000000

False Northing: 0.000000

Latitude of Origin: 0.000000

Linear Unit: Meter

Figure 3 below illustrates the process flow for image proceing and analysis:

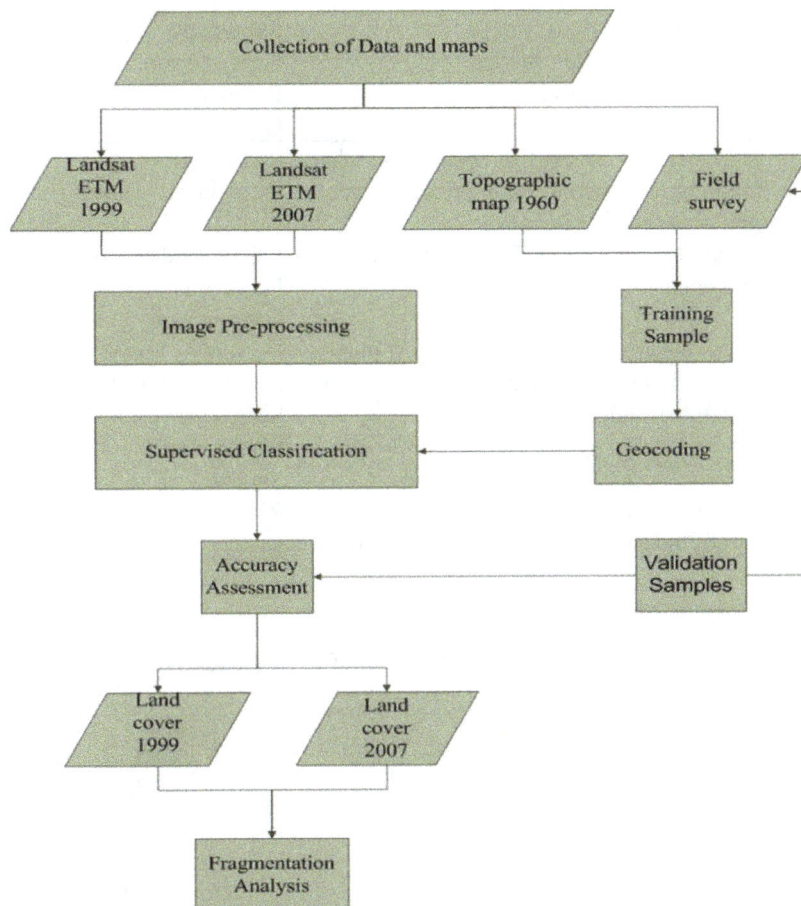

Figure 3. Process flow of the Research Methods

2.3.3 Image Classification

The main aim of image classification was to automatically categorize all pixels in an image into land cover classes (Figures 4a and 4b). The classification legend was made based on spectral characteristics of the multi-spectral dataset acquired. Supervised classification was done by following three stages that included training data sets, classification and output. Training samples were taken for each land use/land cover type to be classified in the image. To obtain true representation of the cover classes, spectra signatures were created by repeatedly selecting training samples, assessing and analysing them by either deleting or merging these spectra.

Using ETM bands 4, 3, 2 for both 1999 and 2007 and three true colour composites, a preliminary land cover map was obtained by visual interpretation. This band combination was chosen because it corresponds to the green reflectance of healthy vegetation and one of the most important bands for vegetation discrimination. In addition, it is also useful for soil-boundary and geological boundary mapping as well as responsive to the amount of vegetation biomass present in a scene. These characteristics of the band 4, 3, 2 combinations make it useful for identifying vegetation types, and emphasize soil-crop and land-water contrasts.

Classification was done by using Maximum Likelihood Classifier (MLC). This option evaluated to which class the pixel most likely belonged, based on the pixel value. The landuse and landcover classification was based on a scheme developed by Berry et al (1994) and further highlighted by Anderson et al. (2002), and was used because it suited the savannah ecosystem of the study area with its five cover classes. A classification scheme was developed of which the following five land use/land cover classes were distinguished: Bare areas, Built-Up area, Crop and Rangelands, Vegetation, and Water (Table 2). Supervised image classification method was employed to extract thematic information on the land covers from 1999 and 2007 satellite images which resulted in the production of the five (5) LULC types.

Table 2. Definition of LULC Classes

LULC Type	Description	Subgroups
Bare	Areas that have neither natural nor artificial vegetal cover	Sand and soils Exposed rock surfaces Gravels
Built-up	Area of land covered by houses or other buildings and increasing in intensity over a period of time	Settlements Other infrastructure
Crop and Rangeland	Areas of land used for cultivation and grassland	Fallow lands Farmlands Grazing land
Vegetation	Areas covered by trees of all ages, plants, and underbrush covering the large area within the study area.	Deciduous Forest Defoliated forest Sacred groves Mixed Forest Shrubs
Water	Includes all land areas holding water most part of the year including the dry season	Rivers Dams

(Source: Anderson et al., 2002)

a) 1999 Landsat TM-RGB: 4,3,2 (30 x 30) m   b) 1999 Landsat ETM-RGB, 4,3,2 (30 x 30) m

Figure 4. Landsat images of the study areas

2.3.4 Fragmentation Analysis

Change in the landscape structure was analysed using FRAGSTATS© version 3.3 software in ArcGIS® environment. FRAGSTATS® is a spatial pattern analysis programme for categorical maps that examines landscape characteristics by quantifying the areal extent and spatial configuration of patches and classes within a landscape. FRAGSTATS® is developed to quantify landscape structure with a comprehensive choice of landscape metrics that can be used to describe the characteristics of individual patches, classes of patches, or the entire landscape (Ampofo, 2008; Frimpong, 2011). This was done for the whole of the study area for the two years, 1999 and 2007.The two maps classified were exported in generic binary 8 bit formats in ERDAS

Imagine® and ran with the FRAGSTATS® program. At the class, the patch and the landscape levels, the following metrics were computed using a cell size of 30 metres: Number of Patches (NP), Patch Density (PD), Largest Patch Index (LPI), Percentage of Landscape (PLAND), Total Class Area (CA) and Mean Patch Size (MPS). Table 3 and Figure 5 illustrate the description of the fragmentation metrics and flowchart of the processes used in the production of the fragmentation metrics respectively.

Table 3. Description of Fragmentation metrics used

Metric	Description	Unit
Number of Patches (NP)	Number of patches (NP) is equal to total number of patches in each class.	None
Patch Density (PD)	Patch Density (PD) which measures the NP per 100 hectares is an index that measures the level of fragmentation within each class.	Number per 100 hectares
Largest Patch Index (LPI)	LPI equals the area (m^2) of the largest patch of the corresponding patch type divided by total landscape area (m^2), multiplied by 100 (to convert to a percentage) or Area of the largest patch in each class as a percentage of the total landscape area.	Percent
Percentage of Landscape (PLAND)	PLAND equals the sum of the areas (m^2) of all patches of the corresponding patch type, divided by total landscape area (m^2), multiplied by 100 (to convert to a percentage)	Percent
Class Area (CA)	CA equals the sum of the areas (m^2) of all patches of the corresponding patch type, divided by 10,000 (to convert to hectares); that is, total class area.	Hectares
Mean Patch Size (MPS)	Average Patch size in each class.	Hectares

The indices of LPI, NP and MPS give the estimation of the degree of fragmentation for different land cover types. NP, PLAND and CA in particular is an excellent measure of the fragmentation of a given class within the landscape; since the landscape size is constant, the greater the number of patches, the greater the degree of fragmentation. The comparison of these spatial characterisations therefore reveals the pattern of land cover modification and transformation for the classified images (Southworth and Munroe, 2004).

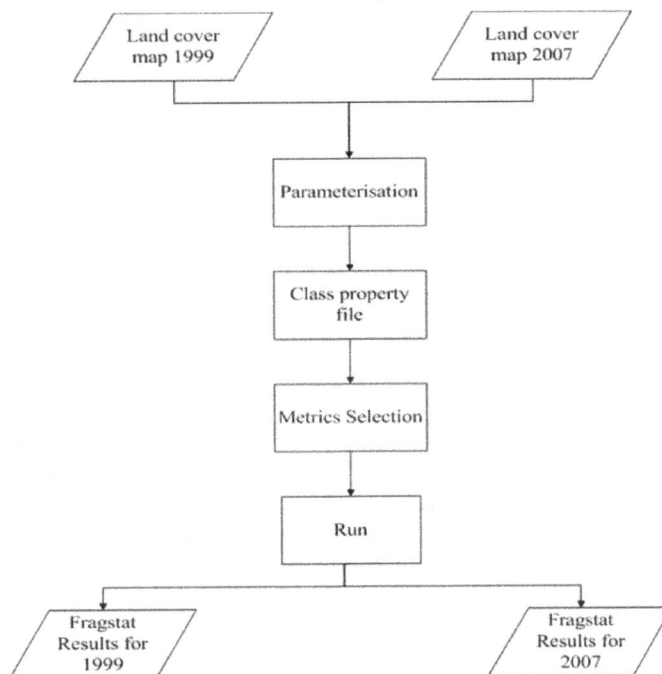

Figure 5. Process flow of Fragmentation analysis

3. Results and Discussions

3.1 Land Cover Classification and Accuracy Assessment

Tables 4 and 5 below illustrate the accuracy of the various LULC in the Talensi and Nabdam Districts Land cover maps for 1999 and 2007 (Figures 6 and 7). To observe the changes in the landscape, accuracy assessment of the classified image is an important step in image classification. The quality of a thematic map from a satellite image is determined by its accuracy. An accuracy assessment performed on the Landsat ETM 1999 and 2007 image resulted in an overall accuracy of 63.03% and 69.39%, respectively.

Table 4. Accuracy Assessment of 1999 Classification

Class Name	Reference Total	Classified Total	Number of Correct	Producers Accuracy (%)	Users Accuracy (%)	Kappa Statistics
Vegetation	25	10	8	32	80	0.7468
Built-up	43	39	34	79.07	87.18	0.7993
Bare Areas	28	49	25	89.29	51.02	0.3595
Crop and Rangeland	19	8	6	31.58	75	0.7025
Water	4	2	2	50	100	1
Total	119	119	87	Overall Accuracy 63.03%	Classification	Overall Kappa Statistics 0.5107

Table 5. Accuracy Assessment for 2007 Classification

Class Name	Reference Total	Classified Total	Number of Correct	Producers Accuracy (%)	Users Accuracy (%)	Kappa Statistics
Vegetation	28	14	13	46.43	92.86	0.9066
Built-up	41	54	35	85.37	64.81	0.4632
Bare Areas	27	22	16	59.26	72.73	0.6472
Crop and rangeland	19	16	13	68.42	81.25	0.7769
Water	4	2	2	50	100	1.0000
Total	119	119	79	Overall Accuracy 66.39%	Classification	Overall Kappa Statistics 0.5530

3.2 Land Use/Land Cover Analysis (LULC)

3.2.1 LULC Analysis in 1999

The spatial extent of the 1999 LULC map after the Supervised Classification yielded five (5) land cover classes (Figures 6) with the Bare areas occupying the highest percentage (49,909.9 ha, 51.62%). The Bare area is fairly distributed in the map with the highest concentration around Tula and Sekoti in the north-east and south-east, respectively (Fig. 6). The next LULC class with the highest area coverage is the vegetation (20,247.8 ha, 20.94%), which is concentrated around the south-eastern and south-western parts of the area with very small patches around the northern part of the study area. Crop and rangelands (11,690.2 ha, 12.09%) are the next in terms of area coverage and are mainly distributed around the central part of the study area and areas closer to the White Volta River to the southern boundary line. The Built-up area has 11,439.7 ha (11.83%) and its area coverage is concentrated at the western and northern part of the study area with some few patches around the southern and central areas. The last is water, covering 3396.78 ha (3.5%) which is also distributed fairly in the study area except the north where there is the Tongo hilly terrain.

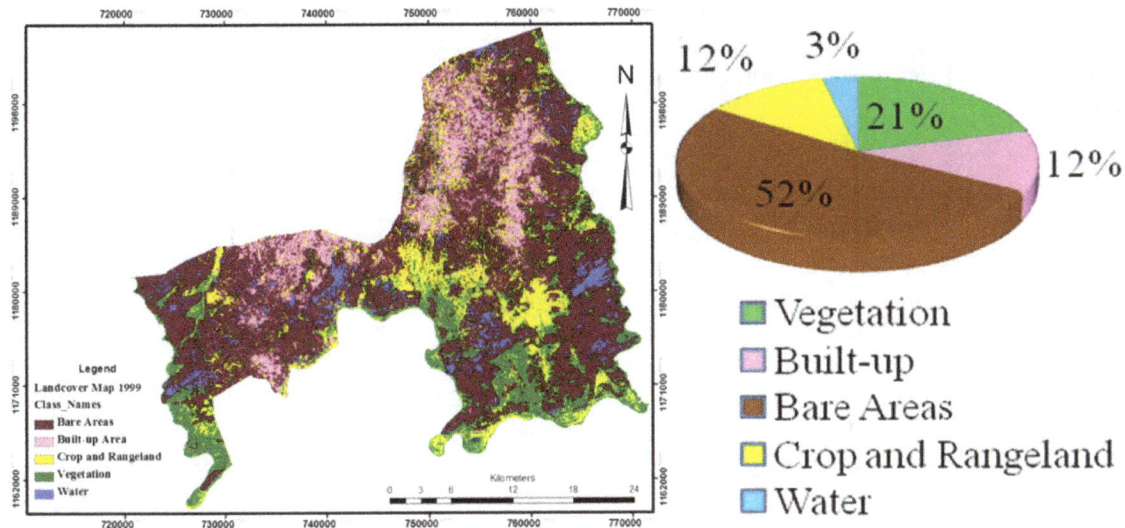

Figure 6. LULC map for 1999 and its cover proportion

3.2.2 LULC Analysis in 2007

The Supervised classification procedures used for the 2007 image yielded five basic land cover types with Bare Area having the largest area coverage of 33,310.89 ha (35%) as compared to the other LULC classes and largely concentrated around the southern and south-eastern part of the study area, with a few around north and western part of the area (Figure 7). Built-up areas constituted the second highest cover type with 24,566.94 ha (25%) and occurred mainly around the northern and western parts, with some few patches at southern and eastern zones. Vegetation, which comes next with 22,882.77 ha (24%), is mainly scattered within the study area with some regular patches at the western part of the study area. Crop and rangeland (14,541.84 ha, 15%) is also scattered across the study area and mixes with patches of vegetation but concentrated at the western and central parts of the study area. Water can be said to be fairly distributed, covering 1,381.86 ha (1.0 %) of the study area.

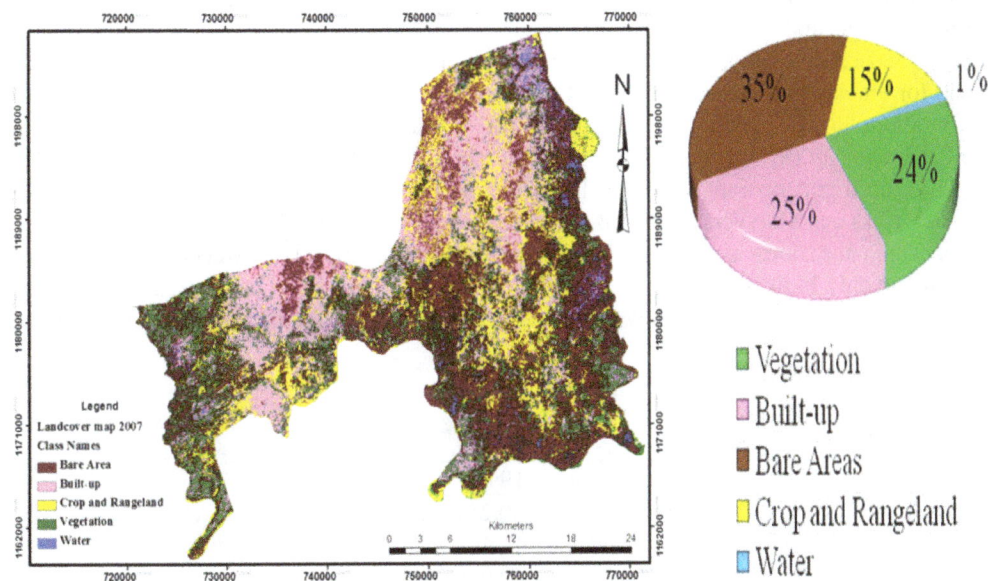

Figure 7. LULC map for 2007 and its cover proportion

3.2.3 Spatial Extent of LULC after Classification

The majority of the LULC changes took place within the Built-up and Bare areas while the remaining classes experienced slight changes over the 8-year period under study. Although, Bare areas covered the highest proportion of the land area in 1999 as in 2007, the results show that much of the Bare areas have been converted into Built-up areas due to settlement expansion. In 2007, crop and rangeland increased by 2,851 ha (20%) which corresponds to the increment in the Built-up areas; i.e., demands for land for housing and consequent settlement expansion as well as parallel demand for agricultural land. Table 6 and Figure 8 show LULC proportions for 1999 and 2007.

Table 6. Proportion of the various LULC types

Class Name	Area (ha) 1999	2007	Area (%) 1999	2007
Vegetation	20247.8	22882.77	20.94	23.66
Built-up	11439.7	24566.94	11.83	25.40
Bare Areas	49909.9	33310.89	51.62	34.45
Agriculture	11690.2	14541.84	12.09	15.04
Water	3396.78	1381.86	3.51	1.42
Total	96684.3	96684.3	100	100

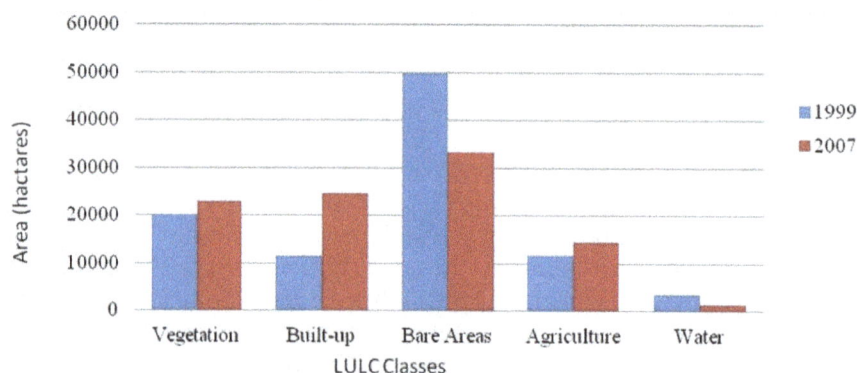

Figure 8. Graph of LULC Proportions in 1999 and 2007

3.3 Fragmentation Statistics of LULC Types within the Talensi and Nabdam Districts

The computed fragmentation statistics of the LULC types showed that significant changes in the structure of the landscape have occurred. The general pattern reveals a continuous change in the landscape structure which is revealed by the changing landscape metrics. From Table 6, it can be observed that within the 8-year period, the number of patches (NP) for all classes increased, with the Crop and rangeland increasing from 13,871 to 37,543 representing a change of about 46.04%. The number of patches for vegetation increased by 17,168 (20.69%) for the same period. The Patch Density (PD), which measures the number of Patches per 100 hectares, is an index which measures the level of fragmentation within each class. The higher the PD, the higher the level of fragmentation within each class and vice versa, and this increased for all classes from 1999 to 2007. However, the largest patch index which measures area of the largest patch in each class as a percentage of the total landscape area increased for only built-up area. This means more patches have been created and that is also reflected in the patch density of the various classes within the 8 years in the vegetation, crop and rangeland and bare areas (Table 7).

Table 7. Fragmentation Statistics for the various LULC types

Metric	LULC Type	1999	2007	Difference
Class Area (CA)	Forest	20,251	22,886	2,635
	Built-up area	11,441	24,568	13,127
	Bare Area	49,910	33,312	-16,599
	Cropland	11,692	14,543	2,851
Percentage Landscape (PLAND)	Forest	8.6	9.8	1.2
	Built-up area	4.9	10.5	5.6
	Bare Area	21.3	14.2	-7.1
	cropland	5.0	6.2	1.2
Number of Patches	Forest	32,902	50,070	17,168.0

(NP)	Built-up area	20,238	36,937	16,699.0
	Bare Area	13,871	37,543	23,672.0
	cropland	29,848	34,527	4,679.0
Patch Density	Forest	14	21	7
(PD)	Built-up area	9	16	7
	Bare Area	6	16	10
	cropland	13	15	2.0
Largest Patch Index (LPI)	Forest	0.5	0.03	-0.47
	Built-up area	0.1	0.76	0.24
	Bare Area	2.0	0.70	-1.30
	cropland	0.1	0.03	-0.07
Mean Patch Size	Forest	0.62	0.46	-0.16
(MPS)	Built-up area	0.57	0.67	0.10
	Bare Area	3.60	0.89	-2.71
	cropland	0.39	0.42	0.03

3.3.1 Class Area (CA), 1999 to 2007

Table 8 below shows that Class area (CA) of forest, Built-up, and Crop and rangelands increased by 2,635 ha, 1,327 ha and 2,852 ha, respectively, whilst Bare areas decreased by 16,599 ha from 1999 to 2007.

Table 8. Class Area metric from 1999 to 2007 for the various LULC types

Class Metric	LULC types	1999	2007	Difference
	Forest	20,251	22,886	2,635
	Built-up area	11,441	24,568	13,127
Class Area (ha)	Bare Area	49,910	33,312	-16,599
	Cropland	11,692	14,543	2,852

3.3.2 Percentage of Landscape (PLAND), 1999 to 2007

From Table 9, Percentage Landscape (PLAND) for only built-up increased, while the PLAND for bare area decreased considerably over the period of the study (Table 9). However, PLAND for crop and rangeland and vegetation experienced a slight increment within the period of study.

Table 9. PLAND metric for 1999 and 2007 for the various LULC types

Metric	LULC Type	1999	2007	Difference
	Forest	8.6	9.8	1.1
	Built-up area	4.9	10.5	5.6
Percentage Landscape (PLAND)	Bare Area	21.3	14.2	-7.1
	Cropland	5.0	6.2	1.2

3.3.3 Patch Density (PD), 1999 to 2007

Table 10 below illustrates the class level metrics for PD of the various LULC classes. PD for bare area, forest and built-up areas witnessed a significant increment; thus from 6, 14 and 9 patches per 100 hectares in 1999 to 16, 21 and 16 patches per 100 hectares in 2007 for bare areas, forest and Built-up areas, respectively. Cropland increased by two (2) patches per every 100 hectares over the 8-year period.

Table 10. Patch Density for 1999 and 2007 for the various LULC types

Metric	LULC Type	1999	2007	Difference
	Forest	14	21	7.0
	Built-up area	9	16	7.0
Patch Density (Patches/100ha)	Bare Area	6	16	10.0
	cropland	13	15	2.0

3.3.4 Largest Patch Index (LPI), 1999 to 2007

Table 11 shows that LPI for forest, bare area and cropland decreased over the 8 year period while built-up area increased from 1999 to 2007. This indicates that, Bare area and Croplands have undergone further fragmentation into smaller patches within the 8-year period.

Table 11. Largest Patch Index for 1999 and 2007 for the various LULC types

Metric	LULC Type	1999	2007	Difference
	Forest	0.5	0.03	-0.47
Largest Patch Index (LPI)	Built-up area	0.1	0.76	0.62
	Bare Area	2.0	0.70	-1.30
	cropland	0.1	0.03	-0.07

3.3.5 Number of Patches (NP), 1999 to 2007

The Number of Patches (NP) increased across all LULC, but a significant increment was observed with Bare Areas having the highest number of patches followed by Forest and Built-up areas and Cropland having the least increment. The increment order are 23,672 ha (63%), 17,168 ha (34%) 16,699ha (45%) and 4,679 ha (14%) for Bare areas, Forest, Built-up area and Cropland, respectively (Table 12).

Table 12. Number of Patch metric for the various LULC types

Metric	LULC Type	1999	2007	Difference
	Forest	32,902	50,070	17,168.0
	Built-up area	20,238	36,937	16,699.0
Number of Patches (NP)	Bare Area	13,871	37,543	23,672.0
	Cropland	29,848	34,527	4,679.0

3.3.6 Mean Patch Size

The Mean Patch Size (MPS) of the areas covered with vegetation and bare areas observed a decrease in the Mean Patch Size, whereas Built-up area witnessed a small increment (Table 13). However, this change in the Crop and rangeland was insignificant. This implies that, Vegetation and Bare areas are undergoing conversion and fragmentation into Cropland and Built-up area.

Table 13. Mean Patch Size for the various LULC types

Metric	LULC	1999	2007	Difference
	Vegetation	0.62	0.46	-0.2
Mean Patch Size (MPS)	Built-up area	0.57	0.67	0.1
	Bare Area	3.60	0.89	-2.7
	Cropland	0.39	0.42	0.0

3.4 LULC Mapping and Changes

The land cover maps (Figures 6 and 7) show the changes that have occurred in the Talensi and Nabdam Districts since 1999. The 1999 and 2007 classified image produced an accuracy of 63.03% and 66.39% with a Kappa Statistic of 0.5107 and 0.5530 respectively (Table 4 and 5). Crop and rangelands had the highest user's accuracy of 91% with Kappa statistic of 0.85, while bare area had the highest producer's accuracy of 71% with the lowest kappa of 0.40. A widely used, acceptable accuracy, which is accepted in LULC classification is 85% (Anderson et al., 2002). The low overall accuracy level is explained in the use of dry season images in mapping vegetation and agricultural cover. For example, a maize field or grassland in the dry season is very bare because there is no chlorophyll matter in the plant stocks which leads to misclassification (Ampofo et al., 2015).

3.5 Landscape Fragmentation Analysis

From 1999 to 2007, increase in vegetation areas was relatively small as compared to the other LULC types within the 8-year period of study, and this resulted in the increase of their Class Area (CA) (Table 6). However, the Number of Patches (NP) and the Patch Density (PD) for vegetation increased significantly, reflecting increased socio-economic activities that are carried out in the areas covered with vegetation, mainly illegal mining, agricultural expansion, settlement expansion and fuel wood harvesting. These activities, for instance, illegal small-scale mining, which require the removal of the top soil, was observed to be taking place under the forest trees. These activities explain the decrease in the Mean Patch Size (MPS) for Vegetation areas.

The CA of Bare areas decreased within the landscape from 1999 to 2007 by being converted to Built-up areas. This is proportional to PLAND and CA, but PD and NP also increased over the period which is attributable to its conversion to built-up area. The trooping of inhabitants to the study area for socio-economic purposes has resulted in built-up expansion over the 8 years.   The CA and the NP of croplands increased considerably over the 8-year period. The increase of croplands within the landscape is due to the population expansion in the study area. Largest Patch Index (LPI) reduced, whereas Mean Patch Size (MPS) of crop and rangelands remained the same from 1999 to 2007, which is an indication that the areas covered with vegetation are being fragmented into crop and rangelands. The LPI for the Built-up area increased considerably within the same period, whilst Largest Patch Index for the other LULC types decreased. This suggests that Built-up is increasingly becoming a dominant factor within the Landscape. The expansion is mainly at the northern part of the study area bordering to the Bolgatanga Municipality, as well as the western area with some few patches around the south-eastern part.

Over short periods (decades or months), natural disturbances, such as forest fires, floods, landslides and windstorms modify and fragment landscapes. In addition, landscapes are naturally fragmented by mountain ridges, rivers and roads (Laverty and Gibbs, 2007). However, landscape fragmentation in the Talensi and Nabdam Districts appears to have been triggered by its endowment with natural resources, such as mineral deposits. The informal sector of the economy of the districts dominates and therefore, there is high dependence on natural resources. As such, subsistence farming, illegal small-scale mining, firewood harvesting as well as charcoal burning are prevalent economic activities in the area. From Table 7, it can be observed that vegetation areas have increased over the period. However, this corresponded to an increase in the level of fragmentation, as depicted by the NP, PD and MPS.

The prevailing land tenure system contributes to the fragmentation of crop and rangelands, as land is constantly fragmented and given to individuals for farming and other purposes, including development of transportation tracks which eventually develop into roads (Seiler and Munroe, 2006). Once a landscape has been fragmented, the size of the remaining patches becomes a critical factor in determining the number and type of faunal species that can survive within them. For all faunal species, large or small, that cannot or will not cross or leave a patch, all requirements to complete their life cycle must be met within the patch; from finding food to mating partners. This is especially important for species with complex life cycles, each with distinct habitat requirements (Laverty and Gibbs, 2007). Fragmentation of landscapes also inhibits modern agriculture, as mechanised farming requires some reasonable patch of land which has not undergone fragmentation for operation.

4. Conclusion

The objective of this study was to analyse the landscape fragmentation of Talensi and Nabdam Districts in the Upper East Region using remotely sensed data and GIS based techniques. The relationships between anthropogenic activities and LULC types as well as the fragmentation of these LULC were investigated and various thematic maps were developed. The main LULC types identified in the Talensi and Nabdam Districts are coded into Bare areas, Vegetation, Built-up areas, Crop and rangeland, and water. The analyses revealed that the composition and configuration of the LULC within the catchment had changed significantly over the 8-year period of study. The Bare area loss and fragmented forest within the area of study could be explained by the LULC conversion for settlement purposes. This loss is attributed to the Built-up expansion in response to population increase. The studies have revealed that there have been significant changes in the LULC types as well as the natural landscape in the Talensi and Nabdam Districts.

References

Alves, D. S., & Skole, D. L. (1996). Characterizing land cover dynamics using multi-temporal imagery. *International Journal of Remote Sensing, 17*(4), 835-839.

Ampofo, S. (2008). *Agricultural expansion and landcover change in the Volta gorge area, Ghana.*, University of Twente, Enchede.

Ampofo, S., Ampadu, B., & Abanyie, S. K. (2015). Landcover change patterns in the Volta gorge area, Ghana: Interpretations from sattelite imagery. *Journal of Natural sciences Research, 5*(4), 71- 82.

Anderson, J. R., Hardy, E. E., Roach, J. T., & WItmer, R. E. (2002). *A Land use and Land Cover Classification System for use with Remote Sensor Data.* Sioux falls)

Berry, M. W., Hazen, B. C., MacIntyre, R. L., & Flamm, R. O. (1996). LUCAS: A system for monitoring land-use change [Electronic Version]. *IEEE Computaional Science and Engineering,*

Bizimana, C., Nieuwoudt, W. L., & Ferrer, S. R. D. (2004). Farm size, land fragmentation and economic efficiency in Southern Rwanda. *Agrekon, 43*(2).

Dwivedi, R. S., & Sreenivas, K. (2005). Land-use/land-cover change analysis in part of Ethiopia using Landsat Thematic Mapper data" *International Journal of Remote Sensing 26*(7), 1285-1287.

Fahrig, L. (2003). Effects of habitat fragmentation on biodiversity *Annual Review of Ecology, Evolution, and Systematics, 34*, 487-515.

Frimpong, A. (2011). *Application of Remote Sensing and GIS for Forest Cover Change Detection (A case study of Owabi Catchment in Kumasi, Ghana).*

Hong, S. K. (1999). Causes and consequence of landscape fragmentation and changing disturbance by socio-economic development in mountain landscape system in South Korea *Journal of Environmental Science, 11*(2), 181-187.

Honnay, O., Piessens, K., Van Landuyt, W., Hermy, M., & Gulinck, H. (2003). Satellite based land use and landscape complexity indices as predictors for regional plant species diversity. *Landscape and Urban planning, 63*, 241-250.

Laverty, M. F., & Gibbs, J. P. (2007). Ecosystem Loss and Fragmentation. *Lessons in Conservation* (I), 72-96.

Liu, Y. L., Jiao, L. M., & Liu, Y. F. (2010). Land use data generalisation indices based on scale and landscape pattern *The International Archives of the Photogrammetry, Remote Sensing and Spatial Information Sciences, 38*(2).

Parent, J., Civco, D., & Hurd, J. (2007). *Simulating future forest fragmentation in a Connecticut region undergoing suburbanisation.* Paper presented at the ASPRS 2007 Annual conference.

Seiler, A., & Folkesson, L. (Eds.). (2006). *Habitat fragmentation due to transportation infrastructure* (Vol. VTI Rapport 530 A). Stockholm: VTI Publications.

Southworth, J. D., & Munroe (2004). Land cover change and landscape fragmentation-comparing the utility of continuous and discrete analyses for a western Honduras region. *Agriculture, Ecosystems and Environment, 101*, 185-205.

Talensi-Nabdam District Planning Coordinating Unit, (D.P.C.U) (2011). *Talensi-Nabdam District Profile.* Tongo: Talensi-Nabdam District Assembly.

Talensi-Nabdam District Planning Coordinating Unit, (D.P.C.U) (2012). *Talensi-Nabdam District Strategic Environmrntal Assessment (SEA).* Tongo: Talensi-Nabdam District Assembly.

Vogt, P., Riiters, K. H., Estreguil, C., Kozak, J., Wade, T. G., & Wickham, J. D. (2007). Mapping spatial patterns with morphological image processing. *Landscape Ecology, 22*, 171-177.

4

Spatial Analysis of Land Surface - Vegetation Relationship in Mountainous Areas of the Tropics Using Srtm-3 Dem

Joel Efiong[1], Opaminola Nicholas Digha[1] & Obianuju Emmanuella Asouzu[2]

[1] Department of Geography and Environmental Science, University of Calabar, Calabar, Nigeria

[2] Department of Geography and Environmental Management, University of Port Harcourt, Port Harcourt, Nigeria

Correspondence: Joel Efiong, Department of Geography and Environmental Science, University of Calabar, P. M. B. 1115, Calabar, Nigeria. E-mail: joel_efiong@yahoo.com

Abstract

Digital elevation models (DEMs) have shown much potential for use in the extraction of land surface parameters and analysis of the relationship between land surface units and vegetation cover. However, there is lack of studies on the use of SRTM-3 DEM in vegetation studies of mountainous regions. This study is therefore an attempt to relate land surface parameters to vegetation cover in the Obudu mountain region using SRTM-3 DEM and Landsat data. Geomorphometric classification of the land surface was done using an unsupervised ISOCLUST algorithm while vegetation cover classification was done using the supervised approach based on the Maximum Likelihood algorithm. The resultant land surface units and vegetation cover maps were then related using grid-based statistic within the geographic information systems. The overall measure of difference between the two maps yielded a chi-square (d.f. = 24) = 1.9154, p > 0.05. This implies that there is no significant difference between the land surface units and the vegetation cover in the study area. This findings support the use of SRTM-3 for land surface and vegetation mapping where there is no higher quality data, or the cost of obtaining one is inhibitive; a situation that is faced by many developing economies like Nigeria. However, this results should be interpreted and used within the context of the uncertainty that is contained in the SRTM-3 DEM.

Keywords: Spatial analysis, land surface-vegetation relationship, Obudu mountain region, SRTM-3 DEM, GIS, geomorphometry

1. Introduction

It has been argued that one of the most important factors for the development of vegetation cover over any land surface is relief (Florinsky & Kuryakova, 1996). Moreover, literature (Hengl & MacMillan, 2009) has shown that information on relief is often sufficient to produce reliable vegetation map, if other factors are held constant. Hence, land surface parameters (elevation, aspect and slope, etc.) are commonly used in vegetation mapping (Florinsky & Kuryakova, 1996). These parameters could be combined to produce land surface units which are then related to vegetation cover. Although this could be achieved, the analysis of land surface parameters and land surface units derived from Shuttle Radar Topography Mission (SRTM)-3 DEM in relation to vegetation mapping in mountainous regions have not been adequately examined and documented.

Three main methodological approaches to analysing relationships between vegetation and environmental site factors are discernible in current Geographical Information Systems (GIS) literature (Höersch, Braun & Schmidt, 2002). These are those that analyse relationships between vegetation and (i) direct influence of environmental site factors, (ii) entire set of environmental site factors, be it direct or indirect, and (iii) direct influence of environmental site factors based on the assumption that such influence can be reveal by land surface parameters. With the third approach, land surface derivatives and maps could provide useful information on the vegetation cover over mountainous regions. Such could assist in land and forest resource management efforts in mountainous areas.

The importance of remote sensing in obtaining data for land cover classification, particularly in difficult terrains has been acknowledged in literature (Lillesand, Kiefer & Chipman, 2008; Mather & Koch, 2011). However, cloud cover has always posed difficulty to the visible and the near infra-red remote sensing in many regions and particularly in the tropics (Langford & Bell, 1997). An alternative to these remote sensors that take measurement within the visible and infra-red bands is radar; since wavelengths emitted by it are not attenuated by cloud cover and rain (Lillesand et al., 2008). Obtaining radar images is cost intensive and may be beyond the reach of many developing nations like Nigeria. However, the SRTM in 2000 has made the radar data available in DEM formats at no cost to the public. Land surface parameters can therefore be extracted from this DEM.   Such parameters can be used to characterise the land surface of an area from where relationships with other environmental variables can be made.

Bolstad, Swank and Vose (1998) noted the importance of the physical environment in determining the spatial diversity of the land surface of mountainous regions. Höersch et al. (2002) and Abbate, Cavalli, Pascucci, Pignatti and Poscolieri (2006) concluded that there is a relationship between the vegetation types and topography in mountain areas. Hence, land surface parameters such as slope, aspect, profile curvature and plan curvature can be considered as important inputs in spatial analysis and estimation of the distribution of vegetation cover in mountainous environments.

In a recent study (Adediran, Parcharidis, Poscolieri & Pavlopoulos, 2004), geomorphometric method that involved multivariate statistical analysis of topographic gradients proposed by Parcharidis, Pavlopoulos and Poscolieri (2001) was used to evaluate the morphological setting that surround each pixel of DEM along the eight azimuth directions in north-central Crete. Ten morphometric classes were isolated using iterative self-organising data analysis techniques (ISODATA) unsupervised classification. A quick estimation of the spatial distribution of similar morphologic units was therefore provided by this approach. The resultant units were then superimposed on land cover types in the study area, where the relative association between the morphologic units and dominant land cover types were determined.

Abbate et al. (2006) adopted the same method described by Parcharidis et al. (2001) in examining the relation between morphological setting and vegetation covers in a medium relief landscape in Central Italy. The results established mutual relations between vegetation types and land cover units through the assessment of corresponding analysis between the results of the classifications of land surface and vegetation cover. Camiz, Papgeorgiou, Poscolieri and Parcharidis (2013) used the same method in landform classification to examine correlation between landforms and ground deformation at Nisyros volcano in Greece.

Several other studies have been carried out in relation to topographic parameters and vegetation. Elumnoh and Shrestha (2000) reviewed many studies (Jones, Settle & Whyatt, 1998;   Janssen, Jaarsma & van der Linder, 1990; Palacio-Pricto & Luna-Gonzalez, 1996;   Cibula & Nyquist, 1987) on the integration of DEM data in land cover classification. Strahler, Logan and Bryant (1979) also used DEM data to describe terrain components in relation to spectral response, all making used of geomorphometric variables such as relief, slope, elevation, aspect and curvature. Civco (1989) was able to normalise Landsat TM data using DEM. Studies (for example, Howard & Mitchel 1985) have shown the strong influence of aspect on vegetation distribution in temperate regions. However, the use of SRTM-3 DEM in achieving this has not been demonstrated in literature, particularly for the tropical environment. A very recent study (Efiong, Eze, Digha & Asouzu, 2015) has however parameterised the land surface of a mountain region in the humid tropical environment using the SRTM-3 DEM, but no relationship of the land surface parameters to vegetation cover were made by the authors. This study therefore bridges this gap.

2. Conceptual framework

A strong relationship between soil and topography (relief) has been established since the work of Dokuchaev (1898), who first recognised the main driving factors of soil formation at a particular place to include, relief, climate, organism, parent materials and time. However, it was only in 1941 that Jenny (Jenny, 1941) was able to translate Dokuchaev observation into a mathematical model (soil catena) expressed as:

$S = f(c, o, r, p, t)$

where,

S   =   soil variable

c   =   climate

o   =   organism

r   =   relief

p   =   parent material, and

t   =   time.

This equation has however been extended (McBratney, Mendonsca-Santos & Minasny, 2003) to include geographic position and the neighbouring soil properties (Dobos & Hengl, 2009). The soil catena concept has also been extended to landform-vegetation studies.   Hengl and MacMillan (2009) re-presented the model as:

S, V = f(c, o, r, p, t),

Where,

V, is vegetation.

While human activities and natural disasters play key roles in the distribution of vegetation cover over any region, the actual vegetation cover in a particular area may be as a result of the interactions between these different components. However, the effect of topography in controlling vegetation distribution is never in doubt (Abbate et al., 2006). The model of vegetation-relief (topography) relationship could therefore be of the form (Modified from Hengl & MacMillan, 2009):

V = f (r, t)

where,

V   =   vegetation

r   =   relief

t   =   time

Hence, when all other factors (e.g. climate, parent materials, organisms etc.) are held constant, reliable vegetation maps could be produced from information on relief of appropriate accuracy.

3. Method of Study

3.1 Study Area

This study was conducted on the Obudu mountain region in Cross River State of Nigeria (Figure 1).   The area is a prolongation of the Cameroon Mountain into the Cross River plains (Ekwueme & Kröner, 2006; Efiong, 2011; Amuyou, Eze, Essoka, Efiong & Egbai, 2013). The area has a rugged topography with steep valleys, highlands and plateau. It extends over 4590km^2 between latitudes 6°00'N and 6°45' N, and longitudes 8°40'E and 9°30'E (Edet & Okereke, 2005). Elevation ranges between 150m and 1600m above the local topography which is over 200m above mean sea level. The plateau has an altitude of 1716m above sea level (Roderkirchen, 2002).

The study area encompasses tropical rainforest, montane and grassland ecosystems with many endemic species of flora and fauna (Rodenkirchen, 2002). It should be noted here that this is one of the two places in Nigeria with significant montane forest, the other being the Mambilla plateau. However, there is an increasing rate of deforestation in the study area.

Part of the plateau has been developed into a ranch resort with the longest cable car in Africa. There are other tourist attractions within the resort itself which makes it a tourist haven. It also hosts a yearly international mountain race competition in November.

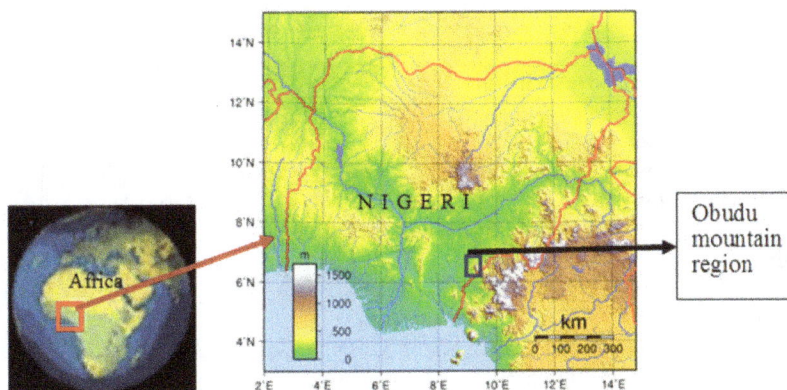

Figure 1.Topographic map of Nigeria showing the study area

Source: en.wikipedia.org

Temperature at the ranch varies between 23°C and 32°C during the dry season and between 4 and 10°C in some months during the wet season. Average rainfall is 4200mm per annum. With the above description, the area is therefore suitable for studies of this nature.

3.2 Data

SRTM-3 DEM and Landsat 8 satellite imagery served as the main data for this study. The SRTM-3 DEM was obtained from http://e4ftl01.cr.usgs.gov/. The specific information regarding the tile that contains the area of interest (AOI) is presented in Table 1. This AOI is the same that was used by Efiong et al. (2015).

Landsat 8 data was obtained from the USGS Earth Resources Observation and Science Center (EROS) via http://glovis.usgs.gov/ free of charge. It has a total of 11 bands: operational land imager (OLI) multispectral bands 1-7 and 9 at 30m spatial resolution, OLI panchromatic band 8 at 15m spatial resolution and the thermal infrared sensors (TIRS) bands 10 and 11collected at 100m but resampled to 30m spatial resolution (http://landsat.usgs.gov/landsat8.php).

Table 1. SRTM DEM data characteristics

Characteristic	Description
Pixel size	3 arc second (90 x 90 m)
DEM output format	HGT, signed 16 bits, in units of vertical metres
Special DN Values	N/A No voids in Version 3
Tile number	N06E009
Area of interest (AOI) (spatial extent)	Top: 714915; Left: 537645; Right: 543585; Bottom: 705195
Columns and rows	66, 108
Areal size	57.7368 km²
Statistics	
Pixel count	7128
Minimum elevation	688
Maximum elevation	1767
Mean	1322.595959596
Standard deviation	225.27588196913

However, bands 1-7 were used in this study since these are the bands where data were collected by the satellite sensor within the visible and near infrared regions of the electromagnetic spectrum. Basic characteristics of the data set of which the area of interest (AOI) was extracted from the stacked layers are shown in Table 2. This AOI has the same spatial extent with the SRTM-3 DEM.

Field work was done for 2 weeks in the month of June 2014 to allow for ground-truthing and collection of data for accuracy assessment. The field exercise involved the identification of the major vegetation cover, collection of elevation data of the land surface and location data for each of the identified vegetation cover type.

Table 2. Satellite image characteristics

Characteristics	Description
Satellite	Landsat 8
Date of acquisition	07/01/2014
WRS-2 Paths/Rows	187/56
Scene information	
ID	LC81870562014007LGN00
Quality	9 Product: OLI_TIRS_L1T
Data type	Unsigned 8-bit
Data format	TIFF
Spatial extent	Top: 714915; Left: 537645; Right: 543585; Bottom: 705195
Pixel size	30 m x 30m
Projection	UTM, zone 32
Spheroid	WGS 84
Datum	WSG 84

3.3 Data Processing

The generalised work-flow regarding the data analysis is presented in figure 2 and discussed in subsequent section.

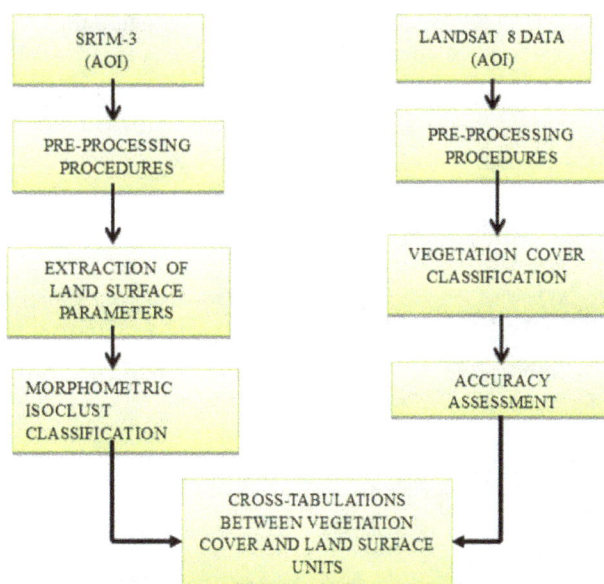

Figure 2. Generalised work-flow

For this study, care was taken in obtaining SRTM data that had been pre-processed to remove voids in the midst of so many versions of the product on the internet. The area AOI was then extracted from the entire tile. The data was re-projected to a projected coordinate system - UTM, zone 32 N with WGS 84 spheroid/datum. It was not possible to filter forest from the SRTM as data on forest height were not available.

Landsat data were also pre-processed before subsequent analysis were performed on it. First, the data were downloaded in bands and stacked. The area of interest (AOI) was then sub-setted from the entire stacked scene. There was no need for mosiaking since the AOI did not overlap another path/row. Haze reduction was performed on the image to reduced the hazy appearance of the image to allow for good visual interpretation. The image was geometrically corerected.

3.4. Data Analysis

3.4.1 Classification of Geomorphometric Units

Both the first (slope and aspect) and second (profile and plan curvatures) derivatives of land surface were extracted from the DEM the simple 3 x 3 window which is dragged across the gridded DEM points to estimate the value of the centre pixel using the neigbouring pixel values (Olaya, 2009)

Slope and aspect was calculated using the very local values (D8: 2-points, 8-direction, maximum-slope), that is, the eight 'Queens's case' neighbours (Horn, 1981). Curvature (profile and plan) were derived using the technique of quadratice surface provided by Zevenbergen and Thorne (1987). The extracted land surface parameters were classified based on a modified Young (1972) angle classification of slopes (Efiong et al., 2015). Curvature, including profile and plan curvatures, were classified into 3 classes based on positive (+), negative)-) or zero (0) curvature.

Since the major intention of this study was to link vegetation cover to geomorphometric units, the extracted land surface paramters were   stacked into an image which served as an input data for geomorphometric classification using an unsupervised ISOCLUST algorithm in 3 iterations with a minimum sample size of 30 pixels per class.

3.4.2 Classification of Vegetation Cover

Vegetation cover classification was done using the supervised approach based on the Maximum Likelihood algorithm on the already pre-processed Landsat data. The classifier was trained to identify four (4) spectral signatures of the major vegetation classes in the study area. Training samples allow for certain characteristics of the statistical nature of the vegetation cover types to be discriminated. For the algorithm that was adopted here, these samples allowed for the mean vector and variance-covariance matrix to be estimated for each class (Mather & Koch, 2011). Since the validity of statistical estimates from training samples depend on the size and how representative the sample is (Mather & Koch, 2011), a total of 120 training samples (30 for each class) was selected from every part of the satellite image. The identification of the classes of these samples was aided by the field data, spectral profiles and visual interpretation of the image. A signature file of these samples was generated and saved in the computer memory and was later used in the supervised classification.

Accuracy of the classified map was then assessed using the confusion matrix. The algorithm for the determination of accuracy of classification of k x k confusion matrix has been dicussed in literature (see Mather & Koch, 2011). In adopting this method, Erdas Imagine 2013 provided the necessary platform for the assessment. Figure 3 is the Erdas Imagine main viewer and the accuracy assessment viewer showing the classified map, class and reference data, together with the eastings and northings.

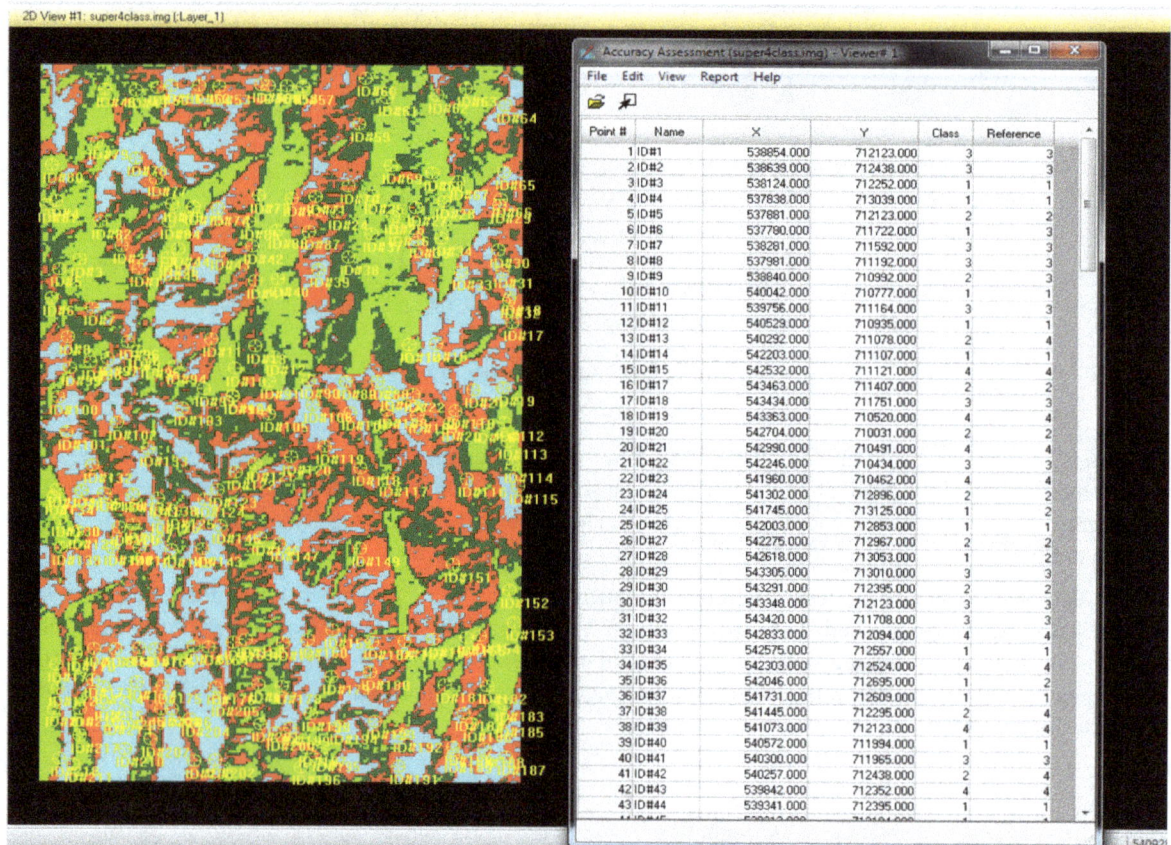

Figure 3. Erdas Imagine and accuracy assessment viewers

The kappa coeffiecient was used to summarised the information provided by the contingency matrix in statistical terms (Bishop, Fienberg & Holland, 1975). Kappa coefficient is literally computed based on the formula:

$$K = \frac{N \sum_i^k x - \sum_{i=1}^k x_i + x + i}{N - \sum_{i=1}^k n_i + x + i}$$

where K =   kappa coefficient

k =   number of classes

N   =   number of reference data samples

$\sum_i^k x =$ sum of the diagonal entries of the confusion matrix

$\sum_{i=1}^k x_i + x + i$ =   sum of the row and column marginal totals

Kappa's coefficient was interpreted based on the standard proposed by Landis and Koch (1977) (Table 3). See also Sim and Wright (2005).

Table 3. Standard for the interpretation of kappa coefficient

Kappa coefficient	Interpretation
≤ 0	Poor
$0.01 - 0.20$	Slight
$0.21 - 0.40$	Fair
$0.41 - 0.60$	Moderate
$0.61 - 0.80$	Substantial
≥ 0.80	Almost perfect

Source: Landis & Koch (1977)

3.4.3 Statistical Analysis

To allow for a grid-based statistical analysis between the geomorphometric units and vegetation cover distribution, the vegetation map was resampled from 30 x 30 m cell size, to 90 x 90 m using the bilinear interpolation method. Since vegeation data are categorical (nominal scale), landform parameters was also scaled down to nominal structure. Such uniformity allows for statistical calculation of correlations between the variables (Höersch, 2003).

There are various grid-based statistics, However, the Crosstab was used in this study. A standard Crosstab allows comparison to be made between two classified images and basically done by the use of Chi-square tests (de Smith, Goodchild & Longley, 2009). Results of cross tabulations are often shown in contingency tables (Abbate et al., 2006; Adediran et al., 2004; Cavalli, Fussili, Pascuici, Pignatti & Poscolieri, 2003; Höersch et al., 2002). Contingency tables are used to compare actual distributions with theoritical ones by means of chi-sqaure (x^2) tests. Chi-Square statistic was computed based on the form:

$$x^2 = \sum \frac{(O-E)^2}{E}$$

Where,

O = Observed frequency

E = Expected frequency

To indicate independence between land surface (geomorphometric) and vegetation cover maps, the chi-sqauare value should be low. Moreover, p should be close to 1, to indicate that the two images are likely similar. Further, two indices of overall (global) similarity (Kappa Index of Agreement and Cramer's V Index), which also are measures of correlation were used to establish the degree of similarity between the two images.

The Kappa Index of Agreement (KIA) is given as:

$$k = \frac{O-E}{1-E}$$

The Cramer's V Index is given as:

$$v = \sqrt{\frac{1}{T} \frac{x^2}{\min\{(M-1),(N-1)\}}}$$

where,

M x N is an array of the source table

These two indices are interpreted in the same way, having typical range of values between 0 and 1; where 1 indicates perfect agreement and 0 indicate chance agreement.

4. Results

The results of the unsupervised ISOCLUST classification of the land surface of the AOI in the Obudu mountainous region is shown in figure 4. From this map, it is clear that 9 land surface clusters were segregated and represented by distinct colours to aid interpretation. The areal coverage of the land surface clusters are presented in Table 4. Cluster 1 has the highest areal coverage of 8.8695 km^2 while cluster 9 has the least areal coverage of 2.9970 km^2.

Figure 5 shows the map of supervised classification of the four dominant vegetation cover types in the Obudu mountain region. This was obtained from the Landsat data with spatial resolution of 30m. The dominant vegetation cover types are forest, scattered trees with shrubs, grasses with shrubs and bare surface / settlements. The classified map was resampled to a spatial resolution of 90 m (Figure 6) to allow for grid-based statistical analysis with the land surface map.

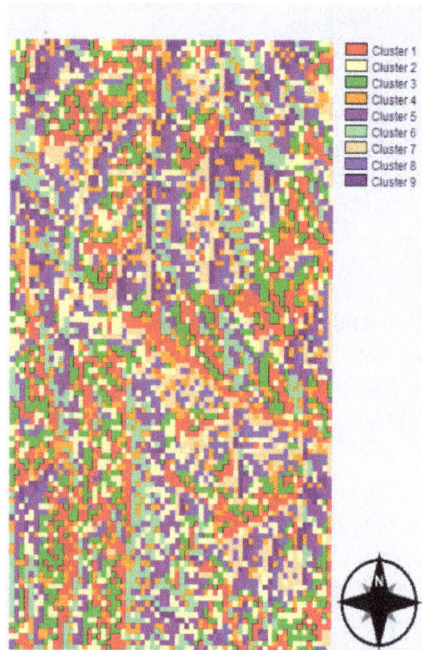

Figure 4. Nine clusters of land surface units

Table 4. Areal coverage of land surface clusters

Land surface cluster	No. of pixels (size = 90m)	Areal coverage (km^2)
1	1095	8.8695
2	1048	8.4888
3	986	7.9866
4	921	7.4601
5	898	7.2738
6	636	5.1516
7	611	4.9491
8	563	4.5603
9	370	2.9970
Total	7128	57.7368

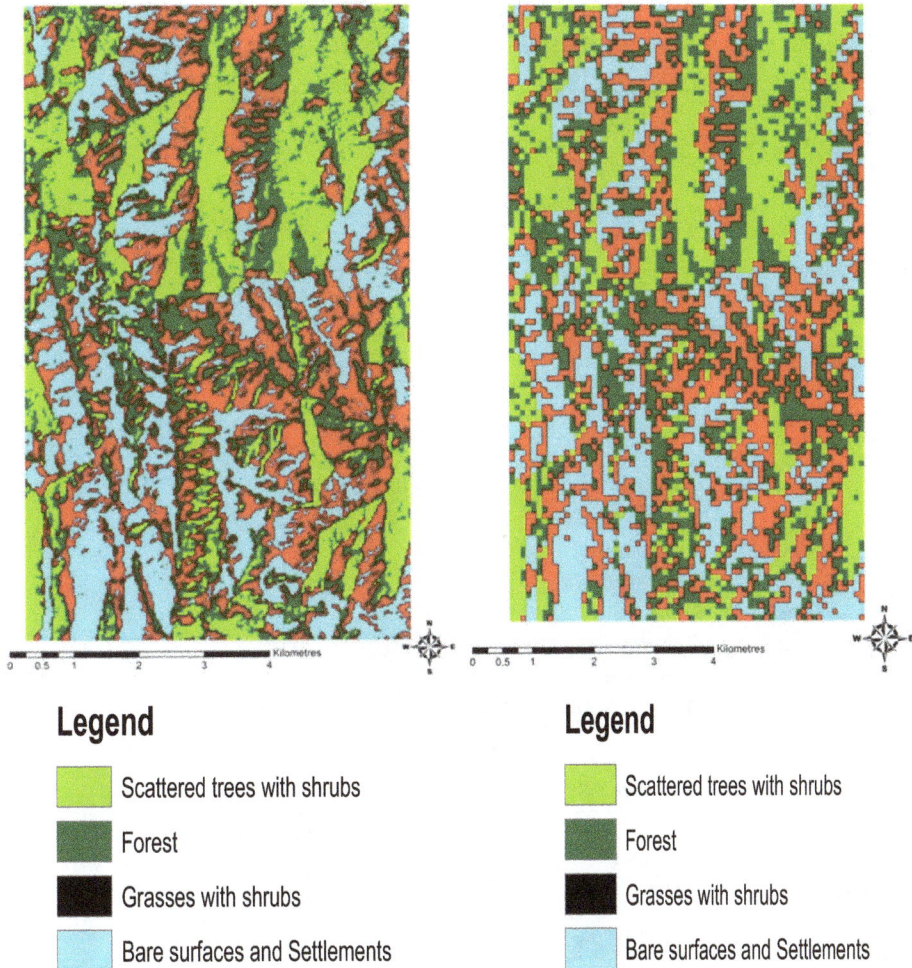

Figure 5. Vegetation cover map based on 30m resolution cover map

Figure 6. Resampled (90m resolution) vegetation cover map

The overall accuracy of the classification of vegetation cover is 81.00% with an overall Kappa statistics of 0.75 The Kappa statistics is interpreted as being substantial (Table 5). The result of the vegetation classification reveals that canopy forest is the most dominant of the four types with total areal coverage and percentage distribution of 16.8075 km^2 and 29.11 per cent while bare surface and settlements has the least with 12.1986 km^2 (21.13 per cent) Table 6 & Figure 7).

Table 5. Vegetation classification accuracy report

a. Error matrix

Classified Data	Unclassi-fied	Scattered trees with shrubs	Forest	Grasses with shrubs	Bare surfaces and settlement	Total Row
Unclassified	0	3	3	2	0	8
Scattered trees with shrubs	0	43	5	2	2	52
Forest	0	4	41	8	5	58
Grasses with shrubs	0	0	3	52	0	55
Bare surfaces and settlement	0	0	0	5	43	48
Column Total	0	50	52	69	50	221

b. Accuracy totals

Class name	Reference totals	Classified totals	Number correct	Producer's accuracy	User's accuracy
Unclassified	0	8	0		
Scattered trees with shrubs	50	52	43	86.00%	82.69%
Forest	52	58	41	78.85%	70.69%
Grasses with shrubs	69	55	52	75.36%	94.55%
Bare surfaces and settlement	50	48	43	86.00%	89.58%
Totals	221	221	179		
Overall classification accuracy				81%	

c. Kappa (K^) Statistics

Conditional Kappa for each Category	
Class Name	Kappa
Unclassified	0.0000
Scattered trees with shrubs	0.7763
Forest	0.6167
Grasses with shrubs	0.9207
Bare surfaces and settlement	0.8654
Overall Kappa Statistics	0.7493

Table 6. Areal and percentage coverage of vegetation cover

Vegetation cover type	No. of pixels covered	Total areal coverage (km²)	% of total area covered
Scattered tress with shrubs	1513	12.2553	21.23
Forest	2075	16.8075	29.11
Grasses with shrubs	2034	16.4754	28.53
Bare surfaces and settlements	1506	12.1986	21.13
Total	7128	57.7368	100.00

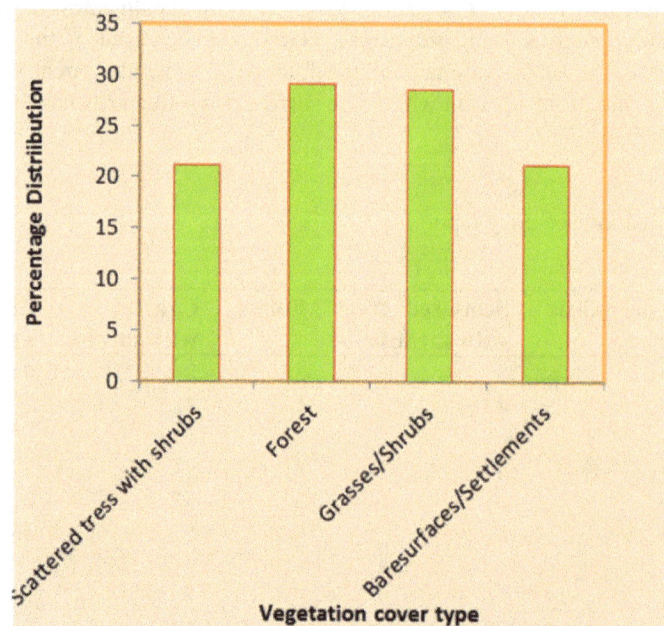

Figure 7. Percentage distribution of vegetation cover

The result of the grid-based cross-tabulation for examining the relationship between land surface units and vegetation cover is presented in Tables 7 and 8. Figure 8 is the map output. Table 7 shows the distribution of the vegetation cover types within the different land surface units. This actually shows the number of pixels for each of the classes in the land surface and vegetation cover maps. For instance, there are 238 pixel classified as grasses with shrubs within land surface unit 1 (concave slopes, facing E - SE - S directions), but 106 pixels classified as scattered tress with shrubs within the same land surface unit.

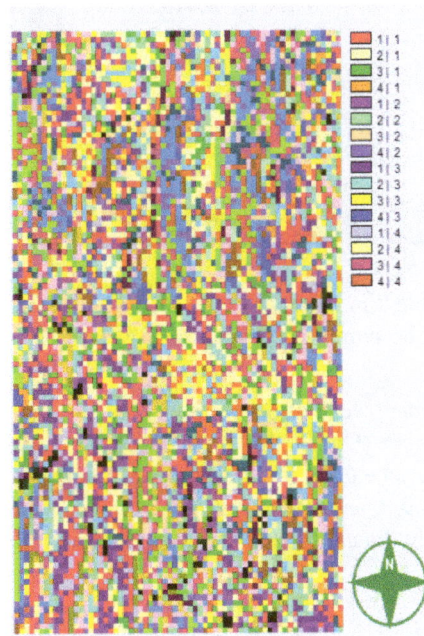

Figure 8. Map output of the cross-tabulation of vegetation cover over land surface units

Table 7. Distribution of vegetation cover over land surface units in terms of number of pixels

Land surface units	Vegetation distribution				
	Scattered trees with shrubs	Forest	Grasses with shrubs	Bare surfaces and settlements	Total
1	106	246	238	505	1095
2	258	278	294	218	1048
3	162	226	322	276	986
4	178	314	233	196	921
5	431	210	172	85	898
6	172	223	143	98	636
7	76	146	277	112	611
8	145	163	168	87	563
9	75	146	111	38	370
Total	1603	1952	1958	1615	7128

Table 8 is the proportional distribution of each vegetation cover type in relation to the land surface units The overall measure of the difference resulted in a chi-square value of 1.91541(df = 24, $p > 0.05$). This means that there is no significant difference between the geomorphometric units (land surface) and the vegetation cover at the 0.05 confidence level. The results of overall (global) similarity between the two maps, which provides a form of correlation measure (de Smith et al., 2009) determined as Cramer's V and Kappa Index of Agreement (KIA) are 0.61 and 0.53 respectively. These are also in support of the fact that the two maps are similar.

Table 8. Proportional distribution of vegetation in relation to land surface units

Land surface units	Vegetation distribution				
	Scattered trees with shrubs	Forest	Grasses with shrubs	Bare surfaces and settlements	Total
1	0.0149	0.0345	0.0334	0.0708	0.1536
2	0.0362	0.0390	0.0413	0.0306	0.1470
3	0.0226	0.0317	0.0452	0.0387	0.1383
4	0.0250	0.0441	0.0327	0.0275	0.1292
5	0.0605	0.0295	0.0241	0.0119	0.1260
6	0.0241	0.0313	0.0201	0.0137	0.0892
7	0.0107	0.0205	0.0389	0.0157	0.0857
8	0.0203	0.0229	0.0236	0.0122	0.0790
9	0.0105	0.0205	0.0156	0.0053	0.0519
Total	0.2248	0.2740	0.2748	0.2264	1.0000

A standardisation of the results in Table 7 in terms of percentage occurrence of each vegetation covers across the land surface units yielded Table 9. The dominant vegetation covers over land surface units are the bold-faced underlined digits.

The criterion for inclusion as dominant vegetation cover within any land surface unit is 33.33% which is one-third of the total vegetation coverage (Udofia, 2011). It is now clearer from Table 9 that scattered trees with shrubs are more prominent in land surface units 5 (convex slopes, facing W - SW and NW directions), forest is typical of land surface units 4, 6 and 9, grasses with shrubs are in 7 (concave slopes, facing NE - N directions) while bare surfaces and settlements are typical of land surface unit 1(concave slopes, facing E - SE - S directions) (See also Table 10).

Table 9. Percentage distribution of vegetation in relation to land surface units

Land surface units	Vegetation distribution (per cent)					Slope direction (Aspect
	Scattered trees with shrubs	Forest	Grasses with shrubs	Bare surfaces and settlements	Total	
1	9.68	22.47	21.69	**46.16**	100	E - SE - S
2	24.62	26.53	28.05	20.80	100	E - NE, W
3	16.43	22.92	32.66	27.99	100	E - SE - S
4	19.33	**34.05**	25.34	21.28	100	E - NE, W
5	**47.97**	23.41	19.15	9.47	100	W -SW - NW
6	27.05	**35.06**	22.48	15.41	100	W - SW
7	12.44	23.88	**45.35**	18.33	100	NE - N
8	25.76	28.95	29.84	15.45	100	NE - N
9	20.27	**39.46**	30.00	10.27	100	NW - N

Table 10. Description of land surface units in Obudu mountain region

Clusters of land surface units	Slope direction (Aspect)	Plan curvature description	Profile curvature description	General description of the land surface
1	E - SE – S	Concave slopes	Concave slopes	Concave slopes, facing E - SE - S directions
2	E - NE, W	Concave slopes	Convex slopes	Concave - convex slopes, facing E - NE and W directions
3	E - SE - S	Convex slopes	Convex slopes	Convex slopes, facing E - SE - s directions
4	E - NE, W	Convex slopes	Concave slopes	Convex - concave slopes, facing E - NE and W directions
5	W -SW - NW	Convex slopes	Convex slopes	Convex slopes, facing W - SW and NW directions
6	W - SW	Concave slopes	Concave slopes	Concave slopes, facing W - SW directions
7	NE - N	Concave slopes	Concave slopes	Concave slopes, facing NE - N directions
8	NE - N	Convex slopes	Convex slopes	Convex slopes, facing NE - N directions
9	NW – N	Concave slopes	Concave slopes	Concave slopes, facing NW - N directions

Source: Efiong et al. (2015)

5. Discussion

The main aim of this study was to relate vegetation cover to land surface units. In the context of spatial analysis, the result of the chi-square test therefore means that there is no significant difference between the land surface units and vegetation cover in the Obudu mountain region. Hence, the two maps (land surface units and vegetation cover) are similar. de Smith et al. (2009) argues that Cramer's V and Kappa Index of Agreement (KIA) provide forms of correlation for grid-based spatial analysis. Where images are similar (showing significant relationships), it is expected that these two indices should be 1 or closer to 1 (de Smith et al., 2009). In the present study, the two indices have values of 0.61 and 0.53 for Cramer's V and Kappa Index of Agreement (KIA) between the two maps respectively. Hence, the two images are similar.

While the results in Tables 7 and 8 do not present a good picture of the vegetation distribution over the land surface units, since they are based on number of pixels, the standardised results in Table 9 gives a better picture of the percentage occurrence of vegetation covers over the land surface units. Several studies including Cavalli et al. (2003) and Abatte et al. (2006) had adopted the standardisation approach to improve the interpretation of their results. In adopting this approach in the present study, each corresponding absolute representative number of pixels was standardised with respect to the total number of pixels of the land surface unit that is considered. The results of this standardisation (Table 9) present a clearer picture of the cross-tabulation, yet there are no sharp boundaries. The non-existence of sharp boundaries is also common in literature. For example, Table 11 presents the results in a similar study by Adediran et al. (2004).

Table 11. Spatial distribution and relationship between the geomorphological units and percent landcover/land use types (Adediran et al., 2004:368)

Landcover types	Geomorphological units									
	1	2	3	4	5	6	7	8	9	10
Bare rocks	0.01	0.00	0.22	0.00	0.03	0.00	0.01	0.05	0.00	0.00
Sclerophyllous vegetations	0.46	0.98	0.45	0.92	0.58	0.45	0.22	0.27	0.36	0.39
Sparsely vegetated areas	10.19	15.82	9.90	9.70	7.31	24.58	5.71	9.13	16.03	21.05
Artificial surfaces	0.08	0.06	0.15	0.03	0.01	0.00	0.00	0.02	0.09	0.00
Transitional woodland/shrub	5.75	5.00	7.26	7.11	13.25	11.34	18.97	12.03	9.83	10.71
Natural grassland	17.19	17.45	17.45	17.68	21.70	25.21	20.56	19.13	19.13	17.01
Mixed forest	0.45	0.41	0.20	0.29	0.43	0.24	1.12	0.47	2.34	1.90
Olive	47.45	43.05	48.61	49.19	41.27	27.84	33.34	38.69	37.35	38.94
Complex cultivation patterns	8.10	10.08	5.97	7.44	4.11	3.32	2.11	3.33	3.98	2.40
Broad leaves forest	0.45	0.97	0.93	1.49	1.25	0.71	0.47	1.07	0.39	0.27
Coniferous forest	9.82	6.16	8.84	6.14	10.06	6.29	17.46	15.79	10.47	7.26

By way of interpreting this results Adediran et al. (2004:368) writes:

"Specifically, (the table) reveals a close spatial association between steeply sloping areas facing NE and sparsely vegetated areas, with this land cover accounting for 24.6% of morphometric unit total coverage. A similar pattern could also be discerned between coniferous forest and steep/average sloping areas facing west, where coniferous forest accounted for 15.8%."

From this standpoint, it has also been demonstrated in this study that there are close spatial associations between some land surface units and vegetation cover; hence some vegetation classes are more typical of particular land surface units than others. For example, scattered trees with shrubs are more typical of land surface unit 5 than in other units while grasses and shrubs are more typical of land surface unit 7. Similarly, forest cover tends to be typical distributed on land surface units 4, 6 and 9 while bare surfaces and settlements are typically located within land surface unit 1.

Moreover, there is an apparent dominance of forest in concave slopes (land surface units 4, 6 and 9) than in other land surface units. Garcia-Aguirre, Ortiz, Zamorano & Reyes (2007) noted the close relationship between slope geometry and fluvial dynamics of overland flow and infiltration. While convex slopes support overland flow, concave slopes on the other hand promote infiltration. It is therefore natural to have higher vegetation density on concave slopes than is found on convex slopes, particularly in the humid tropics where concave slopes reduces fluvial erosion and favours the availability of water for forest growth. On the other hand, scattered trees with shrubs tend to dominate convex slopes (land surface unit 4) where there are less infiltration and more fluvial erosion. The findings of this study agree with those of (Garcia-Aguirre et al., 2007).

This study has shown that some relationships exist between vegetation cover and land surface units in the study area. Hence, classification of land surface units using SRTM-3 can provide a reasonable understanding of spatial distribution vegetation cover of mountainous areas, irrespective of it spatial resolution of 90m. However, application of the findings of this study to other areas should be done with care as entirely different results could be obtained in areas with different profile curvature and plan curvature values. In the present case, the study area is more of a plateau (Efiong et al., 2015).

Again, there are also a number of conceptual problems associated with the use of DEMs (see Fisher, 1993; Wood, 1996; Tate & Wood, 2001). Whilst the issue of scale dependency is not a consideration in this study because the study's interest from the outset has been on the use of the SRTM-3 product which has 90m spatial resolution, it should be noted that the elevation data in SRTM is rather a representation of a digital surface model (DSM) and not a bare-earth model. This implies that built-up areas and dense forest covers are part of the DEM (Nelson, Reuter & Gessler, 2009). This can increase the uncertainty in the results of some kind of analysis done using the SRTM-3 DEM.

The findings of this study can be improved upon firstly, by normalising for topographic effect in the Landsat data for vegetation classification as this could improve the accuracy of the classification. Secondly, the uncertainty of the final outcome of the study could be further reduced if forest covers are filtered from the SRTM-DEM before subsequent extraction of land surface parameters. It is suggested that Wood (1996) algorithm and the self-organizing map as adopted by Eshani and Quiel (2008) could be used in characterising land surface using the SRTM-3 DEM and the results compared with the findings of the present study.

6. Conclusion

The SRTM-3 and Landsat data have provided reasonable understanding of the relationship between land surface and vegetation cover in the Obudu mountain region. The result of the study has shown significant relationship between land surface and vegetation cover in mountainous regions. It is concluded that SRTM-3 DEM can be used to predict vegetation cover in mountainous regions where other data are lacking. However, more studies involving the use of the SRTM publicly available DEM should be conducted on other mountainous areas so as to arrive at a more general relationship. Results from such studies could be used to validate the findings of the present study. Again, the present study could be extended to include field validation of the land surface units to reduce the level of uncertainty in the results.

The study has therefore provided new insights into the use of SRTM-3 in vegetation mapping of mountainous areas which should provoke further research. Moreover, this is the first attempt at characterising the land surface and developing a detailed vegetation map of the major vegetation cover types in the Obudu mountain region. The findings of this study are therefore of immense benefits to land resource managers and forestry agencies. For instance, the cost of reforesting a DEM cell can be estimated using information on slope. Moreover, this study would contribute in solving one of the greatest challenges of today's world which is the conservation of earth's resources for sustainable development. Hence, this study has implications for research, theory and practice.

References

Abbate, G., Cavalli, R. M., Pascucci, S., Pignatti, A., & Poscolieri, M. (2006). Relations between morphological settings and vegetation covers in a medium relief landscape of Central Italy. *Annals of Geophysics, 49*(1), 153-165.

Adediran, A. O., Parcharidis, I., Poscolieri, M., & Pavlopoulos, K. (2004). Computer-assisted discrimination of morphological units on north-central Crete (Greece) by applying multivariate statistics to local relief gradients. *Geomorphology, 58*, 357-370. http://dx.doi.org/10.1016/j.geomorph.2003.07.024

Amuyou, U. A., Eze, E. B., Essoka, P. A. Efiong. J., & Egbai, O. O. (2013). Spatial variability of soil properties in Obudu mountain Region of Southeastern Nigeria. *International Journal of Humanities and Social Sciences, 3*(15), 145-149.

Bishop, Y. M., Fienberg, S. E., & Holland, P. W. (1975). Discrete multivariate analysis: theory and practice. Cambridge: MIT Press.

Bolstad, P. V., Swank, W., & Vose, J. (1998). Predicting Southern Appalachian over story vegetation with digital terrain data. *Landscape Ecology, 13*, 271-283. http://dx.doi.org/10.1023/A:1008060508762

Camiz, S., Papgeorgiou, E., Poscolieri, M., & Parcharidis, I. (2013). *Correlations between landforms and ground deformation at Nisyros volcano (Greece).* Proc. ESA Living Planet Symposium 2013, Edinburg, UK, 9-13. September 2013.

Cavalli, R. M., Fusilli, L., Pascucci, S., Pignatti, S., & Poscolieri, M. (2003). Relationships between morphological units and vegetation categories of Soratte Mount (Italy) as inferred by processing elevation and MIVIS hyperspectral data. In Benes, T. (Ed.), *Geoinformation for European-wide Integration, 573-579.* Rotterdam: Millpress.

Cibula, W. G., & Nyquist, M. O. (1987). Use of topographic and climatological models in a geographical data base to improve Landsat MSS classification for Olympic National Park. *Photogmmmetric Engineering and Remote Sensing, 53*(1), 67-75.

Civco, D. L. (1989). Topographic normalization of Landsat thematic mapper digital imagery. *Photogmmmetric Engineering and Remote Sensing, 55*(9), 1303-1309.

de Smith, M. J., Goodchild, M. F., & Longley, P. A. (2009). *Geospatial analysis: A comprehensive guide to principles, techniques and software tools.* Third edition. UK: Matador.

Dobos, E., & Hengl, T (2009). Soil mapping applications. In Hengl, T and Reuter H, I (Eds.). Geomorphometry: Concepts, Software, Applications, (461-479). Oxford: Elsevier.

Dokuchaev, V. V. (1898). On soil zones in general, and on the vertical zones specifically. *Selected Papers 1949, 3*, 322-329.

Edet, A., & Okereke, C. (2005). Hydrogeological and hydrochemical character of the regolith aquifer, northern Obudu Plateau, southern Nigeria. *Hydrogeology Journal, 13*, 391–415. http://dx.doi.org/10.1007/s10040-004-0358-9

Efiong, J. (2011). Effect of landuse on water discharge in humid regions: an example from southeastern Nigeria. *Global Journal of Social Sciences, 10*(1&2), 53-61.

Efiong, J., Eze, E. B., Digha, O. N., & Asouzu, O. E. (2015). A GIS-based land surface parameterisation of Obudu mountain region using SRTM-3 DEM. *Journal of Basic and Applied Research International, 10*(1), 34- 44.

Ehsani, A. H., & Quiel, F. (2008). Geomorphometric feature analysis using morphometric parameterization and artificial neural network. *Geomorphology, 99*, 1-12. http://dx.doi.org/10.1016/j.geomorph.2007.10.002

Ekwueme, B. N., & Kröner, A. (2006). Single zircon ages of migmatitic gneisses and granulites in the Obudu Plateau: Timing of granulite-facies metamorphism in southeastern Nigeria. *Journal of African Earth Sciences, 44*, 459–469 http://dx.doi.org/10.1016/j.jafrearsci.2005.11.013

Elumnoh, A., & Shrestha, R. P. (2000). Application of DEM data to Landsat Image classification: Evaluation in a tropical wet-dry landscape of Thailand. *Photogrammetric Engineering and Remote Sensing, 66*(3), 297-304.

Fisher, N. I. (1993). *Statistical analysis of circular data. Cambridge*: Cambridge University Press. http://dx.doi.org/10.1017/CBO9780511564345

Florinsky, I, V., & Kuryakova, G. A. (1996). Influence of topography on some vegetation cover properties. *Catena*, 27, 123-141. http://dx.doi.org/10.1016/0341-8162(96)00005-7

Garcia-Aguirre, M. C., Ortiz, M. A., Zamorano, J. J., & Reyes, Y (2007). Vegetation and landform relationships at Ajusco volcano Mexico using a geographic information system (GIS), *Forest Ecology and Management, 239*, 1-12. http://dx.doi.org/10.1016/j.foreco.2006.10.031

Hengl, T., & MacMillan, R. A. (2009). Geomorphometry- a key to landscape mapping and modelling. In Hengl, T and Reuter H, I (Eds.), *Geomorphometry: Concepts, Software, Applications*, (31-63). Oxford: Elsevier.

Höersch, B. (2003). Modelling the spatial distribution of montane and subalpine forests in the central Alps using digital elevation models. *Ecological Modelling, 168*, 267-282. http://dx.doi.org/10.1016/S0304-3800(03)00141-8

Höersch, B., Braun, G., & Schmidt, U. (2002). Relations between landforms and vegetation in alpine regions of Wallis, Switzerland. A multiscale remote sensing and GIS approach. *Computers, Environment and Urban Systems, 26*, 113-139. http://dx.doi.org/10.1016/S0198-9715(01)00039-4

Horn, B. K. P. (1981). *Hill shading and the reflectance map.* Proceedings in IEEE, 69(1), 14-47. http://dx.doi.org/10.1109/PROC.1981.11918

Howard, J. A., & Mitchel, C. W. (1985). *Phyto-geomorphology.* USA: John Wiley & Sons, Inc.

Janssen, L. F., Jaarsma, J., & van der Linder, E. (1990). Integrating topographic data with remote sensing for land-cover classification. *Photogrammetric Engineering & Remote Sensing*, 56(11),1503-   1506.

Jenny, H. (1941). *Factors of soil formation.* New York: McGraw-Hill.

Jones, A. R., Settle, J. J., & Wyatt, B. K. (1998). Use of digital terrain data in the interpretation of SPOT-1 HRV multispectral imagery. *International Journal of Remote Sensing, 9*(4), 669-682. http://dx.doi.org/10.1080/01431168808954885

Landis, J. R., & Koch, G. G. (1977). The measurement of observer agreement for categorical data. *Biometrics, 33*, 159-174. http://dx.doi.org/10.2307/2529310

Langford, M., & Bell, W. (1997). Land cover mapping in a tropical hillsides environment: a case study in the Cauca Region of Colombia. *International Journal of Remote Sensing, 18*(6), 1289-1306. http://dx.doi.org/10.1080/014311697218421

Lillesand, T. M., Kiefer, R. W., & Chipman, J. W. (2008). *Remote sensing and image interpretation* (6th ed.). USA: John Wiley & Sons, Inc

Mather, P. M., & Koch, M. (2011). *Computer processing of remotely-sensed images: an introduction* (4th ed.). UK: Wiley-Blackwell. http://dx.doi.org/10.1002/9780470666517

McBratney, A. B., Mendonca Santos, M. L., & Minasny, B. (2003). On digital soil mapping. Geoderma, 117(1-2), 3-5. http://dx.doi.org/10.1016/S0016-7061(03)00223-4

Nelson, A., Reuter, H. I., & Gessler, P. (2009). DEM production methods and sources. In Hengl, T and Reuter H, I (Eds.), *Geomorphometry: Concepts, Software, Applications*, (65-85). Oxford: Elsevier.

Olaya, V. (2009). Basic land-surface parameters. In T. Hengl, & H. I. Reuter (Eds.). Geomorphometry: Concepts, Software, Applications, (141-169). Oxford: Elsevier.

Palacio-Prieto, J. L., & Luna-GonzAlez (1996). Improving spectral results in a GIS context. *International Journal of Remote Sensing, 17*(11), 2201-2209. http://dx.doi.org/10.1080/01431169608948766

Parcharidis, I., Pavlopoulos, A., & Poscolieri, M. (2001). Geomorphometric analysis of the Vulcano and Nysiros island: clues of the definitions of their volcanic landforms. In Geovannelli, F. (ed.), Proceedings of International Workshop, *"The bridge between Big bang and Biology"*, Stromboli (Messina, It), 13-17 September, 1999, CNR President's Bureau of the National Research Council, Special Volume, pp. 310-320.

Rodenkirchen, K. (2002). *Assessing forests and community water sources of the Obudu Plateau.* Prepared for The Biodiversity Research Programme, Nigerian Conservation Foundation (NCF) – Wildlife Conservation Society (WCS). Retrieved from http://programs.wcs.org/DesktopModules/Bring2mind/.../Download.aspx

Sim, J., & Wright, C. C. (2005). The kappa statistics in reliability studies: use, interpretation and sample size requirements. *Physical Therapy, 85*, 257-268.

Strahler, A. H., Logan, T. L., & Bryant, N. A. (1978). Improving forest cover classification accuracy form

Landsat by incorporating topographic information. *Proceedings of the 12th International Symposium on Remote Sensing of the Environment,* 927-942.

Tate, N. J., & Wood, J. (2001). Fractals and scale dependencies in topography. In Tate, N. J and Atkinson, P. M (Eds.), *Modelling scale in geographical information science,* (35-51), Wiley: Chichester.

Udofia, E. P. (2011). *Applied statistics with multivariate methods.* Enugu: Immaculate Publications Ltd.

Wood, J. D. (1996). *The geomorphologic characterisation of digital elevation models* (Unpublished doctoral dissertation), University of Leicester, UK.

Young, A. (1972). Slopes. Edinburgh: Oliver and Boyd.

Zevenbergen, I. W., & Thorne, C. R. (1987). Quantitative analysis of land surface topography. Earth Surface Processes and Landforms, 12, 47-56. http://dx.doi.org/10.1002/esp.3290120107

5

Morphometric Analyses of Osun Drainage Basin, Southwestern Nigeria

Akinwumiju, A. S.[1] & Olorunfemi, M. O.[2]

[1] Department of Remote Sensing and GIS, Federal University of Technology, Akure, Nigeria

[2] Department of Geology, Obafemi Awolowo University, Ile-Ife, Nigeria

Correspondence: Akinwumiju, A. S., Department of Remote Sensing and GIS, Federal University of Technology, Akure, Nigeria. E-mail: ojhakin@yahoo.com

Abstract

This study evaluated some morphometric parameters with a view to assessing the infiltration potential of Osun Drainage Basin, Southwestern Nigeria. Input data were derived from SPOT DEM using ArcGIS 10.3 platform. The basin has an area extent of 2,208.18 km2, and is drained by 1,560 streams with total length of 2,487.7 km. Drainage Texture (0.52), Stream Number (1,560), Total Stream Length (2,487.7 m) and Main Stream Length (119 m) indicate that larger percentage of annual rainwater would leave the basin as runoff. Infiltration Number increases with increasing Stream Frequency ($r = 0.95$) and Drainage Density ($r = 0.78$). Length of Overland Flow increases with decreasing Drainage Density ($r = -0.83$), Stream Frequency ($r = -0.51$) and Infiltration Number ($r = -0.45$). Regression analysis show that Stream Frequency accounts for 97.43% of the strength of the overall regression model. Thus, Stream Frequency is a strong variable that can solely give meaningful explanation of infiltration potential. However, Basin Perimeter, Length of Overland Flow and Drainage Density also have significant influence on infiltration potential at varying degrees. The overall relationship explains 93.4% of the regression plain. Thus, Stream Frequency, Basin Perimeter, Length of Overland Flow and Drainage Density constitute a set of strong variables that can predict Infiltration Number and consequently, give meaningful explanation to infiltration potential within a basin. The study concluded that infiltration potential is moderate within Osun Drainage Basin as suggested by the mean Infiltration Number.

Keywords: morphometric analysis, infiltration potential, Osun Drainage Basin

1. Introduction

Drainage basin can be defined as a geographically delimited finite area on the earth surface that is drained by a network of streams through a single pore point (Akinwumiju, 2015). Drainage basin is an ideal unit for the interpretation and analysis of fluvial originated landforms where they exhibit an example of open system of operation. Thus, a drainage basin is a fundamental unit of virtually all catchment-based fluvial investigations. The continuous interaction between climate and geology often result to the evolution of landform pattern across a given basin, which can be qualitatively (morphology) and quantitatively (morphometry) analyzed. This topographic expression is known as terrain analysis (Jones, 1999; Obi-Reddy et al., 2002). Terrain analysis is the study of elements relating to the geometric form, the underlying materials, geomorphogenesis and the spatial pattern of landforms (Schmidt and Dikau, 1999). Early studies on terrain analyses were mostly qualitative in approach, which were devoid of numerical analysis of drainage basin (Gregory and Walling, 1973; Ajibade et al., 2010). As a result, detailed understanding of drainage evolution as well as the mechanics of surface runoff was lacking (Ajibade et al., 2010). However, notable scientific approaches to terrain analyses were evident in the literature as far back as 17th Century (Penck, 1894, 1896; Passarge, 1912). Since its introduction by Horton (1940), morphometric analysis has been providing elegant description of basin-scale landscape as well as quantitative parameterization of the earth surface (Easterbrook, 1993; Ajibade et al., 2010). Usually, morphometric analysis is undertaken in many hydrologic investigations such as groundwater potential assessment, pedology, water resource management, flood control, environmental impact assessment and pollution studies among others (Jayappa and Markose, 2011). Furthermore, morphometric analysis could be undertaken with the aim of assessing the impacts of tectonic activities across a drainage basin (Hurtex and Lacazeau, 1999; Sinha-Roy, 2002; Singh, 2008; Walcott and Summerfield, 2008). Thus, morphometric parameters have earlier been observed as crucial indices of surface processes within a given basin. Consequently, these parameters have been determined and analyzed in many geomorphological and surface hydrological studies such as sediment deposition, flood parameterization as well as the evolution of basin morphology (Jolly, 1982; Adejuwon et al., 1984; Anyadike and Phil-Eze, 1989; Lifton and Chase, 1992; Moglen and Bras, 1995; Chen et al., 2003; Haung and Niemann, 2006). More recently, morphometric analyses have been playing a major role in modeling of surface processes such as soil erosion and flooding (Nogami, 1995; Singh et al., 2008; Ajibade et al., 2010; Sumira et al., 2013).

Until recently, scientists usually rely on data garnered from field measurements and or extracted information from existing topographic maps as major inputs in morphometric analyses. Currently, remotely sensed data and Geographic Information System (GIS) have gained recognition as preferred data source and analytical platform for morphometric analyses respectively. For example, multi-resolution Digital Elevation Models have been extensively utilized in various morphometric analyses (Nag and Anindita, 2011; Somashekar and Ravikumar, 2011). Today, many GIS platforms are embedded with various types of morphometric-specific algorithms that enable scientists to determine many morphometric parameters automatically, thereby increasing efficiency as well as reducing rigor and time (Schmidt and Dikau, 1999). Recently, a comprehensive inter-disciplinary-based groundwater potential assessment was undertaken within Osun Drainage Basin, involving terrain analyses. In this study therefore, we present and analyze the adopted morphometric parameters with the aim of evaluating the geomorphometric characteristics; particularly in relation to infiltration potential of the basin.

2. Materials and Methods

2.1 The Study Area

Osun Drainage Basin (ODB) lies within Latitudes 7°35' and 8° 00' north of the Equator; Longitudes 4°30' and 5°10' east of the Greenwich Meridian; in the forested undulating Yoruba Plain of Southwestern Nigeria (Figure 1). Osun Catchment extends from the upland area of Ekiti State to the low lying area of Osun State, covering 21 Local Government Areas with projected population of 6.2 million as at December, 2014 (Akinwumiju, 2015). ODB is a watershed that is drained by a sixth order river network, comprising various perennial rivers that take their courses from Ekiti-Ijesa mountainous region. The basin constitutes the upland northeastern watershed, which is a major donor sub-basin of the much larger Osun-Ogun Drainage Basin in Southwestern Nigeria. Osun-Ogun River Network is one of the few drainage systems in the Southwestern Nigeria that empties its contents directly into the Gulf of Guinea. The climate of the study area is characterized by long rainy season from March to November. The basin lies within the Humid Tropical Climatic Zone that normally experience double maximal rainfall that peaks in July and October. Precipitation is relatively high across the basin (1,500 – 1,700 mm per annum) and the only dry months are January and February. Relative humidity rarely dips below 60% and fluctuates between 75% and 90% for most of the year. In the rainy season, cloud cover is nearly continuous, resulting in mean annual sunshine hours of 1,600 and an average annual temperature of approximately 28oC. The vegetation of the study area is characterized by disturbed rainforest, light forest and patches of thick forest. Experience from change detection analysis showed that the heavily disturbed vegetation has the potential to rejuvenate under sustainable natural resources utilization and management (Akinwumiju, 2015). The study area is underlain by the Precambrian Basement Complex that is characterized by both foliated and non-foliated rocks such as quartzite/quartz schist, amphibole schist, mica schist, migmatite, porphyritic granite, biotite granite, pegmatite, granite gneiss, banded gneiss and charnockite (De Swardt, 1953; Elueze, 1977; Boesse and Ocan, 1988; Oluyide, 1988; Odeyemi et al., 1999; Awoyemi et al., 2005). A unique attribute of Osun Drainage Basin is it's been located at the heart of Ilesa Schist Belt, which is a zone of regional metamorphism that is characterized by notable geological structures such as the Efon (psammite formation) Ridge and Zungeru-Ifewara Mega Fault Line (Akinwumiju, 2015).

2.2 Analytical Procedure

This study relied on the medium resolution Digital Elevation Model (SPOT DEM, 20 m resolution) of Osun Drainage Basin that was acquired from the Office of the Surveyor-General of the Federation in Abuja, Nigeria. Digital spatial data (such as sub-basin and river network maps) were extracted from Akinwumiju (2015). Analyses were undertaken in three stages. The first stage involved the determination of independent morphometric variables such as basin area, basin perimeter, basin relief, stream length, basin length, basin width, maximum order of streams, and number of streams in each order. Thus, automated feature attribute extraction (Add Geometry Attributes) module was adopted to derive the independent morphometric parameters on ArcGIS

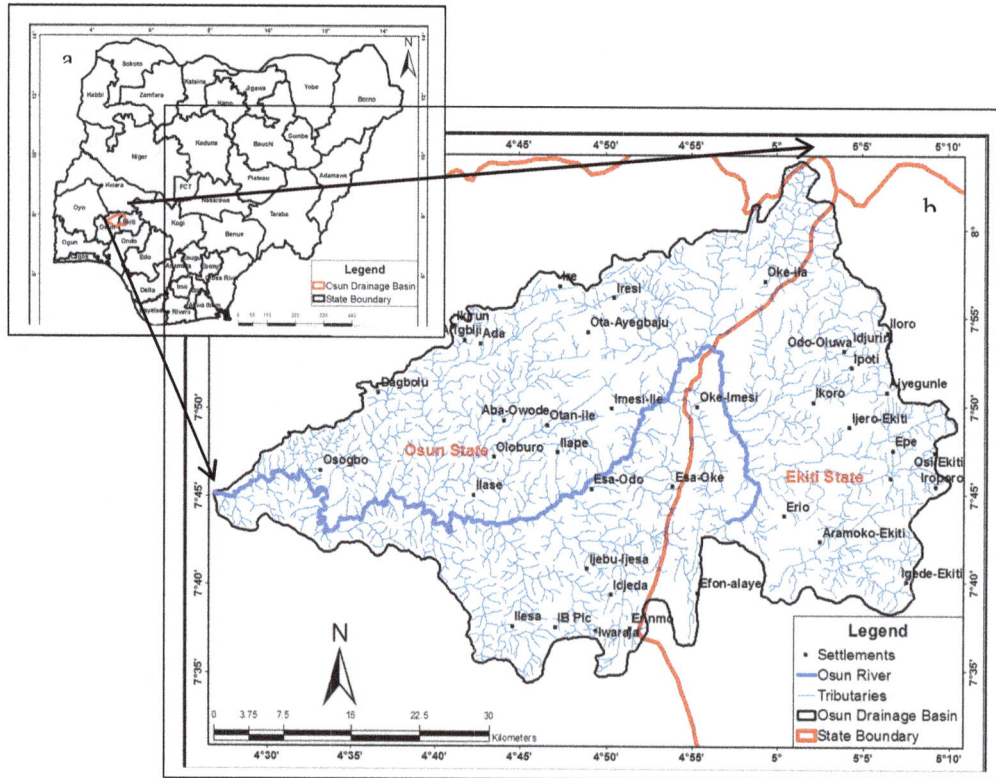

Figure 1. Map of the Study Area showing a): Nigeria's State Boundaries; b): Osun Drainage Basin

platform. The second stage involved the computation of the second level morphometric parameters such as bifurcation ratio, elongation ratio, circularity ratio, Drainage Density, Stream Frequency, Drainage Texture, relief ratio, ruggedness number, length of overland flow, hypsometric integral, topographic traverse symmetry factor, asymmetry factor and Infiltration Number. The third stage involved the modeling of the stream order – stream length ratio curve, stream order – bifurcation ratio curve and the longitudinal profile of the basin's main channel. In the case of hypsometric analyses, the elevation contours were generated from SPOT DEM in ArcGIS 10.3 environment. Thereafter, the study area was delineated into various solid earth surfaces above and below different altitudes as depicted by the elevation contours. The areas of the solid earth surfaces were computed with the aid of automated Measure module of ArcGIS. Subsequently, Hypsometric Integral was computed from the areas of solid earth surfaces and the corresponding altitudes. In order to delve into the influence of tectonic structures and lithology on the main drainage channel, longitudinal profile (involving the plotting of elevation against distance) was constructed for the basin. All the examined morphometric parameters as well as the corresponding formula/procedure are presented in Table 1. The morphometric parameters were directly computed from the DEM-based digital layers of the study area on ArcGIS 10.3 platform. In this study, emphasis is on the morphometric parameters that reflect the infiltration vis-à-vis groundwater potential of the study area. These include basin-scale parameters (Drainage Density, Drainage Texture, Stream Frequency, Length of Overland Flow and Infiltration Number) that rely on information on drainage, topography and geometry of a given basin. Notwithstanding, all the examined parameters represent detail quantitative morphometric analysis of the basin. The drainage network of the study area was comprehensively analyzed. The major analysis undertaken includes river ordering and sub-basin delineation and characterization. The groundwater related parameters were computed for all the sub-basins of the study area and were subjected to correlation and regression analyses with a view to modeling the associations among the variables and the relationship between Infiltration Number (dependent variables) and some selected morphometric parameters (predictor variables). The predictor variables include Drainage Density, Stream Frequency, length of overland flow, Drainage Texture, basin perimeter, stream number, basin area and river order. The regression analysis was computed on SPSS statistical platform.

Table 1. Morphometric Parameters and Formula

S. No.	Parameters	Formula	Reference
1		Linear Morphometric parameters	
1.1	Stream Order (Sμ)	Hierarchical rank	**Strahler (1964)**
1.2	Bifurcation Ratio (Rb)	Rb = Nμ / Nμ +1 Where, Rb = Bifurcation ratio, Nμ = No. of stream segments of a given order and Nμ +1= No. of stream segments of next higher order.	**Schumn (1956)**
1.3	Mean Bifurcation Ratio (Rbm)	Rbm = Average of bifurcation ratios of all orders	**Strahler (1964)**
1.4	Stream Number (Sn)	Sn = Total Number of Stream Segments	
1.5	Stream Length (Lμ)	Length of the stream (kilometers)	**Horton (1945)**
1.6	Mean Stream Length (Lsm)	Lsm = Lμ/Nμ Where, Lμ = Total stream length of order 'μ' Nμ = Total no. of stream segments of order 'μ'	**Strahler (1964)**
1.7	Stream Length Ratio (RL)	RL= Lsm / Lsm-1 Where, Lsm=Mean stream length of a given order and Lsm-1= Mean stream length of next lower order	**Horton (1945)**
1.8	Length of Overland Flow (Lg)	Lg=1/2D Km Where, D=Drainage density (Km/Km2)	**Horton (1945)**
1.9	Basin Perimeter (P)	P=Outer boundary of drainage basin measured in kilometers.	**Schumm (1956)**
1.1	Basin Length (Lb)	Lb=1.312*A0.568	**Gregory and Walling (1973)**
1.11	Standard Sinuosity Index (SS)	SSI = CL/Lv Where, CL = Channel length (Kms) and Lv = Valley length (Kms)	**Muller (1968)**
2		Areal Morphometric parameters	
2.1	Basin Area (A)	Area from which water drains to a common stream and boundary determined by opposite ridges	**Strahler (1969)**
2.2	Drainage Density (Dd)	Dd = Lμ/A Where, Dd = Drainage density (Km/Km2), Lμ = Total stream length of all orders and A = Area of the basin (Km2).	**Horton (1932)**
2.3	Stream Frequency (Fs)	Fs = Nμ/A Where, Fs = Stream frequency. Nμ = Total no. of streams of all orders and A = Area of the basin (Km2).	**Horton (1932)**
2.4	Drainage Texture (Dt)	Dt = Nμ /P Where, Nμ = No. of streams in a given order and P = Perimeter	**Smith (1939) & Horton (1945)**
2.5	Form Factor Ratio (Rf)	(Kms) Rf = A/Lb² Where, A = Area of the basin and Lb = (Maximum) basin length	**Horton (1932)**
2.6	Elongation Ratio (Re)	Re= √A /π / Lb Where, A=Area of the Basin (Km²) Lb=Maximum Basin length (Km)	**Schumm (1956)**
2.7	Circularity Ratio (Rc)	Rc = 4πA/ P² Where, A = Basin Area (Km²) and P= Perimeter of the basin (Km) Or Rc = A/ Ac Where, A = Basin Area (Km2) and Ac = area of a circle having the same perimeter as the basin	**Miller (1953)**
3		Relief Morphometric Parameters	
3.1	Channel Gradient	$C_g = C_c - E_{pp}$ Where, C_c = Channel Crest and E_{pp} = Elevation of Pour Point	**Strahler (1964)**
3.2	Maximum Basin Relief	$R_b = E_h - E_{bm}$ Where, E_b = Highest Elevation of Basin and E_{bm} = Elevation of Basin Mouth	**Horton (1945); Strahler (1964)**
3.3	Relief Ratio	$R_r = R_b/L_b$ Where, R_b = Maximum Basin Relief and L_b = Maximum Length of the Basin	**Schumm (1956)**
3.4	Ruggedness Number	$Rn = R_b D_d$ Where, R_b = Basin Relief and D_d = Drainage Density	**Strahler (1950, 1957)**
4		Tectonic Morphometric Parameters	
4.1	Hypsometric Integral	(h/H):(a/A) Where, h = Lower Interval Elevation – Basin Elevation, H = Basin Relief, a = Area above bottom of Interval and A = Basin Area	**Strahler (1952)**
4.2	TTSF	T = Da/Dd Where, Da = the distance from the main stream channel to the midline of its drainage basin and Dd = the distance from the basin margin (divide) to the midline of the basin	**Cox (1994)**
4.3	Asymmetry Factor	AF = 100 (A_r/A_t) Where, Ar = Area of the basin part to the right of the main drainage channel and At = Area of the entire basin.	**Hare and Gardner (1985)**
4.4	Longitudinal Profile	LP = The Graph of D_c (X axis) and E_c (Y axis) Where, E_c = Elevation Values along main Drainage Channel and D_c = Distance (in kilometer) along main Drainage Channel	

3. Results and Discussion

Osun Drainage Basin has an area extent of 2,208.18 km^2, perimeter of 293.14 km, Axial Length (E-W orientation) of 80.34 km and Axial Width (N-S orientation) of 43.89 km. The basin is drained by 1,560 rivers with total length of 2,487.7 km. The results of the morphometric analysis are discussed below.

3.1 Linear Parameters

Results showed that the study area is drained by a sixth order drainage network comprising 3 fifth order, 14 fourth order, 59 third order, 290 second order and 1,193 first order drainage channels with stream lengths of 119.46 km, 40.76 km, 171.18 km, 272.21 km, 511.21 km and 1,371.70 km respectively. In descending order, the mean stream lengths are 119.46 km, 13.59 km, 12.23 km, 4.61 km, 1.76 km and 1.15 km. In this case, the Mean Stream Length increases with increasing order. The values of Stream Length Ratio are (in descending order) 2. 65 (3rd order), 2.62 (4th order), 1.53 (5th order) 1.11 (2nd order) and 0.12 (1st order). Figure 2 presents the River Order – Stream Length Ratio curve, which reflect a single pick at 3rd order but with values of Stream Length ratio greater than 2.5 for 3rd and 4th order. The interpretation of this is that the channel gradient will be higher for 3rd and 4th order river channels due to relatively steeper slope and undulating topography.

Table 2. The Values of the Linear Parameters

	Parameter	Value
1	Stream Order	6
2	Bifurcation Ratio	3.0 to 4.92
3	Mean Bifurcation Ratio	4.18
4	Stream Number	1 to 1,193
5	Stream Length	40.76 to1,371.70 km
6	Mean Stream Length	1.15 to 119.46 km
7	Stream Length Ratio	0.12 to 2.65
8	Length of Overland Flow	0.44 km
9	Basin Perimeter	293.14 km
10	Axial Length	80.34 km
	Axial Width	43.89 km
11	Standard Sinuosity Index	1.79

The values of Bifurcation Ratio are (in descending order) 4.92 (2nd order), 4.67 (4th order), 4.21 (3rd order), 4.11 (1st order) and 3.0 (5th order). Figure 3 presents the River Order – Bifurcation Ratio curve, reflecting two picks at 2nd and 4th order where the values of Bifurcation Ratio are greater than 4.5. Thus, substantial percentage of 2nd and 4th order drainage channels might likely be structurally controlled. However, the range of Bifurcation Ratio indicates apparently minimal structural control in the drainage development across the study area. The computed value of Standard Sinuosity Index (1.79) indicates that Osun River has a meandering course, suggesting significant influence of geology and topography on channel morphology. The relatively high value of Length of Overland Flow(0.44 km) calculated for the study area also suggest that there would be more time for in situ infiltration of rainwater before the final concentration of runoff into the main stream channels.

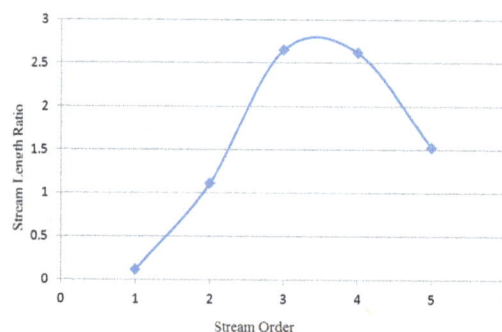

Figure 2. River Order – Stream Length Ratio Curve

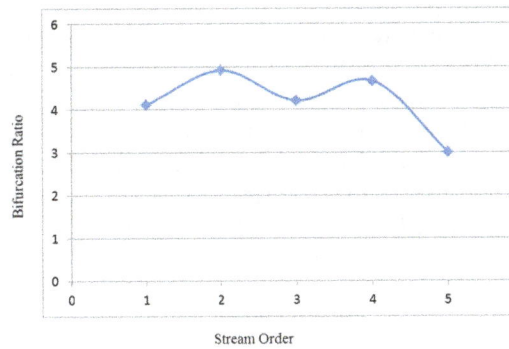

Figure 3. River Order – Bifurcation Ratio Curve

3.2 Shape Parameters

The values of the Form Factor (0.34), Circularity Ratio (0.32) and Elongation Ratio (0.66) reveal that the watershed is elongated and it is at the advanced stage of landform development. Thus, runoff would be easy to manage as peak discharge is expected to be relatively low due to extensive evenly distribution of runoff over the course of the main channel.

Table 3. The Values of the Shape Parameters

	Parameter	Value
1	Form Factor	0.34
2	Circularity Ratio	0.32
3	Elongation Ratio	0.66

3.3 Relief Parameters

The Maximum Basin Relief (450 m), Channel Gradient (2 m/km), Relief Ratio (5.6) and Ruggedness Number (0.10) suggest the occurrence of extreme topographic high and topographic low. Also, these values indicate the prevalence of low lying areas compared to hilly terrain. The relief measures affirmed that the basin is at the advanced stage of landform development.

Table 4. The Values of the Relief Parameters

	Parameter	Value
1	Channel Gradient	2 m/km
2	Maximum Basin Relief	450 m
3	Relief Ratio	5.6
4	Ruggedness Number	0.10

3.4 Landform Evolution and Tectono-Morphometric Parameters

The computed value of the Hypsometric Integral (0.39) shows that 61 percent of the earth materials (above the lowest elevation) have been washed off the basin, either in form of solution or suspension. The basin's topographic Traverse Symmetry Factor (0.36) and the Asymmetry Factor (64.43) indicate that the basin is tilted (N – S direction) and its drainage network is partially structurally controlled.

Table 5. The Values of the Landform Evolution Parameters

	Parameter	Value
1	Hypsometric Integral	0.39
2	Topographic Traverse Symmetry Factor	0.36
3	Asymmetry Factor	64.43

The Longitudinal Profile of the study area is presented in Figure 4. Osun River flows initially towards the north from the southern part of Effon Ridge, then turns southwestward and flows into Asejire Dam in Oyo State. In its course, Osun River loses about 264 (514 - 250) m elevation to various structural displacements in form of faulting and lithological boundaries. Twelve (12) major knick points occur along the river course, which coincide with traversed faults mostly along lithological boundaries. The river course dissects major faults diagonally while reflecting the evidence of significant influence of lithological resistance particularly in the relatively low lying part of the drainage basin. The observed intersection between the Osun River course and the N-S trending lineaments is advantageous in two ways. First, the fault/fracture zones serve as major sources of inflows into some river channels. This probably might have accounted for the all year round base flow of Osun River. On the other hand, the river could be a source of recharge to the fault/fracture zones particularly in the dry season. In this case, the river is said to be effluent. These interrelated ecological through-puts are very crucial to basin-scale environmental sustainability.

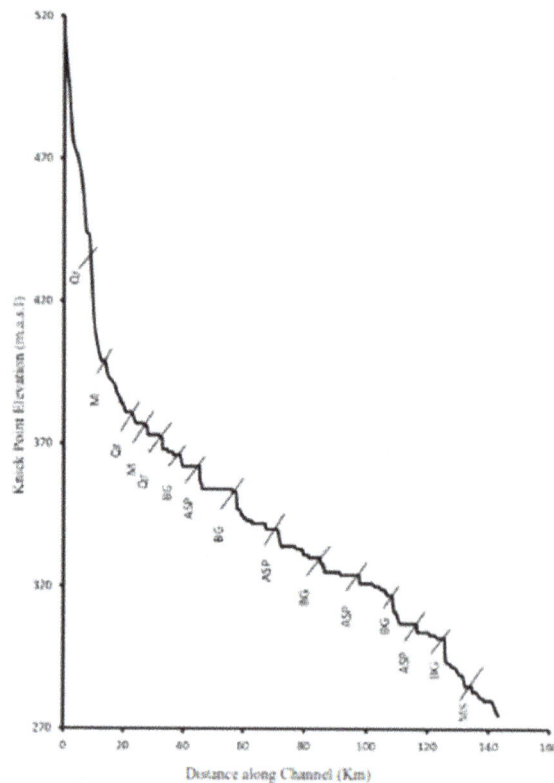

Figure 4. Longitudinal Profile of Osun Main Channel

Qr = Quartzite, M = Migmatite, BG = Banded Gneiss,

ASP = Amphibole Schist interlayer with Pegmatite, MS = Mica Schist

/= Knick Points at Channel – Traversed Fault Intersections

3.5 Infiltration Potential- Related Morphometric Parameters

The map of the sub-basins of the study area is presented in Figure 5 and the statistical summary of the corresponding values of infiltration potential-related parameters are presented in Table 6. The 6th order basin consists of three 5th order sub-basins, six 4th order sub-basins, ten 3rd order sub-basins, forty-two 2nd order sub-basins and forty-six 1st order sub-basins. Areas of sub-basins range from 0.11 km^2 to 416.29 km^2 with a mean of 20.45 km^2. The computed standard deviation (60.13) and coefficient of variation (294.12) indicate that the size of sub-basins vary relatively and absolutely. Perimeters of the sub-basins range from 1.48 km to 223.96 km with a mean of 15.16 km. Computed standard deviation (27.46) and coefficient of variation (181.16) show that basin perimeter is very heterogeneous across the study area. Stream Numbers of the sub-basins vary from 1 to 271 with a mean of 14.44. The computed standard deviation (40.03) and coefficient of variation (277.22) indicate that stream number vary significantly both absolutely and relatively among the sub-basins.

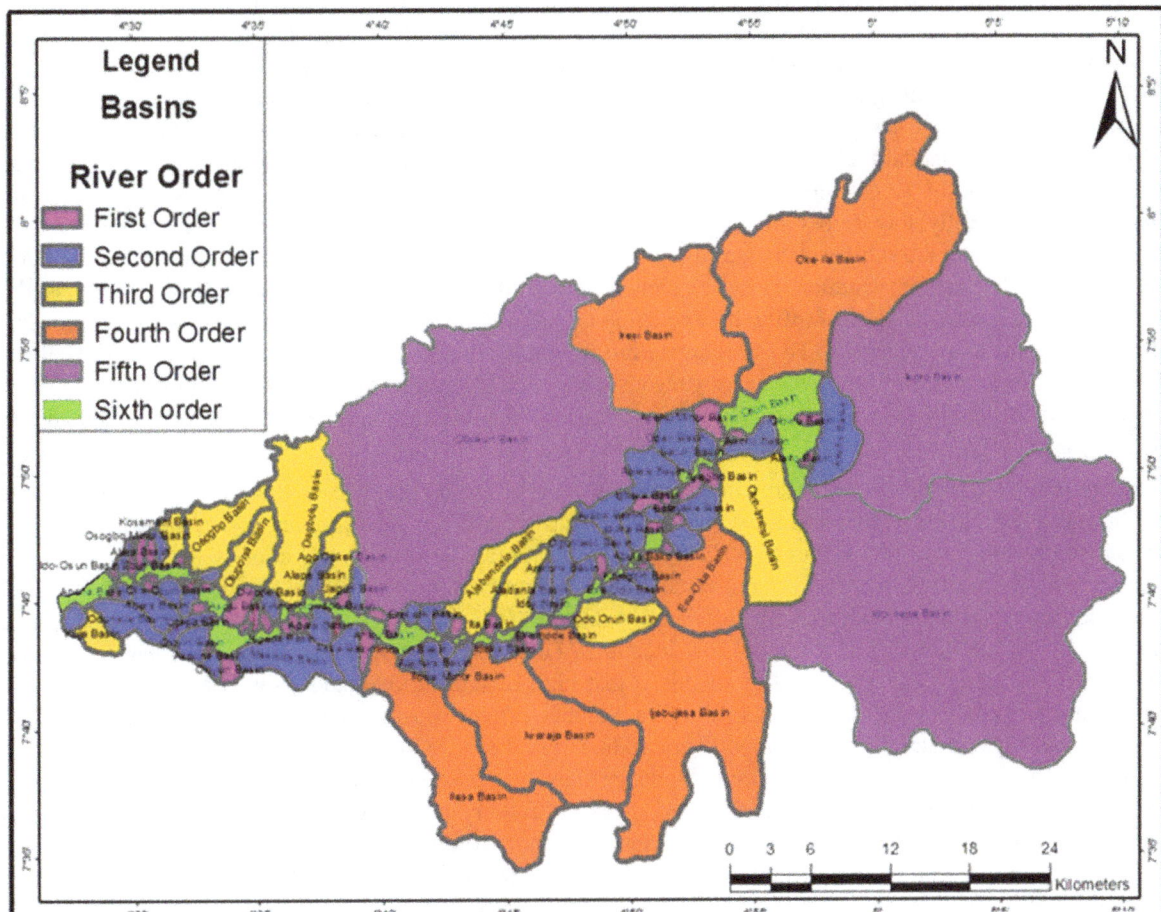

Figure 5. Sub-Drainage Basin Map of the Study Area

Table 6. Spatial Characteristics of Morphometric Parameters

	No.	Minimum	Maximum	Mean	SD	CV
Area	108	0.11	416.29	20.45	60.13	294.12
Perimeter	108	1.48	223.96	15.16	27.46	181.16
SN	108	1.00	271.00	14.44	40.03	277.22
DT	108	0.001	2.54	0.52	0.48	92.31
SL	108	0.36	435.83	23.03	66.07	286.82
LOF	108	0.15	0.86	0.45	0.14	30.79
DD	108	0.58	3.27	1.23	0.36	29.65
Sf	108	0.01	9.09	1.19	1.04	87.49
IN	108	0.01	29.72	1.77	3.06	173.22

SN = Stream Number; DT = Drainage Texture; SL = Stream Length; LOF = Length of Overland Flow; DD = Drainage Density; Sf = Stream Frequency; IN = Infiltration Number

Drainage Textures of the sub-basins range from 0.001 to 2.54 with a mean of 0.52. The values of standard deviation (0.48) and coefficient of variation (92.31) show that Drainage Texture is not homogeneous across the study area. However, Drainage Texture is generally low across the sub-basins, indicating relatively high infiltration potential. The stream lengths of the sub-basins range from 0.36 km to 435.83 km with a mean of 23.03 km. The computed values of standard deviation (66.07) and coefficient of variation (286.86) show that stream length vary heterogeneously across the sub-basins. Lengths of Overland Flow of the sub-basins range from 0.15 km to 0.86 km with a mean of 0.45 km. the computed values of standard deviation (0.14) and coefficient of variation (30.79) revealed that Length of Overland Flow is less heterogeneous across the sub-basins. The interpretation is that runoff would have relatively moderate time-lag to infiltrate before it will be finally confided into main drainage channels. Drainage Density of the sub-basins range from 0.58 km/km2 to 3.27 km/km2 with a mean of 1.23 km/km2. The computed standard deviation (0.36) and coefficient of variation (29.65) indicate that Drainage Density is less heterogeneous across the sub-basins. Thus, infiltration potential is generally moderate in the study area. Stream Frequency of the sub-basins range from 0.01 to 9.09 with a mean of 1.19. The values of standard deviation (1.04) and coefficient of variation (87.49) show that Stream Frequency varies heterogeneously across the sub-basins. However, the computed mean value revealed that Stream Frequency is generally low across the study area, which is an indicator of enhanced infiltration potential. Infiltration Number of the sub-basins range from 0.01 to 29.72 with a mean of 1.77. Values of standard deviation (3.06) and coefficient of variation (173.22) show that Infiltration Number varies significantly across the sub-basins. Analysis indicates that infiltration potential is high in 44% of the sub-basins (with IN < 1) while 32% of the sub-basins was adjudged to be of moderate infiltration potential (with IN ranging from 1 to 2). Analysis showed that infiltration potential was heterogeneously low in 24% of the sub-basins with Infiltration Number ranging from 2 to 30. However, the computed mean indicates that infiltration potential is generally moderate in the study area.

The correlation matrix of the morphometric parameters is presented in Table 7. Results reveal that Basin Order exhibit positive and strong relationship with basin area, basin perimeter, stream number, Drainage Texture and stream length with correlation values of 0.72, 0.81, 0.70, 0.77 and 0.74 respectively at $\alpha = 0.01$. Results showed that Length of Overland Flow exhibit inverse but significant relationship with Drainage Density, Stream Frequency and Infiltration Number with correlation values of -0.83, -0.51 and -0.45 respectively at $\alpha = 0.01$. In this case, when the Length of Overland Flow increases, Drainage Density, Stream Frequency and Infiltration Number will decrease. The interpretation of this is that high Length of Overland Flow is an indicator of high infiltration potential. Results show that Infiltration Number exhibits positive and significant relationship with Drainage Density and Stream Frequency with correlation values of 0.78 and 0.95 respectively at $\alpha = 0.01$. Thus, Infiltration Number increases with increasing Drainage Density and Stream Frequency and decreasing Length of Overland Flow in the study area. This is expected since Infiltration Number is function of Drainage Density and Stream Frequency. Results also showed that Stream Frequency exhibits an inverse but weak relationship with Basin Perimeter and Basin Order with correlation values of -0.23 and -0.20 at $\alpha = 0.05$. Thus, Stream Frequency decreases with increasing Basin Perimeter and Basin Order. However, these associations are weak and might not hold. Results reveal that Length of Overland Flow, Drainage Density, Stream Frequency and Infiltration Number do not have any relationship with Basin Area.

Table 7. Correlation Matrix of Morphometric Parameters

	Area	Perimeter	BO	SN	DT	SL	LOF	DD	Sf	IN
Area	1.00									
Perimeter	0.714	1.00								
BO	0.723	0.809	1.00							
SN	0.981	0.619	0.700	1.00						
DT	0.800	0.501	0.769	0.871	1.00					
SL	0.997	0.739	0.742	0.980	0.811	1.00				
LOF	0.015	-0.013	-0.069	-0.011	-0.176	-0.005	1.00			
DD	-0.099	-0.080	-0.086	-0.076	0.094	-0.084	-0.827	1.00		
Sf	-0.160	-0.234	-0.203	-0.126	0.070	-0.158	-0.509	0.7995	1.00	
IN	-0.107	-0.149	-0.154	-0.088	0.059	-0.105	-0.446	0.781	0.954	1.00

However, it was observed that Stream Frequency and Infiltration Number exhibit inverse but weak relationships with Basin Perimeter at $\alpha = 0.05$. The above facts imply that Infiltration Number is controlled by Stream Frequency, Drainage Density and Length of Overland Flow in the study area. And that it (Infiltration Number) does not depend on basin area and basin order. The relationship between morphometric parameters and Infiltration Number is presented in Table 8 and explained by the equations that follow.

Table 8. Relationship between Morphometric Parameters and Infiltration Number

Model	R	R^2	Adjusted R^2	S.E. of the Estimate	Change Statistics				
					R^2 Change	F Change	df1	df2	Sig. F Change
1	0.954[a]	0.910	0.909	0.92159	0.910	1073.143	1	106	0.000
2	0.957[b]	0.916	0.914	0.89542	0.006	7.287	1	105	0.008
3	0.959[c]	0.919	0.917	0.88123	0.003	4.407	1	104	0.038
4	0.966[d]	0.934	0.931	0.80343	0.014	22.117	1	103	0.000

a. Predictors: (constant), Stream Frequency

b. Predictors: (constant), Stream Frequency, Perimeter

c. Predictors: (constant), Stream Frequency, Perimeter, Length of Overland Flow

d. Predictors: (constant), Stream Frequency, Perimeter, Length of Overland Flow, Drainage Density

$$Y = -1.570 + 2.800X_1 \dots\dots\dots\dots\dots\dots\dots\dots\dots\dots\dots\dots\dots\dots (2)$$
$$(R = 0.95; R^2 = 91.0\%; SE = 0.92)$$

$$Y = -1.767 + 2.854X_1 + 0.009X2 \dots\dots\dots\dots\dots\dots\dots\dots\dots\dots\dots (3)$$
$$(R = 0.96; R^2 = 91.6\%; SE = 0.89)$$

$$Y = -2.599 + 2.964X_1 + 0.010X2 + 1.540X3 \dots\dots\dots\dots\dots\dots\dots (4)$$
$$(R = 0.96; R^2 = 91.9\%; SE = 0.88)$$

$$Y = -7.321 + 2.456X_1 + 0.009X2 + 5.774X3 + 2.810X4 \dots\dots\dots\dots (5)$$
$$(R = 0.97; R^2 = 93.4\%; SE = 0.80)$$

Where, X_1 = Stream Frequency, X_2 = Perimeter, X_3 = Length of Overland Flow, X_4 = Drainage Density

The results of the stepwise regression analysis showed that Stream Frequency accounts for 97.43% of the strength of the overall regression model (eq. 5). The interpretation of this is that Stream Frequency is a strong variable that can solely give meaningful explanation of infiltration potential in the study area. However, basin perimeter, Length of Overland Flow and Drainage Density also have significant influence on infiltration potential at varying degrees. The overall relationship (eq. 5) explains 93.4% of the regression plain, which is quite significant. Thus, it can be affirmed that Stream Frequency, Basin Perimeter, Length of Overland Flow and Drainage Density are strong parameters that can give meaningful explanation of Infiltration Number in Osun Drainage Basin. Therefore, infiltration potential can be predicted based on these parameters.

Table 9 presents the values of some morphometric parameters for the present study area (ODB) and Calabar Drainage Basin in the South-southern Nigeria (Eze and Efiong, 2010). The values of Elongation Ratio, Circularity Ratio and Form Factor computed for the two basins revealed that they are both relatively elongated, which implies that the basin are at advanced stage of landform development. However, based on the classification of Chow (1964), these basins have the tendency of becoming more elongated in the process of time as fluvial processes proceed. Moreover, the values of Area-Perimeter Ratio showed that ODB has higher potential to expand in the process of time.

Table 9. The values of some Morphometric Parameters of Osun Drainage Basin and Calabar Drainage Basin

S/No.	Parameter	Osun Drainage Basin (Source: Authors' Research)	Calabar River Basin [Source: Eze and Efiong, 2010]
1	Basin Area (km^2)	2,208.18	1,514.00
2	Circularity Ratio	0.32	0.34
3	Bifurcation Ratio	4.18	3.57
4	Drainage Density (km/km^2)	1.23	0.34
5	Stream Number	1,560	223
6	Elongation Ratio	0.66	0.64
7	Form Factor	0.34	0.34
8	Stream Frequency	0.71	0.15

9	Basin Length (km)	80.34	62.00
10	Basin Width (km)	43.89	43.00
11	Basin Perimeter (km)	294.14	235.00
12	Total Stream Length (km)	2,487.7	516.34
13	Main Stream Length (km)	119	68
14	Relief Ratio	5.6	0.014
15	Length of Overland Flow (km)	0.44	1.47
16	Drainage Texture	0.52	0.05
17	Area-Perimeter Ratio	7.53	6.44

The values of Drainage Density, Stream frequency and Length of Overland Flow showed that infiltration potential is higher in Calabar Drainage Basin compared to Osun Drainage Basin. This is expected as the former is located within the sedimentary environment while the latter is located within the Basement environment. The values of Relief Ratio suggest that the basins are located in environments of contrasting topographic characteristics. While the relief of Calabar Drainage Basin is observed to be relatively gentle, the relief of ODB is characterized by extreme topographic high and topographic low. Consequently, infiltration potential would be higher in Calabar Drainage Basin as surface runoff would have more time to infiltrate compared to ODB where surface runoff is relatively rapid. In the same vein, the values of Drainage Texture, Stream Number, Total Stream Length and Main Stream Length recorded for the basin indicate that larger percentage of annual rainfall would infiltrate within Calabar Drainage Basin while contrastingly, larger percentage of annual rainwater would leave ODB as river discharge as a result of the basin's relatively low infiltration potential.

4. Conclusion

This study has attempted to examine the morphometric characteristics of Osun Drainage Basin, Southwestern Nigeria, with a view to assessing its infiltration potential. Several parameters were determined and analyzed in order to have in-depth knowledge of the geomorphometric features as well as the infiltration potential of the study area. The study shows that the drainage network of the study area is partially structurally controlled. ODB tilts southwestward and the meandering main channel reflects the evidence of geological disturbance along its course. Results reveal that the basin is at advanced stage of landform development with the tendency to become more elongated in the process of time. Except for Length of Overland Flow and Drainage Density, other parameters (Basin Area, Basin Perimeter, Stream Number, Drainage Texture, Stream Length, Stream Frequency and Infiltration Number) vary heterogeneously across the sub-basins. Basin Order, Basin Area, Basin Perimeter, Stream Number, Drainage Texture and Stream Length exhibit positive and significant associations with one another. Infiltration potential-related parameters (Length of Overland Flow, Drainage Density, Stream Frequency, and Infiltration Number) do not exhibit significant association with other basin-scale morphometric parameters in the study area. Stream Frequency exhibits weak association with Basin Perimeter and River Order. The study shows that Stream Frequency is the strongest variable that influences infiltration potential. Basin Perimeter, Length of Overland Flow and Drainage Density also have significant influence on infiltration potential at varying degrees. Thus, Stream Frequency, Basin Perimeter, Length of Overland Flow and Drainage Density constitute a set of strong variables that can give meaningful explanation of infiltration potential. Analysis reveals that larger percentage of annual rainwater would leave ODB as runoff discharge as a result of its relatively low infiltration potential. Finally, results of the correlation statistics show that Infiltration Number increases with increasing Stream Frequency and Drainage Density; and Length of Overland Flow increases with decreasing Drainage Density, Stream Frequency and Infiltration Number.

The study concluded that the basin's infiltration potential is moderate as suggested by the value of Infiltration Number. However, there is the need to examine the characteristics of the basin's vadose zones as well as the aquifers, which are the major determinant factors of groundwater percolation and accumulation.

Acknowledgement

We acknowledge the Office of the Surveyor-General of the Federation, Nigeria for providing the Digital Elevation Model. The authors are grateful to the anonymous reviewers whose comments and suggestions have significantly improved the quality of this article.

References

Adejuwon, J. O., Jeje, L. K., & Ogunkoya, O. O. (1984). Hydrological Response Patterns of some Third Order Streams on the Basement Complex of Southwestern Nigeria. *Hydrological Science Journal, 28*(3), 377 –

391.

Akinwumiju, A. S. (2015). *GIS-Based Integrated Approach to Groundwater Potential Assessment of Osun Drainage Basin, Southwestern Nigeria.* An Unpublished Ph.D. Thesis Submitted to the Institute of Ecology and Environmental Studies, Obafemi Awolowo University, Ile-Ife, Nigeria. 316pp

Awoyemi, M. O., Onyedim, G. C., Arubayi, J. B., & Ariyibi, E. A. (2005). Influence of Lithology and Geological Structures on Drainage Patterns in Part of the Basement Complex Terrain of Southwestern Nigeria. *Ife Journal of Science, 7*(2), 291-296

Ayandike, R. N. G., & Phil-Eze, R. C. (1989). Runoff Response to Basin Parameters in Southwestern Nigeria. *Annals of Geography, 71A*(1&2), 75 – 84

Boesse, T. N., & Ocan, O. O. (1988). Geology and Evolution of the Ife-Ilesa Schist Belt, Southwestern Nigeria. *Symposium on Benin-Nigeria Go-traverse of Proterozoic Geology and Tectonics of High Grade Terrains.,* 87-107.

Chen, Y. C., Sung, Q., & Cheng, K. (2003). Along-Strike Variations off Morphotectonic Features in the Western Foothills of Taiwan: Tectonic Implications based on Stream-Gradient and Hypsometric Analysis. *Geomorphology, 56,* 109 – 137

Cox, R. T. (1994). Analysis of Drainage-basin Symmetry as a Rapid Technique to identify Areas oof Possible Quaternary Tilt-block Tectonic: An Example from the Mississippi Embayment. *Ame.Soc.Geol.Bull, 106,* 571-581

De Swardt, A. M. J. (1953). The Geology of the Area around Ilesa. *Geol.Surv.Nig.Bulletin, 23.*

Easthernbrook, D. J. (1993). Surface Processes and Landforms. Macmillian Publishing Co., New York. 325pp.

Elueze, A. A. (1977). *Geological and Geochemical Studies in the Ilesa Schist Belt in relation to the Gold Mineralization.* An Unpublished M. Phil. Thesis. University of Ibadan.

Eze, E. B., & Efiong, J. (2010). Morphometric Parameters of the Calabar River Basin: Implication for Hydrologic Processes. *Journal of Geography and Geology, 2*(1), 18-26.

Gregory, K. J., & Walling, D. E. (1973). Drainage Basin Form and Process: A Geomorphological Approach. Edward Arnold, London. 456 pp.

Hare, P. W., & Gardner, T. W. (1985). *Geomorphic Indicators of Vertical Neotectonism along Converging Plate Margins, Nicoya Peninsula, Costa Rica.* In: Morisawa, M. and Hack, J. T. *Tectonic Geomorphology Symposium,* Allen and Unwin, Boston, 90-104.

Haung, X. J., & Niemann, J. D. (2006). Modeling the Potential Impacts of Groundwater on Long-term drainage Basin Evolution. *Earth Surface Processes and Landforms, 31,* 1802-1823.

Horton, R. E. (1932). Drainage basin characteristics. *American Geophysical Union of Transactions,13,* 350-361.

Horton, R. E. (1945). Erosional Development of streams and their drainage basins, Hydrophysical approach to quantitative morphology. *Geol.Soc.Amer.Bull, 56,* 275-370.

Hurtrez, J. E., & Lacazeau, F. (1999). Lithological Control on Relief and Hypsometry in the Herault Drainage Basin (France), Comptes Rendues Academie des Sciences de la terra et des Planets. *Earth and Planetary Sciences, 328*(10), 687-694.

Jain, V., & Sinha, R. (2003). Evaluation of Geomorphic Control on Flood Hazard through Geomorphic Instantaneous Unit hydrograph. *Current Science, 85*(11), 26-32.

Jayappa, K. S., & Markose, V. J. (2011). Hypsometric Analysis of Kali River Basin, Karnataka, India, using Geographic Information System. *Geocarto International, 26*(7), 553-568.

Jones, J. A. A. (1999). Global Hydrology: Processes, Resources and Environmental Management. Longman. 399pp.

Lifton, N. A., & Chase, C. G. (1992). Tectonic, Climatic and Lithological Influences on Landscape Fractal Dimension and Hypsometry: Implications for Landscape Evolution in the San Gabriel Mountains, California. *Geomorphology, 5,* 77-114.

Miller V. C. (1953). A Quantitative Geomorphic Study of Drainage Basin Characteristics in the Clinch Mountain Area, Virginia and Tennessee. Department of Geology, Columbia University.

Moglen, G. E., & Bras, R. L. (1995). The Effect of Spatial Heterogeneities on Geomorphic Expression in a

Model of Basin Evolution. *Water Resources Research, 31*, 2613-2623.

Morisawa, M. E. (1957). Measurement of Drainage Basin Outline Form. *Journal of Geology, 66*, 86-88.

Morisawa, M. E. (1959). Relation of Morphometric Properties to Runoff in the Little Mill Creek, Ohio Drainage Basin. Technical Report 17. Office of Naval Research. Project NR 389-042.

Morisawa, M. E. (1962). Quantitative Geomorphology of some watersheds in the Appalachian Plateau. *Bulletin of American Society of Geology, 73*, 1025-1046.

Nag, S. K., & Anindita, L. (2011). Morphometric Analysis of Dwarakeswar Watershed, Baukura District, West Bengal, India, using Spatial Information Technology. *IJWREE, 3*(10), 212-219.

Nogami, M. (1995). *Geomorphometric measures for Digital Elevation Models* – Z.Geomorph. N. F., Suppl. Bd., 101, 53-67, Berlin, Stuttgart.

Obi-Reddy, G. E., Maji, A. K., & Gajbhiye, K. S. (2002). GIS for Morphometric Analysis of Drainage Basins. *GIS India, 11*(4), 9-14.

Odeyemi, I .B.; Anifowose, Y. B., & Asiwaju-Bello, Y. A. (1999). Multi-Technique Graphical Analyses of Fractures from Remote Sensing Images of Basement Regions of Nigeria. *J.Min.Geol., 35*(1), 9-21.

Oluyide, P. O. (1988). Structural Trends in the Nigerian Basement Complex. In Oluyide, P. O., Mbonu, W. C., Ogezi, A. E., Egbuniwe, I. G., Ajibade, A. C., and Umeji, A. C. (Eds.): *Precambrian Geology of Nigeria.* Geol.Surv.Nigeria, 99-102.

Orimoogunje, O. O. I., Oyinloye, R. O., & Momodou, S. (2009). Geospatial Mapping of Wetlands Potential in Ilesa, Southwestern Nigeria. *FIG Working Week, Surveyors Key Role in Accelerated development,* Eilat, Israel, 2009.

Passarge, S. (1912). Physiologische Morphologie. *Mitt.Geogr.Ges.Hamburg, 26*, 135-337.

Penck, A. (1894). Morphologie der Erdoberflache-Stuttgart.

Penck, A. (1896). Die Geomorphologie als genetische Wissenschaft: eine Einleitung zur Diskussion uber geomorphologische Nomenklatur – Comptes Rendas 6. Int.Geogr.Congress, London, Section C., 735-752.

Pitlick, J. (1994). Relations between Peak Flows, Precipitation and Physiography for Five Mountainous Regions in Western U.S.A. *Journal of Hydrology, 158*, 219 – 240.

Schmidt, J., & Dikau, R. (1999). Extracting Geomorphometric Attributes and Objects from Digital Elevation Models – semantics, Methods, Future Needs. In: Dikau, R. and Saurer, H. (Eds.). GIS for Earth Surface Systems. Gebruder Borntraeger, D – 14129, Berlin. D-70176 Stuttgart, 152-173

Schumm, S. A. (1956). The Evolution of Drainage Systems and Slopes in Badlands at Perth Amboy,New Jersey, *Geol.Soc.Amer.Bull, 67*, 597-646.

Singh, O., Sarangi, A., & Sharma, M. C. (2008). Hypsometric Integral Estimation Methods and its Relevance on Erosion Status of North-Western Lesser Himalayan Watersheds. *Water Resource Management*, 22, 1545-1560.

Singh, T. (2008). Hypsometric Analysis of Watersheds developed on Actively Deforming Mohand Anticlinal Ridge, Northwestern Himalaya. *Geocarto International, 23*, 417-427.

Sinha-Roy, S. (2002). Hypsometry and Landform Evolution: A Case Study in the Banas Drainage Basin, Rajasthan, with Implications for Aravalli Uplift. *Journal of Geological Society of India*, 60, 7-26.

Smith G. H. (1939). The Morphometry of Ohio: The Average Slope of the Land. *Annals of the Association of American Geographers, 29*, 94.

Smith, K. G. (1950). Standard for grading Texture of Erosional Topography. American Journal of Science, 248, 655-668.

Somashekar, R. K., & Ravikumar, P. (92011). Runoff Estimation and Morphometric Analysis for Hesaraghatta Watershed", A Remote Sensing and GIS Approach. *Journal of Indian Society of Remote Sensing, 39*(1), 95-106.

Strahler A. N. (1950). Equilibrium Theory of Erosional Slopes, approached by Frequency Distribution Analysis. *Am. Jour. Sci., 248*, 800-814.

Strahler, A. N. (1952). Hypsometric (area-altitude) Analysis of Erosional Topography. *Geol.Soc.Amer.Bull., 63*, 117-142.

Strahler A. N. (1957). Quantitative Analysis of Watershed Geomorphology. *American Geophysical Union Transactions, 38*, 912-920.

Strahler, A. N. (1964). Quantitative Geomorphology of Drainage Basins and Channel Networks. In: V. T. Chow (Ed.). *Handbook of Applied Hydrology.* McGraw-Hill, New York, pp.439-476.

Strahler A. N. (1969). Quantitative geomorphology of Drainage Basin and Network. Er. Hand book of Applied Geomorphology, Van Te Chow (Ed), New York.

Sumira, R., Pandit, A. K., Wanganeo, A., & Skinder, B. M. (2013). Drainage Basin Characteristics and Soils Erosion Intensity of Lidder Watershed in Lidder Valley, Kashmir Himalaya, India. *IJMAES, 1*(2), 47-58.

Tali, P. A., Kanth, T. A., & Wani, R. A. (2013). Morphometric Analysis of Upper Jhelum Catchment usinf Geographic Information System. *IJRSG, 2*(3), 70-75.

Walcott, R. C., & Summerfield, M. A. (2008). Scale Dependence of Hypsometric Integrals: An Analysis of Southeast African Basins. *Geomorphology, 96*, 174-186.

6

Mineralogy and Geochemistry of the Weathering Profiles above the Basement Rocks in Idi- Ayunre and Akure Districts, Southwestern Nigeria

Adewole J. Adeola[1] & Abisola M. Oyebola[1]

[1] Department of Geology and Mineral Sciences, Crawford University, Igbesa, Nigeria

Correspondence: Adewole J. Adeola, Department of Geology and Mineral Sciences, Crawford University, Igbesa, Nigeria. E-mail: kiwoleadeola@yahoo.com

Abstract

Idi-ayunre and Akure areas are part of the basement complex of southwestern Nigeria and are predominantly consisted of gneisses, granite and migmatite with some minor quartz veins and pegmatite. These rocks have been greatly weathered to form clay, laterite and soils.

Chemical analysis were carried out on basement rocks and exposed profiles. The weathering profile was subjected to X ray diffraction (XRD) analysis to determine mineralogical compositions whereas Chemical Index of Alteration (CIA) was calculated from the elemental concentrated data.

Weathering of basement rocks in Idi-Ayunre and Akure districts resulted in the formation of soil layer which ranged 0-0.4m, laterite layer 1.2-2.2m, and clayey zone 3.8-6.6m. Quartz, plagioclase, microcline, and biotite were the main minerals in parent rocks. Some of the primary minerals such as biotite and K-feldspar have been weathered to form kaolinite. Quartz, kaolinite and goethite formed the dominant minerals revealed by X-ray diffraction on decomposed granite sequences. The results from chemical analysis showed that Al and Fe have been enriched in weathering profiles of banded gneiss, migmatite gneiss and porphyritic granite whilst on the other hand Ca, Mg, Mn, Na, K, Ti were reported to be depleted.. Silica was relatively stable from basement to the topsoil in the profile. The CIA generally ranged between 80 - 99

The lateritic profiles over banded gneiss, granite and porphyritic granite of Idi-Ayunre and Akure areas varied with the composition of the parent rocks. The thick clayey layers could be of great economic importance for the production of ceramics wares and for constructional purposes.

Keywords: basement, weathering, kaolinite, laterite, Nigeria

1. Introduction

The majority of the soils present in the tropics are reddish soils called lateritic soils or laterite soils according to Anupam S. and Rajamani V. (2000). Lateritic soils are the commonest tropical soils and they are widely distributed soils in southwestern Nigeria in an area is characterized by tropical climate. Lateritic soils are essentially product of tropical or subtropical weathering of various crystalline igneous rocks, sediments, detrital deposits and volcanic ash. The climatic conditions favorable for laterization include: the warm and humid atmospheric conditions, abundant and regular rainy seasons with well defined dry and wet seasons,

The Precambrian basement complex of southwestern Nigeria consists predominantly of gneisses, schists and quartzites with emplacement of granitic and basic rocks. Some of the basement rocks have been greatly weathered to form residual soils. In southwestern Nigeria, extensive occurrences of such residual bodies have been reported by various scientists. Emofurieta et, al. (1995) examined the secondary geochemical and mineralogical dispersion patterns associated with lateritization process in Ile-Ife, southwestern Nigeria. Kehinde-Phillips et. al. (1995) described the mineralogy and geochemistry of the weathering profiles over amphibolites, anthophylite and talc schists in the Ilesha schist belt, southwestern Nigeria. According to Bolarinwa (2006) examined the mineralogy and geochemistry of the weathering profiles above basement rocks in Ibadan and concluded that the weathering profiles are composed of quartz, kaolinite, goethite and limonite. Ige et al. (2005) studied the mineralogy and geochemistry of lateritic weathering profiles on ultramafic rock bodies around Mokuro in Ile-Ife area, southwestern Nigeria and observed that the weathering is towards lateritization. According to Adegbuyi et al. (2015), the weathered products in Akure, Isua-Akoko areas are predominantly clay materials. These clay deposits are kaolinitic clay and plastic kaolin and they are found to be very good raw materials for the industrial production of ceramics, paints and refractory, low alumina chemical pharmaceutical products and as additives in cement industry.

Investigation of weathered profiles reveals erratic variations in the major oxide composition and pronounced dissimulation in mineralogical characteristics in line with the parent rock and chemical environments. Hence, translocation and redistribution of the weathering products through groundwater and percolating rain water under appropriate Eh and pH conditions resulted in the lateritization or duricrust formation (Krauskopf, 1985).

Metallic ores and residual clays of economic values often occur within lateritic profiles above parent rocks in tropical regions due to supergene enrichment and leaching of labile components. Previous studies in Idi Ayunre and Akure district, were mainly on geochemistry of the basement rocks with little attention on the residual profiles over other rock types. The studying of the removal of mobile elements by meteoric water from the crystalline rocks, the resulted chemical variation along the profile and subsequent concentration of stable weathering products is the focus of present investigation

2.Geologic Setting

The basement rocks in Akure and Idi-Ayunre have been weathered to produce the residual soils. Ola (1983) defined lateritic soil as a residual soil derived essentially from chemical weathering of igneous and metamorphic rocks under tropical climate. During weathering, decomposition of rock and formation of soil occur simultaneously and are indistinguishable from each other. Thus, soil formation can be considered as an advanced stage of weathering. The unique characteristic of soil is the organization of its constituents and properties into layers that are related to the present-day surface and that change vertically with depth. Consequently the essential process in soil formation is the transportation of material from one level in the weathered debris to another resulting in different compositional horizons which may range in thickness from a few centimeters to several meters. The vertical sequence or soil profile generally comprises three principal horizons identified from the surface downward as A, B and C. A-horizon is the soil layer from which down-ward moving water has removed much of the soluble material, the B-horizon is the intermediate layer in which some of the soluble and colloidal materials are deposited and the C-horizon is the zone of fragmented but still largely unaltered rock that grades into underlying bedrock. Laterites commonly occur within the B-horizon below the top-soil. The parent rock, climate, topography, biological activity and time are the main factors that influence the development and the nature of a soil profile.

Idi-Ayunre and Akure areas lie in southwestern Nigeria (Figure1). The investigated residual profiles in Idi-ayunre are at the outskirt of Ibadan along the Abanla road. The areas of study are easily accessible due to the presence of secondary and primary roads, road cuts and recently blasted outcrops.

The major rock associations of Idi Ayunre and Akure areas are part of the Basement complex of Nigeria. Idi-Ayunre is underlain by banded gneiss and granite gneiss The minor rocks such as quartz vein and pegmatite occur as intrusions in the major rock types (Rahaman 1984). The entire western part of Idi-Ayunre is underlain by banded gneiss while the eastern part is predominantly made up of granite gneiss (Figure 2). Mineralogically, it is consisted of plagioclase, quartz, and biotite. Banded gneiss is strongly foliated and the foliation is defined by the alternation of mafic and felsic bands. The mafic bands contain mostly biotite while the felsic layer are made up of quartzofeldspartic minerals. The banded gneiss is highly susceptible to weathering due to its exposed planes of weakness.

According to Oyinloye (2007), Akure area is underlain by porphyritic granite, charnockite, banded gneiss and migmatite. (Figure 3) The porphyritic granite occurs mostly as massive outcrops around Oba-Ile area and along Owo road, Akure. It consists of phenocryst of feldspar mineral embedded in the matrix of feldspar and quartz. The felsic minerals consist of feldspar and quartz while the mafic mineral consists of biotite and other accessory minerals. The mafic minerals are randomly distributed within the porphyritic granite. The migmatite which is made up of leucocratic components alternating with basic components is located at the central part of Akure. The leucocratic component ranges from fine-grained granitic gneisses to medium grained banded gneisses especially around Shagari estate, Akure (Figure 3).

3. Material and Methods

Samples of rocks, clay, laterite and soil were obtained from fresh road cuts at specific intervals from top soil down to fresh basement rocks. Sixty samples comprising 15 each of top-soil, laterite, clay and rock were collected from three lateritic profiles above granite gneiss, migmatite, and porphyritic granite in the Idi-Ayunre and Akure areas. Petrographic study was carried out on thin sections of the rock samples. Clay, laterite and soil mineralogy was determined using X-ray Diffraction (XRD) at Activation Laboratory in Canada. The X-ray diffraction analysis was performed on a Panalytical X'Pert Pro diffractometer, equipped with a Cu X-ray source and an X'celerator detector, operating at the following conditions: voltage - 40 kV; current - 40 mA; range - 5-80 $^{\circ}2\theta$; step size: 0.017 , $^{\circ}2\theta$; time per step: 50s; divergence slit: fixed, angle 0.5°. The crystalline mineral phases were identified by X'Pert HighScore Plus software using the PDF-4 ICDD database. The quantities of the crystalline minerals were determined using the Rietveld method. The Rietveld method is based on the calculation of the full diffraction pattern from crystal structure information. Elemental compositions of the rocks and clay samples were determined using Inductively Coupled Plasma-Mass Spectrometer (ICP-MS) at Activation Laboratory, Canada . For ICP-MS, microwave high pressure/temperature decomposition of samples (230°C, 7.0MPa; Paar Physical Multiwave sample preparation system) using Merck Suprapurs grade reagents (HF, $HCIO_4$, HNO_3 and HCl). All measurements were made on a Sciex/Perkin-Elmer ELAR 6000 ICP-MS. The Chemical Index of Alteration (CIA = Al_2O_3 /Al_2O_3 + CaO + Na_2O + K_2O) X 100, where all components expressed in molecular proportions was calculated to measure the Intensity of Chemical Weathering (ICW) of soils.

4. Results and Discussion

4.1 Field Work and Macro-Petrography

Profile 1 is located above the banded gneiss at Idi- Ayunre via Ibadan. A 6m thick profile over banded gneiss is exposed in a quarry and is essentially residual. The presence of relic structures of parent rock and angular quartz grains in the profile strongly supported its insitu nature. There is no evidence of large scale movement of weathered materials. Four distinct layers were identified based on colour, texture along with relic structures in the saprolitic zone. It is noticed on the field that the color gets brownish towards the top. (Figure4a)

The upper horizon which is the topsoil is generally light brown in colour and is about 0.2m thick. It contains rootlet of plants and is characterized by the presence of organic matter (humus). This layer is friable and contains some pebbles of quartz. Below this layer is a reddish and partly consolidated laterite which is about 1.2m thick. The lateritic layer grades gradually into the underlying clayey horizon. In this layer, the organic content has almost completely disappeared. The clay layer is underlain by saprolitic zone characterized by some whitish and reddish spots and this graded into the bedrock.

Profile 2 is located above porphyritic granite in Akure. A 7m thick profile was exposed around Fiwasaye Girls Grammar School, Akure (Figure 4b). The weathering profile over granite gneiss consists of four distinct layers. The layers are characterized based on their texture, colour and relic structures in the saprolitic zones. The topsoil layer is dark grey in colour and is about 0.3m thick. It contains rootlets of plants, organic matter, decomposed leaves and angular quartz pebbles. Below this horizon is a reddish brown laterite with a thickness of about 3m,organic matter has completely disappeared. Laterite grades gradually into the clay layer which is about 3m thick. The clay horizon is brownish in colour and organic material in this layer has disappeared completely. The saprolite characterized by relic structures grades to the fresh parent rock. The fresh bedrock could not be reached due to the thickness of the weathered profile.

Profile 3 is located above migmatite gneiss opposite the Federal University of Technology Akure (FUTA). A 10m thick profile was exposed at the back of the filling station about it is about 300m opposite the FUTA second gate. (Figure 4c). In the field, there is no evidence of soil movement or transportation, hence, they are in-situ. Three layers are recognized and they are characterized by their different colours and textures. The upper layer is dark grey in colour and it is about 0.5m thick, it contains rootlets of plants and angular pebbles of quartz which proves that the profile is residual. The topsoil grades gradually into the underlying reddish brown laterite which is about 3m thick. Laterite grades gradually into clay horizon which is brown in color with an average thickness of 8m. It is noticed on the field that the clay gets finer towards the bottom and shows colour laminations of the mineralogical banding of the parent rock. The basement rock could not be reached due to thick overburden.

4.2 Mineralogy

The X-ray diffractograms of soil, laterite and clay on the banded gneiss are shown in Figure 5. Conspicuous peaks of quartz, microcline and albite are recorded by the X-ray chart of the soil sample (Figure 5), while quartz, kaolinite and goethite are the major minerals reflected in the difractograms of the laterite and clay. Quartz is apparently present in all the profiles because of its high resistance to weathering. Biotite and other ferromagnesian minerals in the fresh rock have probably weathered to generate kaolinite in the laterite. While microcline which is present in the soil has also altered by chemical weathering to kaolinite and equally leached into laterite horizon.

On the migmatite gneiss, only the peaks of quartz, microcline and albite are recorded by the X-ray chart of the soil sample. The presence of microcline and albite show that weathering is still at incipient stage (Figure 6), while quartz, kaolinite and goethite are the major minerals reflected in the diffractograms of the laterite and kaolinite and quartz dominated the clay charts.

The X-ray diffractograms on the porphyritic granite show that the soil is predominantly made of quartz and kaolinite, microcline, albite. Kaolinite and quartz are the major minerals reflected in the diffractograms of the laterite, while kaolinite and quartz dominated the clay charts (Figure 6).

4.3 Chemical Compositions

The average chemical data of weathering profile over banded-gneiss at Idi-Ayunre are presented in Table 1. The Table shows the concentrations of major and trace elements of banded gneiss. The average compositions of SiO_2 are 66.04%, 59.25%, 57.74% and 67.05% in the parent rock, clay, laterite and soil respectively. This shows that relative to parent rock, there has been depletion in the SiO_2 in the weathering profile (Figure8). The enhanced value of 67.05% in the top soil may be due to relative depletion of MnO, MgO, CaO, Na_2O and K_2O in the soil horizon. Hence, free quartz, SiO_2 is present in silicate minerals and their weathering and dissolution apparently led to the enrichment of SiO_2 and Fe_2O_3 in the top soil. The average concentration values of Al_2O_3 increase from 15.85% in the parent rock to 22.53% in the clay with 24.63% in laterite and slightly reduced to 17.21% in the top soil. This indicates that there is a significant enrichment of Al_2O_3 in the clay and laterite profile compared to the parent rock (Figure 8). From the chemical data, the relative enrichment could be explained by the removal of MgO and weathering of Al_2O_3 bearing minerals such as biotite in the banded-gneiss. Fe_2O_3 has average of 4.10%, 4.72%, 6.73% and 7.79% in the parent rock, clay, laterite and soil respectively. This indicates a strong enrichment of Fe_2O_3 in the weathering profile relative to the parent rock. This is typical for the weathering and lateritization of felsic rocks. The values of the average Fe_2O_3 concentration show that the iron bearing minerals such as biotite in the parent rock have probably been affected by chemical weathering to release iron oxides/hydroxide in the weathering horizons (Kehinde-Phillips et al. 1995). The values of CaO, Na_2O, K_2O, P_2O_5 and MnO and reduce up the profile indicating strong leaching. The progressive dissolutions and mobility of Mg, Ca, K, Na led to their depletions in the weathering profiles (Aleva 1994, Kehinde-Philips 1991, Bolarinwa 2006).

Trace element data in Table 1 show enhancement of Ba, Zr, Y, Sc in clay, laterite and soil compared to the parent rock, however, Sr is depleted from 780 to 363ppm in clay subsequently being reduced to 115ppm in laterite. The silica-sequioxide ratio (S.R) of the rock (3.31), clay (3.35), laterite (2.04), soil (2.22) are stated in Table (1), the alumina-iron (A.R) are (3.86), (4.77), (6.60), (6.17) for rock, clay, laterite and soil. The relatively high S.R and A.R strongly suggest that true laterites are produced from the weathering of iron bearing minerals in the banded gneiss (Aleva, 1994). The total values of (MgO + CaO) are 6.33 in the rock and 1.05, 0.71 and 1.73 in the clay, laterite and soil respectively while the total values of (Na_2O + K_2O) is 5.21 in the rock, 0.93, 5.39 and 4.85 in clay, laterite and soil. Relatively high Na_2O and K_2O are mainly contributed by the weathered plagioclase and muscovite minerals while low MgO and CaO is probably due to the low amount of mafic minerals in the parent rock, this trend observed is similar to the granite gneiss at Ibadan area by Bolarinwa 2001

The average concentrations of major and trace elements in the weathering profile over migmatite are presented in Table 2. The average compositions of SiO_2 are 68.49%, 47.43%, 42.64% and 44.37% in the parent rock, clay, laterite and soil respectively. This shows that relative to the parent rock, there has been depletion in the SiO_2 in the weathering profile. The average concentration values of Al_2O_3 increase from 15.53% in the parent rock to 36.01% in the clay with 28.59% in laterite. This indicates that there is a significant enrichment of Al_2O_3 in the clay and laterite profile compared to the parent rock. Such enrichment could be explained by the removal of MgO and weathering of Al_2O_3 bearing minerals such as biotite in the migmatite gneiss (Figure 9). Fe_2O_3 has average of 2.85%, 2.46%, 15.03% and 14.25% in the parent rock, clay, laterite and soil respectively. This indicates strong enrichment Fe_2O_3 in the weathering profile relative to the parent rock. This is typical of the weathering and lateritization of felsic rocks. The values of the average Fe_2O_3 concentration show that the iron bearing minerals (e.g biotite) in parent rock have probably been affected by chemical weathering to release iron oxides/hydroxide in the weathering horizons.

The depletion of MgO, CaO, Na_2O, K_2O, P_2O_5, and MnO are particularly pronounced and they are removed at a faster rate. The corresponding values for MgO are 1.18%, 0.05%, 009.% and 0.02% CaO are 4.34%, 0.01%, 0.02% and 0.37%, Na_2O are 4.95% 0.01%, 0.01%, 0.02%, K_2O are 1.31%, 0.03%, 0.53%, 0.21%, P_2O_5 are 0.14%, 0.05% 0.09%, 0.19%, MnO are 0.05%, 0.01%, 0.01%, 0.14%, Cr_2O_3 are 0.23%, 0.02%, 0.07 in the rock, clay, laterite and soil respectively. This progressive dissolutions and mobility of magnesium, calcium, potassium, sodium has led to their depletions in the weathering profiles (Kehinde – Philips 1995, Bolarinwa 2006). According to Aleva (1994), the decrease in CaO, Na_2O, K_2O, P_2O_5, MnO and Cr_2O_3 contents of the weathering profiles relative to the parent rock is due to leaching of these oxides.

Trace element data in Table 2 show enhancement of Zr, Y, Sc in clay, laterite and soil compared to the parent rock, however, in Ba and Sr exhibit depletion. The silica-sequioxide ratio (S.R) of the rock 3.77 , clay 1.22, laterite 0.99, soil 1.23 are stated in table (2) the alumina-iron (A.R) are 5.39, 14.63, 1.90, 1.59 for rock, clay, laterite and soil respectively. The relatively high S.R and A.R strongly suggest that true laterites are produced from the weathering of iron bearing minerals in the pegmatite (Aleva, 1994). The total values of alkalis (MgO + CaO) are 5.52 in the rock and 0.06, 0.11 and 0.06 in the clay, laterite and soil respectively while the total values of (Na_2O + K_2O) is 6.26 in the rock, 0.44, 0.06 and 0.26 in clay, laterite and soil. The silica- sequioxide ratio has been used to quantify the degree of weathering and lateritization (Martins and Doyne, 1927).

The average concentrations of major and trace elements in the weathering profile over porphyritic granite at Akure are presented in Table 3. The mean compositions of SiO_2 are 68.86%, 50.11%, 46.47% and 50.12%in the parent rock, clay, laterite and soil respectively. This shows that relative to the parent rock, there has been depletion in the SiO_2 in the weathering profile. The enhanced value of 50.12% in the top soil may be due to relative depletion of MnO, MgO, CaO, Na_2O, K_2O, in the soil horizon. Hence, apart from the occurrence of free quartz, SiO_2 is present in silicate minerals such as biotite, amphiboles and pyroxene. The weathering and dissolution of these silicate minerals apparently led to the enrichment of SiO_2 and Fe_2O_3 in the top soil. The average concentration values of Al_2O_3 increase from 15.39% in the parent rock to 32.32% in the clay with 28.95% in laterite. This indicates that there is a significant enrichment of Al_2O_3 in the clay and laterite profile compared to the parent rock. From the chemical data, the relative enrichment could be explained by the removal of MgO and weathering of Al_2O_3 bearing minerals such as biotite in the pegmatite. Fe_2O_3 has average weight percentage of 3.3%, 3.45%, 10.02% and 8.34% in the parent rock, clay, laterite and soil respectively. This indicates strong enrichment Fe_2O_3 in the weathering profile relative to the parent rock. This is typical of the weathering and lateritization of felsic rocks. The corresponding values for CaO are 3.38%, 0.02%, 0.03% and 0.33%, Na_2O are 3.15% 0.01% 0.02%, 0.18%, K_2O are 1.75%, 0.37%,0.58%, 2.28%, P_2O_5 are 0.14%, 0.09%0.34%,0.20%, MnO are 0.05%,0.01%, 0.02%, 0.08% Cr_2O_3 are 0.01%, 0.01%, 0.007% in the rock, clay, laterite and soil respectively. This progressive dissolutions and mobility of magnesium, calcium, potassium, sodium has led to their depletions in the weathering profiles (Kehinde – Philips 1995, Bolarinwa 2006). According to Aleva (1994), the decrease in CaO, Na_2O, K_2O, P_2O_5, MnO and Cr_2O_3 contents of the weathering profiles relative to the parent rock is due to leaching of these oxides. Trace element data in Table 3 show enhancement of Sr, Zr,Y,Sc in clay, laterite and soil compared to the parent rock. As stated in Table 3, the silica-sequioxide ratio (S.R) of the rock, clay, laterite and soil are 7.84%, 1.80%, 1.19%, 1.80% respectively while alumina-iron (A.R) are 4.63%, 9.32%, 2.60%, 2.69% for rock, clay, laterite and soil. The relatively high S.R and A.R strongly suggest that true laterites are produced from the weathering of iron bearing minerals in the migmatite gneiss (Aleva, 1994). The total values of (MgO + CaO) are 4.45% in the rock and 2.47%, 0.60% and 2.47% in the clay, laterite and soil respectively while the total values of (Na_2O + K_2O) is 6.06% in the rock, 0.59%, 0.12% and 0.59% in clay, laterite and soil. This is similar to the trend observed in banded gneiss and migmatite.

5. Mineralogical and Geochemical Trends in the Soil.

Weathering of banded gneiss, porphyritic granite, and migmatite in Idi-Ayunre and Akure areas resulted in the formation of soil layer (0 - 0.5m), laterite layer (0.9-1.8m), and clayey zone (3.9-7.6m). The maximum depth of the profile observed in the field is 10m.

The rocks samples are grey or light coloured due to their felsic nature. Petrographic study carried out on thin sections showed that quartz, plagioclase feldspars, and biotite were the essential minerals in the granitic rocks Whole rock mineralogy determined using the X-ray diffraction showed that quartz and kaolinite were the dominant minerals in the decomposed rocks. In most of the soil horizon, prominent peaks of quartz, kaolinite and microcline were recorded while kaolinite, quartz, and goethite peaks are more conspicuous in the laterite layer. The concentration of goethite in the laterite horizon supported leaching from the top-soil and supergene enrichment of the iron oxides in the underlying laterite horizon (Kehinde-Phillips 1995). The clayey zone, which is 3.9m, 4.8m and 8m thick on banded gneiss, porphyritic granite and migmatite respectively, showed conspicuous peaks of kaolinite. Bauxite mineral (gibbsite) was conspicuously absent in the profile, the trend of weathering in all the profiles is towards iron enrichment (ferralitization) rather than aluminium accumulation (bauxitization).

The general trend in all the profile is a decrease in SiO_2 content from bedrock to the lateritic zone. Loss of silica up the profiles on granitic rocks could be due to the break down of silicate minerals due to weathering and leaching of silica as silicic acid $SiO(OH)_2$, in the solution along with Ca, Na, K and Mg. The Fe_2O_3 content increased from the bedrock to the lateritic zone due to free ions produced during weathering which combined with air and water to form oxide or hydroxide. The increase could also be due to the development of clay complexes from intense weathering of feldspar to form clay. It could also result from hydration of hematite near the surface to produce hydrate iron oxide or hydrate limonite [$FeO(OH). nH_2O$], which formed a secondary coating in most soil particles. This is similar to the trends reported in the weathering profiles of Older granite and granite gneiss in Ibadan, Abeokuta, Ilesha and Ile-Ife areas (Emofurieta et al., 1995, Ige et. al. 2005 Bolarinwa, 2006). The Na_2O, K_2O, CaO, P_2O_5 content increased down the profile from the lateritic zone to the bedrock. Decrease in their concentrations upward towards A horizon is due to decomposition of silicate minerals (feldspar) during weathering to form clay minerals and leaching of these elements.

Chemical Index of Alteration (CIA) = $AL_2O_3 /AL_2O_3 + CaO + Na_2O + K_2O$) X 100, where all components expressed in molecular proportions is commonly used as a measure of the Intensity of Chemical Weathering (ICW) of soils. (This parameter was derived by Nesbitt and Young 1982, 1984). Nesbitt and Young (1984), indicated that CIA values of 30 to 55 is an indication of weathering at incipient zone, while CIA values ranging from 51 to 85 could be considered as intermediate zone of weathering. Weathering that is at the advanced stage will have CIA values greater than 85. Tables 1,2 and 3 show the summary of the average values of the calculated Chemical Index of Alterations (CIA) for the different weathering profiles in Idi-Ayunre and Akure areas. Based on the CIA values recorded in Tables, it is observed that all weathered profiles have experienced moderate to high intensity of chemical weathering. The average CIA values of banded gneiss are 57.13 and 81.39 for rock and laterite respectively. These values showed increasing trends of weathering that have reached the moderate stage. The average values of CIA of migmatite are 51.05, 99.29, 99.72 and 98.36 in the rock, clay, laterite and soil respectively while, the CIA values for porphyritic profile from rock to weathered layers, varied from 59.11 (rock) to 97.20 (laterite). The values show progressive chemical weathering that has reached a matured stage with high intensity in the Akure area (Figure 11), this is similar to the intensity of weathering in gneissic rocks observed by Nesbitt and Young, (1984) and Anuspam and Rajamani, (1999). The clay mineralogy and CIA values of banded gneiss, migmatite and porphyritic granite showed that nearly all the primary minerals have decomposed to form secondary minerals such as kaolinite. This trend is similar to the weathering of the Ife, Ibadan and Abeokuta banded and granitic rocks by Emorforieta et al. (1995), Kehinde-Phillips and Tietz (1995) and Bolarinwa and Elueze (2004).

The silica to sequioxide (S.R) ratio is used to determine the types of soil formed from the chemical weathering of the parent rocks. True laterite is assigned a ratio of 1.33; lateritic soil 1.33 to 2.0 and non lateritic if the value is greater than 2.00 (Nesbitt and Young 1984, 1989, Adeyemi and Ogundero 2001, Ola 1982 and Martin and Doyen 1927). Results from Tables 1, 2 and 3 showed that laterites above migmatite and porphyritic granite with average S.R of 0.99 and 1.19 can be categorized as true laterite. The soils formed from banded gneiss with S.R ratios of 2.04 are non lateritic soils due to their high S.R values.

The ternary plots of SiO_2 - Al_2O_3 - Fe_2O_3 (Gardner and Walsh 1996) of the different profiles showed that the trend of weathering inclines towards lateritization rather than bauxitization (Figure 12). The dominance of iron bearing minerals such as goethite n the lateritic profiles and the high Fe_2O_3 content in their geochemical data proved that the weathering is towards iron enriching (ferralitization) rather than aluminum accumulation (bauxitization).

6. Conclusions

Lateritization is the dominant process of chemical weathering in tropical and subtropical regions of the world. This present case-study reveals that the weathering of banded gneiss, migmatite, and porphyritic granite has reached advanced stage and the dominance of iron bearing minerals in the lateritic profiles shows that the weathering is towards lateritization rather than bauxitization. The leaching process involving the alkali and alkali earth metals resulted in a residue enriched in quartz, Fe-oxides and kaolinite clays which vary in proportion along the profiles. This study also shows that laterites above migmatite and porphyritic granite can be categorized as true laterite while the soils formed from banded gneiss are non lateritic soils due to their high S.R values

Acknowledgement

The authors wish to acknowledge the Geology Department of Crawford University for permission to use the laboratory for petrographic analyses..

References

Adegbuyi, O., Ojo, G. P., Adeola, A. J., & Alebiosu, M. T. (2015) Compositional and industrial assessment of Isua Akoko, Akure. Ayadi and Ode Aye clay deposits of Ondo State, Nigeria. *Global Journal of Pure and Applied Sciences, 21,* 38-46

Adeyemi, G. O. (1992). Highway geotechnical properties of laterised residual soils in the Ajebo–Ishara geological transition zone of southwestern Nigeria. Unpublished Ph.D Thesis. Department of Geology, Obafemi Awolowo University, Ile – Ife, Nigeria.

Adeyemi, G. O., & Ogundero, O. C. (2001). Some Geotechnical properties of soil developed over migmatite-gneiss in Oru-Ijebu, southwestern Nigeria. Journal of Applied Sciences. 4(3) 2130-2150. Aleva, G.J. 1994. Laterites; Concepts, Geology, Morphology and Chemistry, ISRIC, Wageningen. 169p

Anupam S., & Rajamani, V. (2000). Weathering of gneissic rocks in the upper reaches of Cauvery river, south India: Implications to neotectonics of the region. *Chemical Geology, 166,* 203-223.

Bolarinwa, A. T. (2006). Mineralogy and geochemistry of the weathering profiles above basement rocks in Ibadan, southwestern, Nigeria. *Global Journal of Geological Sciences*, *4*(2), 183-191.

Bolarinwa, A. T. (2001). Compositional characteristics and economic potentials of the lateritic profiles over basement and sedimentary rocks in Ibadan-Abeokuta area, southwestern Nigeria, Unpublished Ph.D. Thesis, University of Ibadan. 255p.

Elueze, A. A., & Bolarinwa, A. T. (1994). Assessment of function, application of lateritic clay bodies in Ekiti environs, south western Nigeria. *Journal of Mining and Geology, 31,* 79-83.

Elueze, A. A., Ekengele, N. L., & Bolarinwa, A. T. (2004). Industrial assessment of the residual clay bodies over gneisses and schists of Younde area, southern Cameroon. *Journal of Mining and Geology, 40*(1), 9-15.

Emofurieta, W. O., Ogundimu. T. O., & Imeokparia, E. G. (1944): Mineralogical, Geochemical and Economic Appraisal of some clay and shale deposits in Southwestern and northeastern Nigeria. *Jour. Mineral. Geol.,* 30(2), 151-159.

Gardner, R., & Walsh, N. (1996). Chemical weathering of metamorphic rocks from low elevations in southern Himalaya. *Chemical Geology, 127,* 161-176.

Ige, O. A., Durotoye, B., & Oluyemi, A. E. (2005). Mineralogy and geochemistry of lateritic weathering profiles on ultramafic rock bodies around Mokuro in Ile-Ife area, southwestern Nigeria. *Journal of Mining and Geology, 41*(1), 11-18.

Kehinde-Phillips, O. O. (1991). Compositional variations within laterite profiles over mafic and ultramafic rock units of the Ilesha schist belt, Southwestern Nigeria *PhD. Thesis* university of Ibadan, Unpubl 201p.

Kehinde-Phillips, O. O., & Tietz, G. F. (1995). The mineralogy and geochemistry of theweathering profiles over amphibollite and Talk-schists in the Ilesha scist belt, southwestern. *Nigeria Journal of Mining and Geology, 31*(1), 53-62.

Martins, F. J., & Doyne, F. C. (1927): Laterite and lateritic soils in Sierra Leone. *Int. Jour, Agric sci., 17,* 530-55.

Nesbitt, H. W., & Young, G. M. (1984). Prediction of some weathering trends of plutonic and volcanic rocks based on thermodynamic and kinetic. *Geochim. Cosmochin. Acta, 48,* 1523-1534.

Ola S. A. (1982). Geotechnical properties and behavior of some Nigeria lateritic soils. In Ola S.A. (Ed), *Tropical soils of Nigeria in Engineering Practice* (pp. 61-63).

Oyinloye, O. A. (2007). Geology and geochemistry of some crystalline besement rocks in Ilesha area, southwestern Nigeria: Implications on provenance and evolution. *Pakistan Journal of Scientific and Industrial Research, 50*(4), 233-231.

Rahaman, M. A. (1984). Recent advances in the study of the basement complex of Nigeria. In Oluyide P.O. et al. (Eds.), *Precambrian Geology of Nigeria* (pp 241-256).

Appendix

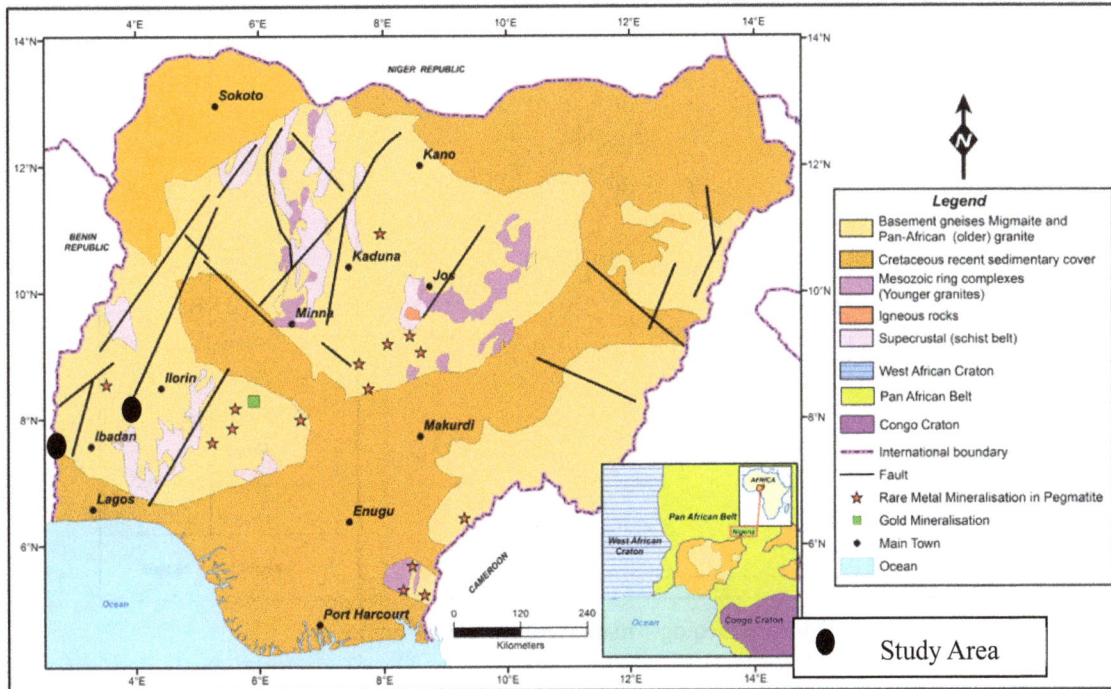

Figure 1. Map of Nigeria showing the location of the study area

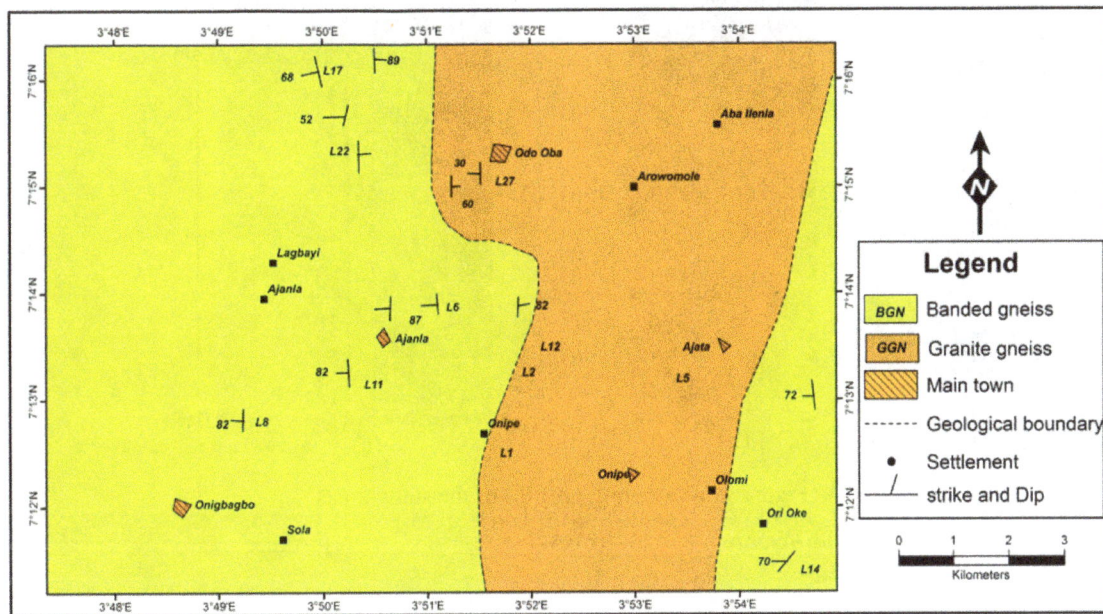

Figure 2. Geological map of Idi-Ayunre area and its' environs

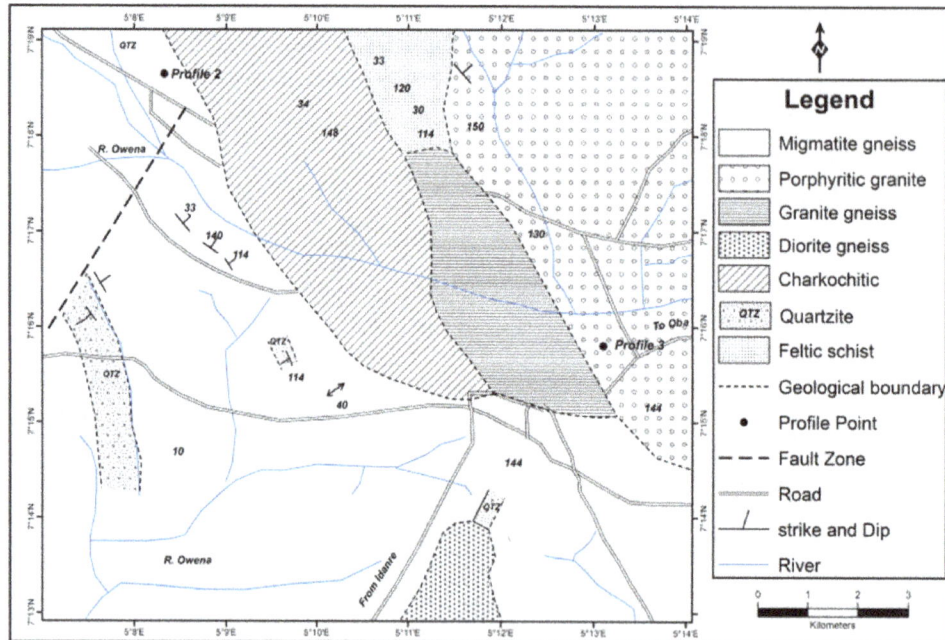

Figure 3. Geologic map of Akure and its' environs

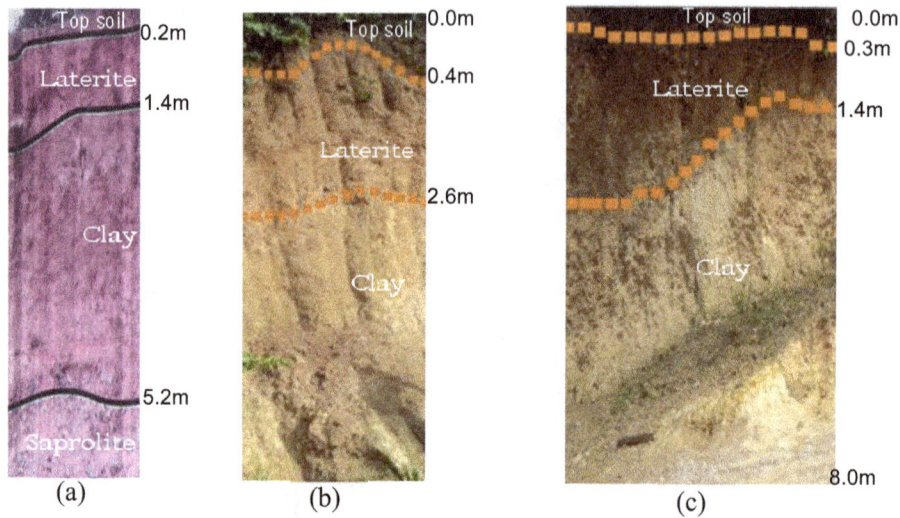

Figure 4. Weathering profiles of the study areas

(a)- Idi-Ayunre   (b)-Akure,   (c)- Akure

Mineralogy and Geochemistry of the Weathering Profiles above the Basement Rocks...

79

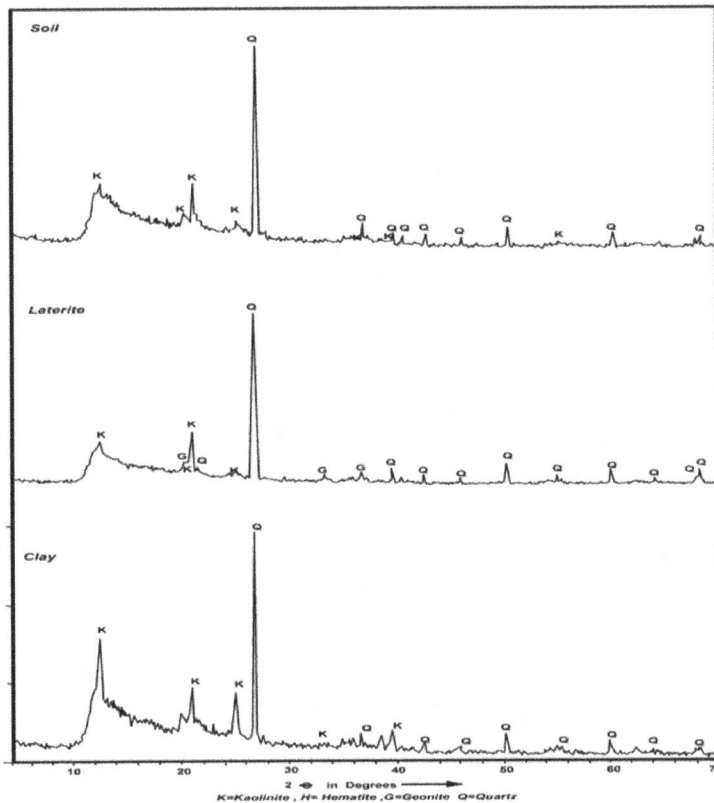

Figure 5. X-ray diffractograms of the weathered profiles above banded gneiss at Idi-Ayunre, Ibadan

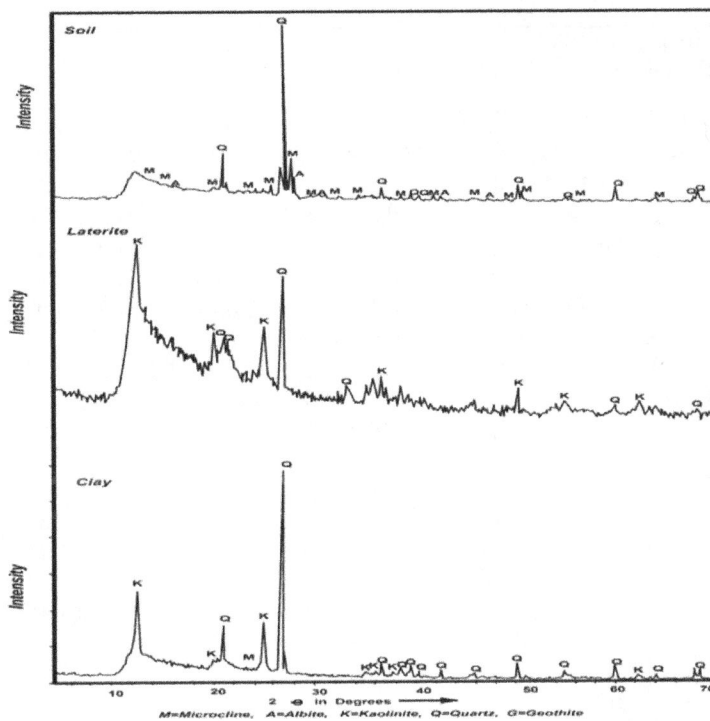

Figure 6. X-ray diffractograms of the weathered profiles above migmatite gneiss at Akure

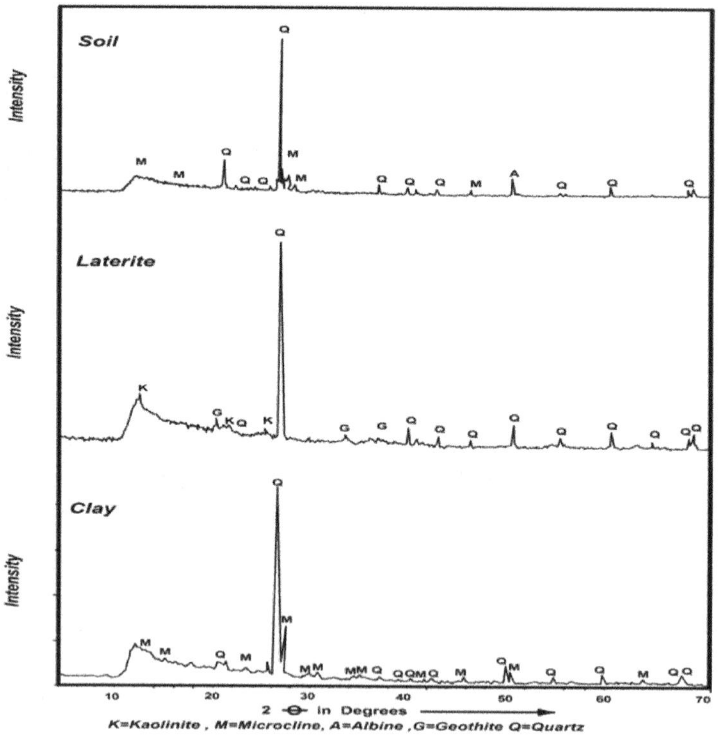

Figure 7. X-ray diffractograms of the weathered profiles above porphyritic granite at Akure

Table 2. Average chemical composition of major (%) and trace element (ppm) over banded gneiss at Idi-Ayunre area

Oxides	Rock Mean	Rock Range n=5	Clay Mean	Clay Range n=5	Laterite Mean	Laterite Range n=5	Top soil Mean	Top soil Range n=5
SiO_2	68.49	67.23-69.75	47.43	46.55-47.79	42.64	38.77-49.54	44.37	43.43-44.96
Al_2O_3	15.35	14.99-15.17	36.01	35.71-36.20	28.59	24.67-31.07	22.12	19.69-23.34
Fe_2O_3	2.85	2.56-3.14	2.46	1.81-3.76	15.03	13.98-16.36	14.25	12.73-17.05
MgO	1.18	1.02-1.34	0.05	0.04-0.08	0.09	0.09-0.10	0.23	0.22-0.25
CaO	4.34	3.98-4.71	0.01	0.01-0.01	0.02	0.01-0.03	0.37	0.34-0.41
Na_2O	4.95	4.72-5.18	0.01	0.01-0.01	0.01	0.01-0.01	0.02	0.01-0.04
K_2O	1.31	0.19-1.72	0.025	0.12-0.49	0.05	0.03-0.07	0.21	0.18-0.26
TiO_2	0.20	0.17-0.23	0.84	0.68-0.94	1.62	1.54-1.75	3.42	3.28-3.57
P_2O_5	0.14	0.12-0.17	0.05	0.03-0.09	0.09	0.05-0.13	0.20	0.17-0.23
MnO	0.05	0.05-0.06	0.01	0.01-0.01	0.02	0.01-0.02	0.14	0.12-0.15
LOI	0.50	0.050-0.51	12.75	12.44-12.98	11.68	9.90-12.62	14.35	13.96-14.87
Total	99.36		99.65		99.85		99.68	
Trace Element (ppm)								
Ba	610.5	622-1130	11	5-22	69.3	37-136	217.67	193-261
Sr	719.5	665-780	4	2-8	29.3	13-45	39.3	52-73
Zr	107	4-7	504	677-827	473	433-524	970	930-1058
Y	19	3-7	22.3	22-23	13	13-15	28.67	25-31
Nb			41.67	18-54	35	28-51	1.67	43-63
Sc	13	1-1	10.67	12-18	22	22-32	28.3	26-30
Silica – Aluminum Ratio								
SR	3.77	3.97-3.57	1.22	1.18-1.24	0.99	0.84-1.28	1.23	1.18-1.26
AR	5.39	5.30-6.10	14.63	9.50-19.95	1.90	1.76-2.10	1.59	1.15-1.80
MgO+ CaO	5.52	5-6.05	0.6	0.56-0.66	0.11	0.10-0.13	0.06	0.05-0.09
Na_2O+ K_2O	6.26	6.09-6.44	0.44	0.20-0.81	0.06	0.04-0.08	0.26	0.13-0.50

CIA   57.13   56.20- 58.31   79.67   74.57- 82.23 81.39   81.07- 81.64   71.52 71.00-71.54

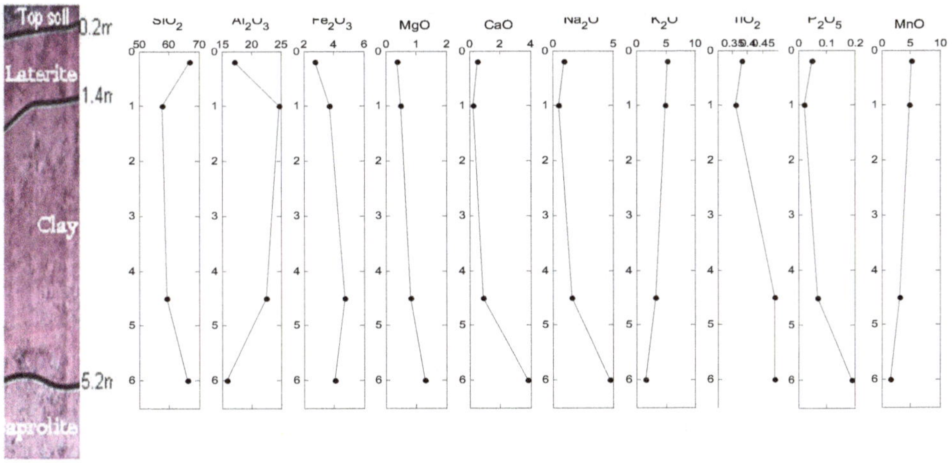

Figure 8. Chemical variation along the weathering profile over biotite gneiss in Idi-Ayunre area

Table 2. Average chemical composition of major (%) and trace element (ppm) of migmatite gneiss at Akure

Oxides	Rock Mean	Range n=5	Clay Mean	Range n=5	Laterite Mean	Range n=5	Top soil Mean	Range n=5
SiO$_2$	68.49	67.23-69.75	47.43	46.55-47.79	42.64	38.77-49.54	44.37	43.43-44.96
Al$_2$O$_3$	15.35	14.99-15.17	36.01	35.71-36.20	28.59	24.67-31.07	22.12	19.69-23.34
Fe$_2$O$_3$	2.85	2.56-3.14	2.46	1.81-3.76	15.03	13.98-16.36	14.25	12.73-17.05
MgO	1.18	1.02-1.34	0.05	0.04-0.08	0.09	0.09-0.10	0.23	0.22-0.25
CaO	4.34	3.98-4.71	0.01	0.01-0.01	0.02	0.01-0.03	0.37	0.34-0.41
Na$_2$O	4.95	4.72-5.18	0.01	0.01-0.01	0.01	0.01-0.01	0.02	0.01-0.04
K$_2$O	1.31	0.19-1.72	0.025	0.12-0.49	0.05	0.03-0.07	0.21	0.18-0.26
TiO$_2$	0.20	0.17-0.23	0.84	0.68-0.94	1.62	1.54-1.75	3.42	3.28-3.57
P$_2$O$_5$	0.14	0.12-0.17	0.05	0.03-0.09	0.09	0.05-0.13	0.20	0.17-0.23
MnO	0.05	0.05-0.06	0.01	0.01-0.01	0.02	0.01-0.02	0.14	0.12-0.15
LOI	0.50	0.050-0.51	12.75	12.44-12.98	11.68	9.90-12.62	14.35	13.96-14.87
Total	99.36		99.65		99.85		99.68	
Trace Element (ppm)								
Ba	610.5	622-1130	11	5-22	69.3	37-136	217.67	193-261
Sr	719.5	665-780	4	2-8	29.3	13-45	39.3	52-73
Zr	107	4-7	504	677-827	473	433-524	970	930-1058
Y	19	3-7	22.3	22-23	13	13-15	28.67	25-31
Nb			41.67	18-54	35	28-51	1.67	43-63
Sc	13	1-1	10.67	12-18	22	22-32	28.3	26-30
Silica – Aluminum Ratio								
SR	3.77	3.97-3.57	1.22	1.18-1.24	0.99	0.84-1.28	1.23	1.18-1.26
AR	5.39	5.30-6.10	14.63	9.50-19.95	1.90	1.76-2.10	1.59	1.15-1.80
MgO+ CaO	5.52	5-6.05	0.6	0.56-0.66	0.11	0.10-0.13	0.06	0.05-0.09
Na$_2$O+ K$_2$O	6.26	6.09-6.44	0.44	0.20-0.81	0.06	0.04-0.08	0.26	0.13-0.50

| CIA | 51.25 | 50.09-51.72 | 99.29 | 99.20- 99.80 | 99.72 | 98.01 – 99.79 | 97.36 | 96.30-97.78 |

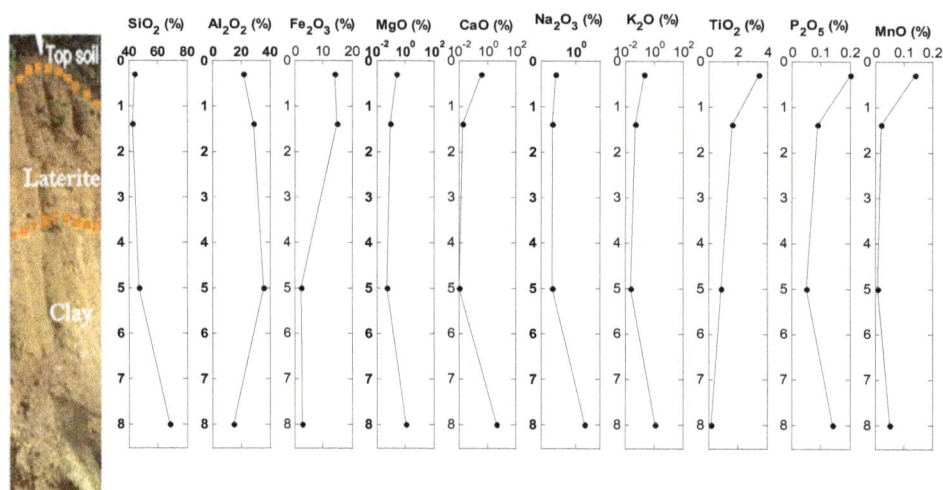

Figure 9. Chemical variation along the weathering profile over migmatite in Akure

Table 3. Average chemical composition of major (%) and trace element (ppm) over porphyritic granite in Akure.

	Rock		Clay		Laterite		Top soil	
Oxides	Mean	Range	Mean	Range	Mean	Range	Mean	Range
SiO_2	68.86	68.05-69.44	50.11	49.43-51.32	46.47	45.12-47.38	51.12	45.39-62.09
Al_2O_3	15.08	14.89-15.08	32.34	31.71-32.67	28.95	27.18-31.91	22.48	16.47-25.57
Fe_2O_3	3.30	2.76-3.79	3.45	2.93-4.43	10.02	9.31-10.43	8.34	5.52-9.92
MgO	1.07	0.73-1.53	0.03	0.02-0.04	0.09	0.08-0.10	0.78	0.24-0.28
CaO	3.38	3.13-3.67	0.02	0.02-0.02	0.03	0.01-0.04	0.33	0.31-0.36
Na_2O	4.30	4.07-4.45	0.01	0.01-0.01	0.02	0.01-0.05	0.18	0.01-0.53
K_2O	1.75	1.27-2.05	0.04	0.01-0.09	0.58	0.18-0.81	2.28	0.33-6.18
TiO_2	0.42	0.31-0.40	2.05	0.77-2.70	1.28	1.04-1.40	1.54	1.50-1.61
P_2O_5	0.12	0.11-0.14	0.09	0.07-0.11	0.34	0.06-0.48	0.20	0.08-0.27
MnO	0.05	0.36-0.38	0.01	0.01-0.01	0.02	0.01-0.03	0.08	0.02-0.13
LOI	0.37		11.76	11.47-11.93	11.84	11.64-12.13	13.00	
Total	98.39		96.47		99.65		100	
Trace Element (ppm)								
Ba	916	622-1130	145.3	82-178	1102	85-1627	963	460-1967
Sr	704	665-780	57.3	24-75	482	22-720	207	199-213
Zr	5	4-7	451	267-566	342.67	310-390	1119	648-2024
Y	5	3-7	29	11-38	68	23-91	36	26-42
Nb			50	23-65	28	22-32	44.3	42-49
Sc	5	136-163	23	19-25	23	18-26	17	8-22
Silica – Aluminum Ratio								
SR	7.84	7.42-8.21	1.80	1.28-2.82	1.19	1.09-1.26	1.80	1.28-2.82
AR	4.63	4.05-5.39	9.37	7.16-11.14	2.66	2.63-3.43	2.69	2.58-2.98
MgO+ CaO	4.45	3.86-4.29	2.47	0.34-6.71	0.60	0.19-0.86	2.47	0.34-6.71
Na_2O+ K_2O	6.06	5.34-6.45	0.59	0.57-0.61	0.12	0.11-0.12	0.59	0.59-0.6

CIA	59.11	59.02-59.40	99.78	99.59-99.88	97.20	89.14-99.74	88.95	88.52-97.50

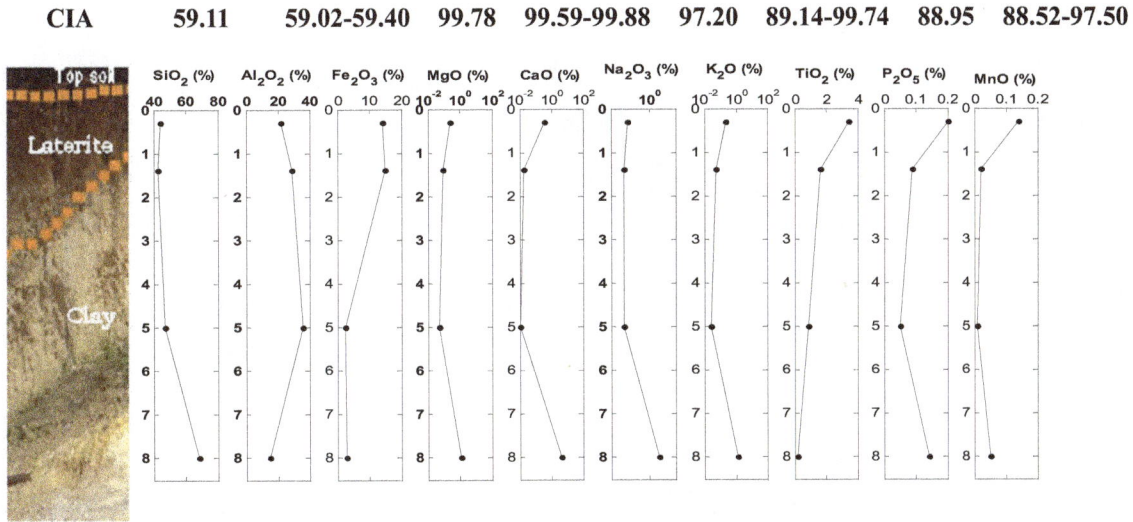

Figure 10. Chemical variation along the weathering profile over migmatite in Akure

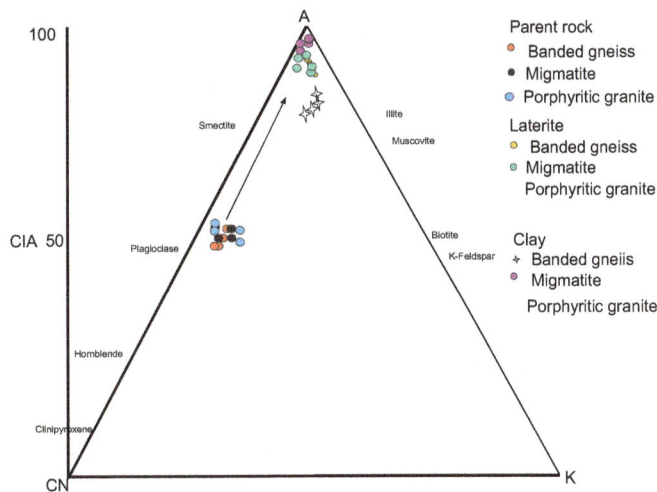

Figure 11. A-CN-K diagram (A=Al$_2$O$_3$, CN = CaO + Na$_2$O, K = K$_2$O), all in moles when A-CN-K are recalculated to 100) of weathering profiles from Idi-Ayunre banded gneiss and Akure migmatite and Phorphyritic granite (After Nesbitt and young, 1984)

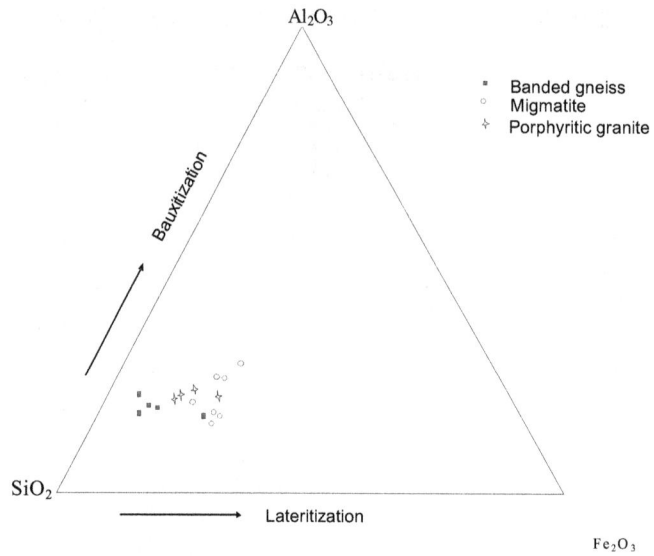

Figure 12. Ternary plot of SiO$_2$-Al$_2$O$_3$-Fe$_2$O$_3$ o lateritic profiles showing iron enrichment (lateritization) after (Gardner and Walsh 1996)

7

Using topographic map interpretation methods to determine Tookany (Tacony) Creek erosion history upstream from Philadelphia, Pennsylvania, USA

Eric Clausen[1]

[1] Jenkintown, PA. USA

Correspondence: Eric N. Clausen, 100 West Ave D-17, Jenkintown, PA, 19046, USA. E-mail: eric2clausen@gmail.com

Abstract

Topographic map interpretation methods are used to determine erosional landform origins in and adjacent to the Tookany (Tacony) Creek drainage basin, located upstream from and adjacent to Philadelphia, PA. Five wind gaps notched into the Tookany-Wissahickon Creek drainage divide (which is also the Delaware-Schuylkill River drainage divide), a deep through valley crossing the Tookany-Pennypack Creek drainage divide, a Tookany Creek elbow of capture, orientations of Tookany Creek tributary valleys, a narrow valley carved in erosion resistant metamorphic bedrock, and the relationship of a major Tookany Creek direction change with a Pennypack Creek elbow of capture and a Pennypack Creek barbed tributary are used along with other evidence to reconstruct how a deep south oriented Tookany Creek valley eroded headward across massive southwest oriented flood flow. The flood flow origin cannot be determined from Tookany Creek drainage basin evidence, but may have been derived from a melting continental ice sheet, and originally flowed across the Tookany Creek drainage basin region on a low gradient topographic surface equivalent in elevation to or higher than the highest present day Tookany Creek drainage divide elevations with the water flowing in a complex of shallow diverging and converging channels that had formed by scouring of less resistant bedrock units and zones. William Morris Davis, sometimes referred to as the father of North American geomorphology, spent much of his boyhood and several years as a young man living in the Tookany Creek drainage basin and all landforms discussed here were within walking distance of his home and can be identified on a topographic map published while he was developing and promoting his erosion cycle ideas. Davis never published about Tookany Creek drainage basin erosion history, but he developed and promoted uniformitarian and erosion cycle models that failed to recognize the significance of Tookany Creek drainage basin erosional landform features providing evidence of the immense floods that once crossed present day drainage divides and eroded the Tookany Creek drainage basin.

Keywords: drainage divide, Pennypack Creek, Tookany Creek, William Morris Davis, wind gap, Wissahickon Creek

1. Introduction

1.1 What is the William Morris Davis Relationship with the Tookany Creek Drainage Basin?

William Morris Davis (1850-1934) played a major role in shaping the science of geomorphology, and is sometimes referred to as the "father of North American geomorphology." Davis is best known for descriptive landform interpretations and his cycle of erosion or geographical erosion cycle. Davis first developed his erosion cycle ideas in the 1880s and subsequently he, his students, and their students enhanced and promoted the concept so the Davis erosion cycle interpretation became the dominant geomorphology paradigm used throughout much the early 20th century (Orme, 2007). Important to the Davis erosion cycle idea was the uniformitarianism paradigm applied by Davis when describing landform evolution. While most researchers today consider the Davis erosion cycle interpretations obsolete the underlying Davis uniformitarianism paradigm still influences many landform evolution studies.

Tookany Creek drains most of Cheltenham Township, located in the Upper Piedmont Section of the Piedmont Province in Montgomery County, PA (USA) and adjacent to the City of Philadelphia (shown as Philadelphia County in figure 1). In Cheltenham Township the stream drains an area slightly larger than 20 square kilometers.

After leaving Cheltenham Township Tookany Creek enters Philadelphia where its name changes to Tacony Creek and it flows in a southwest and then south direction separating northeast city neighborhoods from the main city to the west before reaching the southwest oriented Delaware River as Frankford Creek. West and north of the Tookany Creek drainage basin is the Wissahickon Creek drainage basin. East of the Tookany Creek drainage basin is the Pennypack Creek drainage basin.

Davis spent much of his boyhood and several years as a young adult living in the Tookany Creek drainage basin (Clausen, 2015). His home was located near the Tookany-Wissahickon Creek (also Delaware-Schulykill River) drainage divide. By walking 1 or 2 kilometers to the east young Davis could watch Tookany Creek water flow to the Delaware River or by walking 2 to 4 kilometers to the west he could watch Wissahickon Creek water flow to the Schuylkill River. Near his family's home were headwaters of a northeast oriented barbed tributary (Mill Run) flowing to a southeast oriented Tookany Creek valley segment, which a short distance downstream turned to become a southwest oriented Tookany Creek valley segment. Upstream were five wind gaps notched into the Tookany-Wissahickon Creek drainage divide and a well defined through valley across the Tookany-Pennypack Creek drainage divide. Both Tookany Creek and Wissahickon Creek have deep valleys with steep walls carved in erosion resistant metamorphic bedrock. A rolling upland on which the Davis home was located separates the two valleys.

While living in the Tookany Creek drainage basin Davis had no reason to believe the study of landforms would become important in his future career. His father was President of the Barclay Coal Company (Chorley, R. J., Beckinsale, R. P, & Dunn, A. J., 1973) and there may have been expectations young Davis would take an interest in coal mining. These expectations may be why young Davis studied mining engineering while a student at Harvard. But Davis also had interests in astronomy, meteorology, and entomology and immediately after graduation took a job as a meteorologist at an Argentine observatory where he spent his free time observing local insects. After three years in Argentina he returned to the family home and worked in the Barclay Coal Company office until being invited to teach as an instructor at Harvard University. It was only while trying to secure his shaky Harvard position that his landform origin hypotheses developed (Chorley et al, 1973).

Davis claimed ideas for his geographic erosion cycle hypotheses originated while working in the Montana Crazy Mountains (Chorley et al, 1973), but observations made in the Tookany Creek drainage basin as a boy and young man must have played an important role. Topographic maps of the Tookany Creek drainage basin were first published after Davis secured his Harvard University faculty position. The United States Geological Survey (USGS) 1:62:500 scale Germantown PA map (from which figure 1 is taken) was published in 1896 when Davis was actively developing and promoting his geographical erosion cycle and uniformitarianism hypotheses (e.g. Davis, 1899). Davis apparently had access to topographic maps almost as soon as they were issued as shown by dates of his publications (e.g. Davis, 1889a), but no evidence was found that any of his numerous publications specifically discussed Germantown map evidence.

Whatever Davis as a boy and young man observed in the Tookany Creek drainage basin apparently did not conflict with his (1895) uniformitarianism paradigm:

"The deepening of a valley by its stream is a slow process; the widening of the valley by the wasting of its slopes is still slower; the development of subsequent streams by headward erosion, the accompanying migration of divides, and the resulting rearrangement and adjustment of waterways are slowest of all. The deepening of a canyon is a rapid process compared to the creeping of a divide. …Here, if anywhere, the slow processes of uniformitarianism are justified, and the hurried processes of catastrophism are completely at fault. To attempt to substantiate principles so widely accepted as those of uniformitarianism may seem to some an unnecessary task. It might be compared to adducing new evidence in support of the law of gravitation."

This uniformitarianism concept was not unique to Davis and at that time also reflected the view of many geologists. For example, Bascom, F., Clark, W. B., Darton, N. H., Knapp, G. N., Kuemmel, H. B., Miller, B. L., and Salisbury, R. D. (1909) who mapped geology in the Germantown and adjacent quadrangles describe a regional geologic history consistent with the Davis uniformitarianism paradigm.

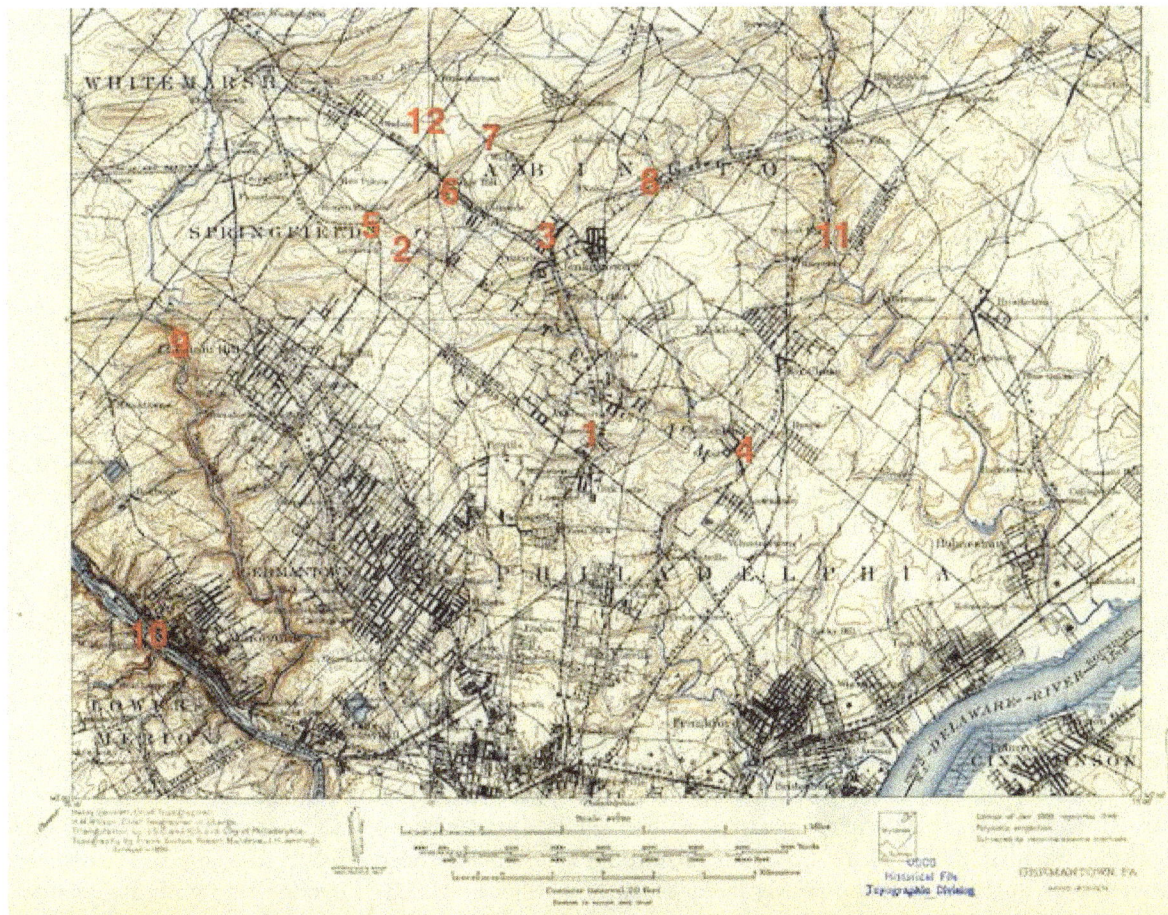

Figure 1. Section of the USGS 1896 Germantown 1:62,500 topographic map (20-foot contour interval) showing the Tookany Creek and adjacent drainage basins. Number 1 identifies the Davis family home, 2–Tookany Creek headwaters, 3–Jenkintown elbow of capture, 4–Cheltenham abrupt direction change, 5–Seminary wind gap, 6– Edge Hill wind gap, 7-Weldon wind gap, 8–Jenkintown-Bethayres through valley, 9–Wissahickon Creek gorge, 10–Schuylkill River, 11– Pennypack Creek, and 12–Chester Valley. Tookany and Pennypack Creeks flow to the southwest oriented Delaware River

1.2 Why is the Davis Uniformitarianism Paradigm Important Today?

The Davis uniformitarianism paradigm implies landscapes evolve slowly over extremely long periods of time measured in millions if not tens of millions of years. While Thornbury (1969, p 29) argues, "Little of the earth's topography is older than Tertiary and most of it is no older than the Pleistocene" Bishop (2007) claims the Thornbury conclusion is based on "measured erosion rates that were almost certainly too high because of anthropogenic disturbance" and even "when more 'reasonable' long-term erosion rates were used... the Davisian erosion cycle required between 10 and 25 Myr to run its full course." Newer research approaches to studying landform evolution that replaced the Davis developed erosion cycle models do not challenge the Davis concept that landscapes evolve slowly over long periods of time. For example Hack's (1957) dynamic equilibrium scheme for landscape evolution, while quite different from the Davis erosion cycle concept also implies slow processes of landscape evolution. Likewise quantitative geomorphology introduced by Horton (1945) and Strahler (1952) does not challenge the Davis uniformitarianism model.

Bishop in his 2007 review paper summarizes more recent work related to long-term landscape evolution, which the plate tectonics revolution and the "development of numerical models that explore the links between tectonic processes and surface processes" stimulated. Bishop points out how breakthroughs in analytical and geochronological techniques are enabling researchers to confirm, "more sophisticated Davisian-type numerical models of slope lowering under conditions of tectonic stability (no active rock uplift), and... will indicate that the Davis and Hack models are not mutually exclusive." In other words the Bishop (2007) paper suggests researchers using the newly developed analytical and geochronological techniques are in many cases indirectly making use of the Davis uniformitarianism paradigm and rejecting the Thornbury (1969) proposed Pleistocene erosion paradigm. But, what if Davis overlooked significant evidence when developing his erosion cycle models and as a result his uniformitarianism paradigm is flawed? Is it possible Davis pointed geomorphologists in an unproductive direction and the geomorphology research community needs to study, perhaps for the first time, landform features Davis overlooked?

1.3 How was the Tookany Creek Drainage Basin Eroded?

The Davis uniformitarianism paradigm, while permitting wind gaps, through valleys, barbed tributaries, and elbows of capture, implies such features are rarely formed. Yet in and surrounding the Tookany Creek drainage basin such landform features are common. The Hack dynamic equilibrium paradigm also fails to provide a good explanation for abundant wind gaps, through valleys, barbed tributaries, and elbows of capture. But could a paradigm defined by headward erosion of deep valleys into an upland surface on which massive floods are flowing in a complex of diverging and converging channels explain the formation of wind gaps, through valleys, barbed tributaries, and elbows of capture? J H. Bretz in the 1920s proposed in a series of papers that an immense flood eroded the channeled scabland in eastern Washington State (e.g. Bretz, 1923). The geology research community of that time, which W. M. Davis had strongly influenced, overwhelmingly rejected the Brez's hypothesis (Baker in 1981 prepared a book containing some of the original papers published during the ensuing channeled scabland debate). Immense floods flowing in giant anastomosing complexes of diverging and converging channels are now commonly used to explain eastern Washington State landscape features, but are rarely used to explain landform evolution in other regions, especially in southeast Pennsylvania where the Tookany Creek drainage basin is located.

The study described here interpreted topographic map evidence to determine how the wind gaps, through valleys, barbed tributaries, elbows of capture, valley orientations, and similar landform features found within and near the relatively small Tookany Creek drainage basin (and near the former Davis boyhood home) were formed. The goal was to use these erosional landforms like pieces of a picture puzzle to reconstruct a coherent sequence of erosion events that would explain all of the observed landform evidence. Since Davis should have been intimately familiar with Tookany Creek drainage basin landforms the reconstructed Tookany Creek drainage basin erosion history should also answer the question of whether or not Davis overlooked significant landform features when developing and promoting his erosion cycle and uniformitarianism paradigms and if so whether or not his still surviving uniformitarian paradigm is now causing modern day geomorphologists to overlook the importance of similar landform features.

2. Method

Topographic maps used in this study were obtained from the United States Geological Survey Historical Map Collection website and Pennsylvania Department of Conservation and Natural Resources (DCNR) Interactive Map Resources website. The DCNR Interactive Map Resources website also provided access to a digital Pennsylvania geologic map used to identify bedrock units underlying the Tookany Creek drainage basin. Topographic maps available on the DCNR Interactive Map Resources website were particularly useful as they could be scrolled in any direction, however the USGS website identified the Frankford and Germantown (and to a lesser degree the Ambler and Hatboro) 1:24,000 scale topographic maps as being the specific maps from which the DCNR website map data had been obtained. Elevations on all maps used in this study were shown in feet and for that reason all elevations given in this paper are also in feet.

This study began by identifying landforms of interest along the Tookany-Wissahickon Creek (also the Delaware-Schuylkill River) divide and on the Tookany-Pennypack Creek divide. The Tookany-Wissahickon Creek divide is at the southern end of the Delaware-Schuylkill River divide and potentially provides information as to how those two much larger drainage basins evolved. Specifically identified were five identifiable wind gaps notched into the Tookany-Wissahickon divide and a deep through valley crossing the Tookany-Pennypack Creek divide. Two Tookany Creek abrupt direction changes, a narrow valley downstream from Jenkintown, and a less obvious through valley across the Pennypack-Tookany Creek divide and barbed tributaries were also observed. The through valleys and wind gaps were interpreted to be evidence water once crossed present day drainage basin divides at multiple locations.

The identified wind gaps and through valleys were interpreted to be evidence a completely different drainage system had crossed the region prior to Tookany Creek drainage basin erosion. The nature of this earlier drainage system was determined by looking at the locations and number of wind gaps and through valleys and also at the orientations of Tookany Creek (and some Wissahickon and Pennypack Creek) valley segments and tributaries, which were interpreted to reflect previous drainage system channel orientations and locations. The wind gaps, through valleys, and valley orientations suggested a large number of drainage channels had once crossed the Tookany Creek drainage basin region and that at least some of the earlier drainage channels had converged and diverged within or near the present day Tookany Creek drainage basin. These interpretations strongly suggested that prior to Tookany Creek drainage basin erosion a large complex of diverging and converging channels had crossed the entire region.

Next the barbed tributaries and elbows of capture were considered to be evidence a deeper Tookany Creek valley had eroded headward across this anastomosing complex of channels and the barbed tributaries had formed when water on downstream ends of beheaded channels had reversed to move toward the newly eroded and deeper Tookany Creek valley. Reversal of flow in beheaded channels was interpreted to only have been possible if the beheaded channel had been a low gradient channel suggesting the anastomosing channel complex had formed on a low gradient topographic surface. Erosion of the barbed tributary valleys was interpreted to have been by water captured from diverging and converging channels that Tookany Creek valley headward erosion had not yet beheaded. Valley depths and the erosion resistant nature of bedrock into which the Tookany Creek valley has been eroded were considered to be evidence that headward erosion of a deep Tookany Creek valley into the region required immense water volumes. These considerations suggested Tookany Creek valley headward erosion had been across an anastomosing channel complex through which massive volumes of water were flowing on a low gradient topographic surface at least as high or higher than the highest Tookany Creek drainage basin elevations today.

3. Results

Topographic map interpretation results for the identified Tookany Creek drainage basin wind gaps, through valleys, barbed tributaries, and elbows of capture or major valley direction changes are described below. While recent 1:24,000 scale topographic maps with 10- or 20-foot contour intervals were used all landform features described here are located within 6 kilometers of the former Davis home and are also recognizable on the 1896 Germantown topographic map with its 20-foot contour interval.

3.1 Laverock and Seminary Wind Gaps

Figure 2 with a 10-foot contour interval shows two closely spaced wind gaps both notched into the Tookany-Wissahickon (Delaware-Schuylkill) drainage divide. The red four-lane highway passes through the Seminary wind gap (next to Westminster Theological Seminary) while the Laverock wind gap is located north of the word "Laverock". Both wind gaps appear to have had natural floor elevations of between 350 and 360 feet. The Seminary wind gap is notched into a northeast oriented ridge known as Edge Hill. The grade of an abandoned railroad can be seen extending in a north direction through the gap. That railroad is shown on the 1896 Germantown map, but was used primarily for freight traffic so Davis probably never rode on it.

The Edge Hill ridge according to the Pennsylvania DCNR Interactive Map Resources digital geologic map is composed of Cambrian Chickies Formation with quartzite the dominant rock type and marks the boundary between the Piedmont Upland Section (south) and Piedmont Lowland Section (Potter, 1999). Felsic gneiss of Precambrian age underlies the region south of the Edge Hill ridge while north of the ridge is the carbonate floored Chester Valley with Cambrian or Ordovician Conestoga Formation limestone being adjacent to the Edge Hill ridge. Cambrian Elbrook Formation (calcareous shale) and Cambrian Ledger Formation (dolomite) underlie unseen Chester Valley areas north of figure 2. Camp Hill, which is a linear ridge like Edge Hill, forms the unseen Chester Valley north boundary and north of the Camp Hill ridge Triassic age sedimentary rocks underlie the Piedmont Province Gettysburg-Newark Lowland Section.

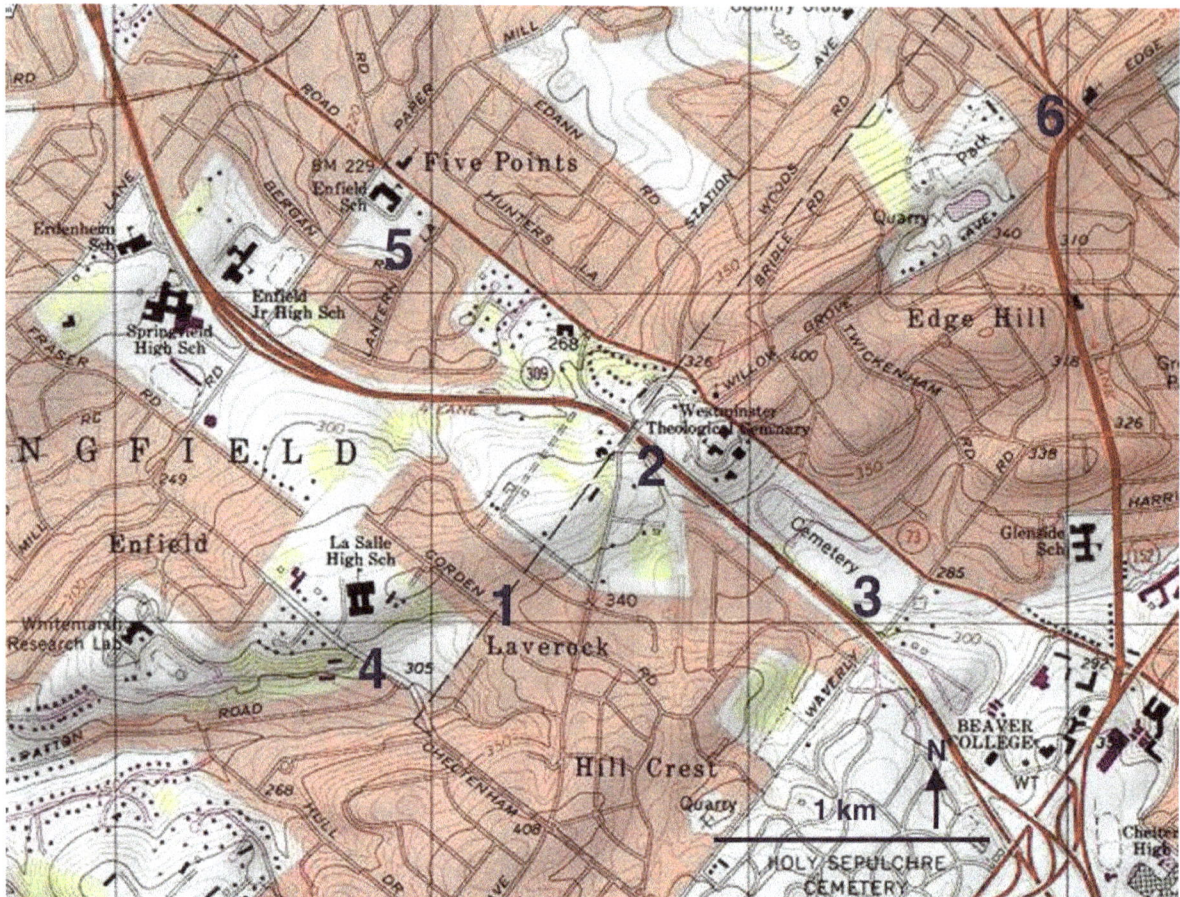

Figure 2. USGS 1:24,000 topographic map (with a 10-foot contour interval) taken from Pennsylvania DCNR Interactive Map Resources website showing Laverock wind gap (1), Seminary wind gap (2), east oriented Tookany Creek headwaters (3), west oriented Wissahickon Creek tributary headwaters (4), a northwest oriented tributary (5) to an unseen west oriented Wissahickon Creek tributary in the Chester Valley, and the Edge Hill wind gap (6). La Salle High School is located on the Edge Hill ridge

East oriented Tookany Creek headwaters are located east of the Laverock wind gap just north of Beaver College (now Arcadia University). West oriented headwaters of a Wissahickon Creek tributary are located west of the Laverock wind gap. The wind gap is evidence water once flowed one way or the other through the gap and dismemberment of that flow required one of the two streams now flowing in opposite directions to have reversed its flow direction. A flow direction reversal could only have occurred if headward erosion of a much deeper valley beheaded a low gradient channel. However, the two streams now flowing in opposite directions from the Laverock wind gap today have steep gradients as they both flow into deep valleys. Somehow after the earlier channel had been dismembered water continued to flow into the reversed channel so as to erode the deep valley seen today.

A saddle notched into the Edge Hill ridge north of the "L" in the word "Laverock" may have been used by south oriented water flowing to the Laverock wind gap location that helped to erode the deep reversed flow valley. The possibility of north and east oriented flow is rejected because the Tookany and Wissahickon drainage basins are today both south oriented drainage basins. If south oriented water did flow to the Laverock wind gap location then at that time Chester Valley elevations (north of Edge Hill) must have been greater than 380 feet, or significantly higher than they are now. Figure 3 illustrates how the Laverock and Seminary wind gaps are located midway between south oriented Wissahickon Creek (west) and the south oriented Tookany Creek valley downstream from the Jenkintown rail junction elbow of capture. Prior to being beheaded and reversed by Tookany Creek valley headward erosion water flowed in a west direction from the Jenkintown rail junction elbow of capture across the Laverock wind gap to join Wissahickon Creek just north of the present day Wissahickon gorge (seen in the southwest corner of figure 3). Note how the west oriented Wissahickon Creek tributary originating near the Laverock wind gap turns to flow through a water gap south of Erdenheim and joins a southwest oriented Wissahickon Creek tributary, which is much easier to explain if the water in that tributary valley always flowed in a west direction.

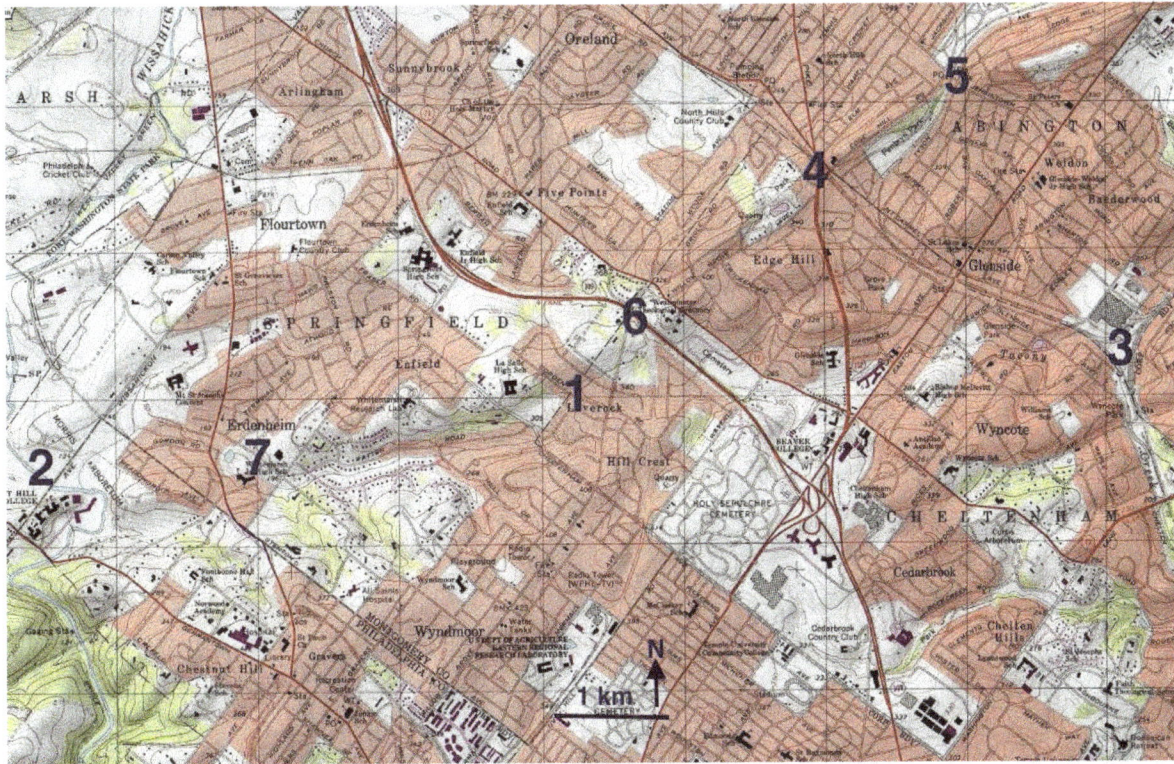

Figure 3. USGS 1:24,000 topographic maps (with a 10-foot contour interval) showing relationship of Laverock wind gap (1) to Wissahickon Creek (2), Jenkintown elbow capture (3), Edge Hill wind gap (4), Weldon wind gap (5), and Seminary wind gap (6). The west oriented Wissahickon tributary originating west of the Laverock wind gap crosses the Edge Hill ridge in a water gap (7) to enter the Chester Valley before joining a southwest oriented stream that flows to Wissahickon Creek near point 2

The Seminary wind gap is located further east on the Edge Hill ridge and its natural floor elevation was probably 30-40 feet lower than the floor elevation of the saddle directly north of the Laverock wind gap (which suggests south oriented water moved through it after flow across the higher elevation saddle ceased). The gap links the east oriented Tookany Creek valley with the valley of a northwest and west oriented Wissahickon Creek tributary (on the Chester Valley floor). The Seminary wind gap location is such that water flowing through it would have moved to the east oriented Tookany Creek headwaters valley confirming the hypothesis that the Tookany Creek headwaters flow direction was reversed during the dismemberment of a west oriented flow channel leading to the deep south oriented Wissahickon Creek gorge (seen in figure 3 southwest corner). However, prior to the beheading and reversal of flow that created the east oriented Tookany Creek headwaters it is possible a west oriented diverging channel led in a northwest direction through the Seminary wind gap to converge with a west oriented channel located in what was at that time the yet to be eroded Chester Valley north of the Edge Hill Ridge.

Evidence seen in figures 2 and 3 documents dismemberment of a west oriented flow channel (south of the Edge Hill ridge) that was moving water to the south oriented Wissahickon gorge. Since the original flow direction was to the west the evidence also suggests the west oriented flow was eroding the deep west oriented Wissahickon Creek tributary valley headward toward the Laverock wind gap location, which means the south oriented Wissahickon gorge existed prior to flow dismemberment. Further the figure 2 and 3 evidence documents that following the flow dismemberment elevations in the Chester Valley (north of the Edge Hill ridge) were high enough that enough water could spill from the Chester Valley in a south direction into the reversed flow channel to erode a deep east oriented valley while at the same time there was enough west and southwest oriented water moving north of the Edge Hill ridge toward the south oriented Wissahickon Creek valley to significantly lower the Chester Valley floor.

3.2 Edge Hill, Weldon, and Highland Wind Gaps

Figure 4 illustrates in detail the Edge Hill wind gap located where the railroad near the figure southwest corner goes under a red 4-lane highway (also seen figures 2 and 3), the Weldon wind gap located at the unnamed railroad station north of the "A" in the word "ABINGTON", and the Highland wind gap located near the Highland School (in northeast corner of figure 4). The Weldon gap was known as Tyson's Gap when Davis lived in the region (Camburn, 1977). Railroad and highway construction has not significantly altered the Weldon gap elevation, where a spot elevation of 326 feet is shown, or the Highland wind gap floor (elevation between 380 and 390 feet), but has deepened the Edge Hill wind gap. Today the highway bridge over the railroad is built at approximately the original wind gap floor elevation of between 320 and 330 feet while the railroad passes underneath in a deeper railroad excavated cut. The railroad through the Edge Hill gap was constructed in 1854 and had a station near the former W.M. Davis home.

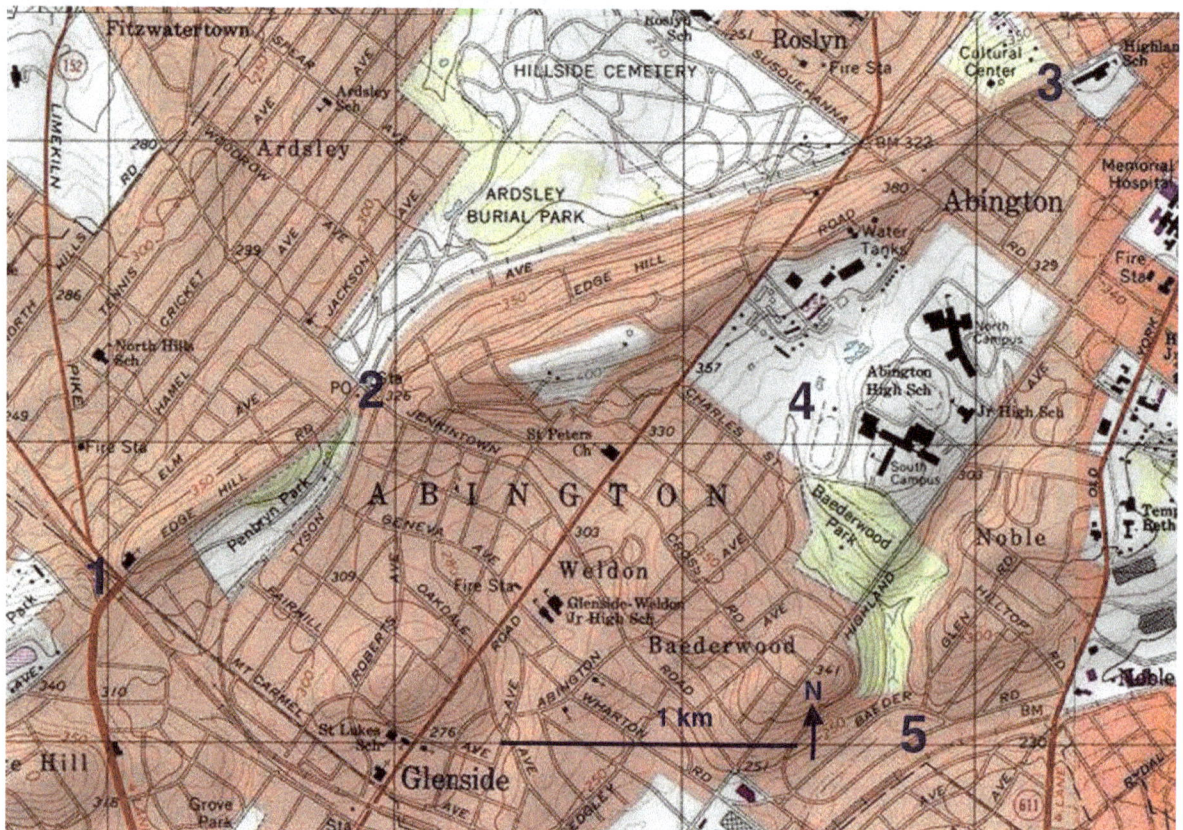

Figure 4. USGS 1:24,000 topographic map taken from the Pennsylvania DCNR Interactive Map Resources website (with a 10-foot contour interval except along the lower 90% of the eastern edge where the contour interval is 20 feet) showing the Edge Hill wind gap (1), Weldon wind gap (2), Highland wind gap (3), headwaters of south and southwest oriented Baederwood Creek (4), and western end of the Jenkintown-Bethayres through valley (5).

The Edge Hill ridge elevation exceeds 410 feet southwest of the Edge Hill wind gap (near the Westminster Seminary) and exceeds 420 feet between the Weldon and Highland gaps, but at multiple locations is slightly lower. Lower ridge elevations at various locations suggest considerable water flowed across the ridge and that the water was not confined to the wind gap channels. North of the Edge Hill wind gap are headwaters of a west oriented Wissahickon Creek tributary (seen in figure 3) and the Edge Hill wind gap orientation suggests prior to Tookany Creek valley headward erosion a diverging channel (diverging from a west oriented channel south of the present Edge Hill ridge) had flowed in a northwest direction to converge with a west oriented channel in what at that time was the yet to be eroded Chester Valley. If so, headward erosion of the much deeper Tookany Creek valley beheaded and reversed flow through the Edge Hill wind gap until Chester Valley lowering beheaded the then south oriented flow that had been moving into the newly eroded Tookany Creek valley. The Weldon and Highland wind gap orientations suggest southwest oriented flow moved to west oriented channels south of the present day Edge Hill ridge before Tookany Creek valley headward erosion captured that flow. Today streams flowing along segments of these routes are incorporated into the Glenside sewer system, but are shown on the 1896 Germantown map (figure 1).

3.3 Summary of the Wind Gap Evidence

Map evidence (only partially seen here) documents three west oriented streams draining the Chester Valley (north of the Edge Hill ridge) to south oriented Wissahickon Creek. This evidence strongly suggests that while water was spilling across the Edge Hill ridge and eroding the Seminary, Edge Hill, Weldon, and Highland wind gaps (and perhaps lowering the Edge Hill ridge at other locations) water was also eroding three west oriented valleys headward into the carbonate bedrock underlying the Chester Valley floor. Further, previously discussed evidence from the Laverock wind gap demonstrates that prior to the flow reversal that created the east oriented Tookany Creek headwaters water also flowed in a west direction south of the Edge Hill ridge. The Seminary and Edge Hill wind gap orientations suggest that for a time channels may have diverged from the west oriented channel south of the present day ridge and then converged with west oriented channels in the yet to be eroded Chester Valley. Weldon and Highland wind gap orientations suggest water diverged from west oriented channels in what was then the yet to be eroded Chester Valley and converged with west oriented channels south of the Edge Hill ridge. These diverging and converging channels appear to have been components of a west or southwest oriented anastomosing channel complex through which massive volumes of water flowed across the entire region.

This west or southwest oriented complex of diverging and converging channels must have initially developed on a low gradient regional topographic surface equivalent in elevation (or higher than) the highest Edge Hill elevations seen today. Headward erosion of the deep south-oriented Wissahickon Creek valley into this former topographic surface captured the southwest and west oriented flow and also significantly lowered base level permitting the massive southwest oriented flow to erode the Chester Valley carbonate rock floor headward in an east-northeast direction. At the same time headward erosion of the deep south-oriented Tookany Creek valley beheaded and reversed a west oriented flow channel moving water to the newly eroded Wissahickon Creek valley south of the present day Edge Hill ridge, captured southwest oriented flow channels that crossed what is now the Edge Hill ridge and also beheaded and reversed diverging northwest oriented flow channels.

The west oriented Laverock gap channel valley eroded headward from the actively eroding south oriented Wissahickon valley head while the Wissahickon valley was still eroding headward across the Chester Valley with the result that elevations south of the Edge Hill ridge were lowered as Chester Valley lowering was just beginning. Chester Valley lowering proceeded from west to east and water ceased to flow through the Seminary and Edge Hill wind gaps while water still flowed through the Weldon and Highland gaps. The Highland gap, being near the Chester Valley east end, in spite of its higher floor elevation was probably the last of the south oriented channels moving water across the Edge Hill ridge to be beheaded by headward erosion of west oriented (Chester Valley) Wissahickon Creek tributary valleys. Headward erosion of the deep south-oriented Pennypack Creek valley seen in figures 1 and 5 ended all west and southwest oriented flow to both the newly eroded Tookany Creek drainage basin and the Chester Valley.

3.4 Jenkintown Rail Junction Elbow of Capture and the Jenkintown-Bethayres through Valley

East of the Tookany Creek elbow capture near the Jenkintown rail junction and the east oriented Tookany Creek headwaters (and upstream from the south oriented Tookany Creek valley) is an east-northeast oriented through valley seen in figure 5 that extends from the elbow of capture to south oriented Pennypack Creek. The through valley is drained by east-northeast oriented Meadow Brook to Pennypack Creek and a southwest oriented Baederwood Creek segment (the stream flows south before entering the through valley where it turns to flow to Tookany Creek at the elbow of capture). Tookany Creek crosses the 200-foot contour line downstream from its elbow of capture while a benchmark near where Meadow Brook joins Pennypack Creek has an elevation of 145 feet. The Tookany-Pennypack Creek drainage divide on the through valley floor is located near the number 1 and has an elevation of between 220 and 230 feet.

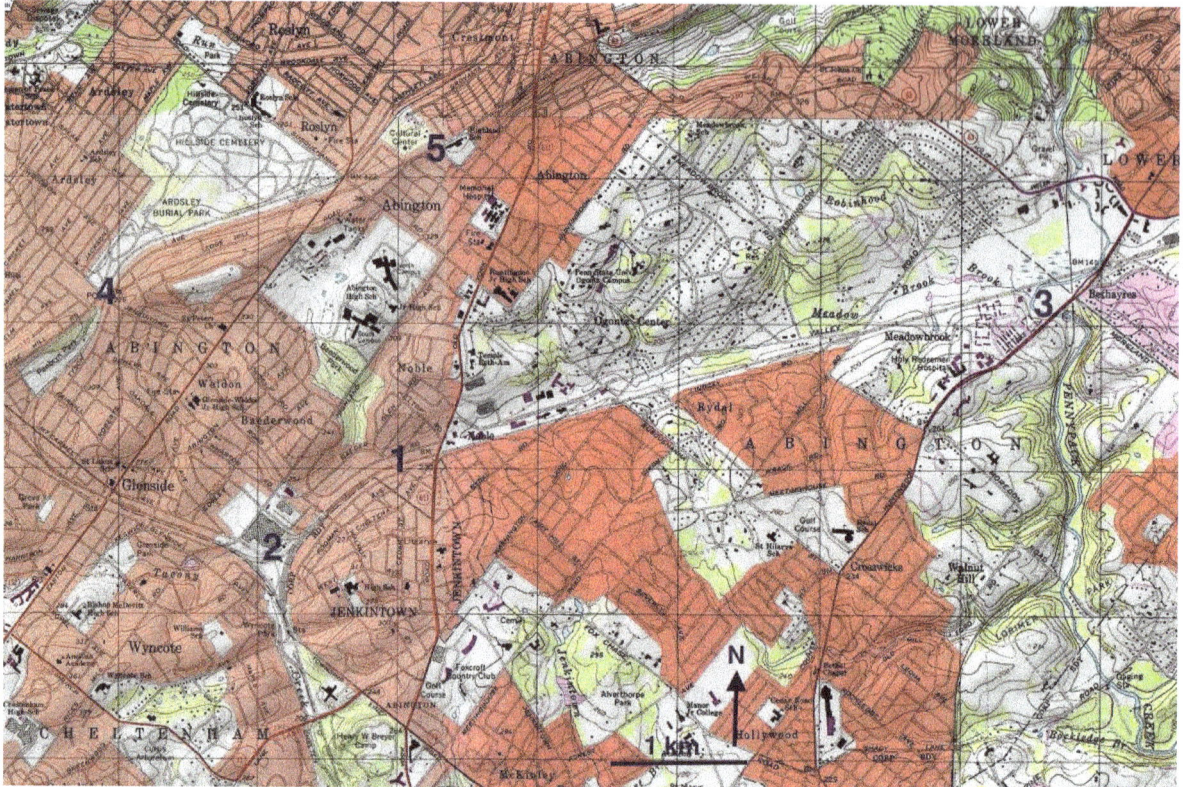

Figure 5. USGS 1:24,000 topographic maps (with 10-foot contour intervals in the west half and along the north edge and a 20-foot interval elsewhere) taken from Pennsylvania DCNR Interactive Map Resource website show the Jenkintown-Bethayres through valley and its drainage divide at (1), Tookany Creek at Jenkintown elbow of capture (2), Pennypack Creek (3), Weldon wind gap (4), and Highland wind gap (5)

The digital geologic map available at Pennsylvania DCNR Interactive Map Resources website shows the Jenkintown-Bethayres through valley to be located along the boundary between Precambrian felsic gneiss (north) and the Wissahickon Formation or schist of probable lower Paleozoic age (south). Near the Pennypack Creek valley a thin wedge of carbonate rock is located between the two metamorphic rock units. The contact is mapped as the Huntingdon Valley fault and small earthquakes in 1980 may have been associated with the fault (Bischke, 1980). Elevations to the north are generally more than 300 feet with still higher elevations found along the Edge Hill ridge. Elevations to the south are somewhat lower and only exceed 300 feet in the Jenkintown area. For its entire length the through valley floor is at least 100 feet lower than surrounding uplands on both sides and the valley provided a logical route for construction of the railroad line. The railroad was built in 1876 shortly before W.M. Davis left Pennsylvania to assume his Harvard position and he must have been aware of the railroad plans and construction.

The same low gradient west-southwest and west oriented channel that eroded the Laverock wind gap initiated the through valley erosion. At that time the channel floor elevation must have been at least as high as the Laverock wind gap floor elevation today or more than 120 feet higher the present day Tookany-Pennypack Creek divide (on the through valley floor). Tookany Creek valley headward erosion captured the west-southwest oriented flow when it also beheaded and reversed the channel further to east to create the east oriented Tookany Creek headwaters. Large volumes of west-southwest oriented water then flowed along the through valley alignment and significantly lowered the through valley floor. Southwest and then south oriented flow that had crossed the Edge Hill ridge at the Highland wind gap location entered the through valley along the present day Baederwood Creek alignment. Large volumes of water that were using the Seminary, Edge Hill, and Weldon, wind gaps from what was at that time the yet to be eroded Chester Valley crossed the Edge Hill ridge and joined the west-southwest oriented flow at the present day Tookany Creek elbow of capture. The combined flow at times must have been very large and it deepened and enlarged the Tookany Creek valley downstream from the Jenkintown rail junction elbow of capture.

Flow in the low gradient west-southwest oriented channel on the through valley alignment was next beheaded and reversed by headward erosion of the deeper south-oriented Pennypack Creek valley. A reversal of flow occurred when at least some water was still pouring across the Edge Hill ridge into the Tookany Creek drainage basin, especially at the Highland wind gap location (at that time Pennypack Creek valley headward erosion had not progressed far enough to behead the west or southwest oriented flow moving into the actively eroding Chester Valley) and for a time that remaining flow moved in a southwest and south direction along the present day Baederwood Creek alignment to enter the through valley where some water moved in an east direction to the new Pennypack Creek valley while the remaining water moved in a west direction to the Tookany Creek elbow of capture and then south in Tookany Creek. The east oriented flow eroded the present day east-northeast oriented Meadow Brook valley on the through valley floor. Headward erosion in the Chester Valley of west oriented Wissahickon tributary valleys and (to the east) of the Pennypack Creek valley beheaded all flow routes to the Tookany Creek drainage basin (the last being on the Baederwood Creek alignment) so as to create the through valley drainage divide.

3.5 Cheltenham through Valley

The Cheltenham through valley, unlike the Jenkintown-Bethayres through valley, is not easily seen on the ground or even on topographic maps, although railroad builders in 1906 used it when constructing a low-grade railroad line across the divide between Tookany and Pennypack Creeks. That rail line is seen in figure 6 (with a 20-foot contour interval) and extends in a northeast direction from near the figure 6 southwest corner to near the figure northeast corner. Even a close look at the map may suggest evidence for the through valley's existence to be less than convincing. However, other types of evidence suggest southwest oriented water did flow across the Pennypack-Tookany Creek divide at this location.

The Cheltenham through valley links an elbow of capture where a southwest oriented Pennypack Creek segment turns to flow in an east-northeast direction with a Tookany (Tacony) Creek abrupt direction change where Tookany Creek turns from flowing in a southeast direction to flow in a southwest direction. Note how upstream from its southwest oriented segment Pennypack Creek flows in a southeast direction and how downstream from its east-northeast oriented segment Pennypack Creek flows in a south direction. Also note how a northeast oriented (or barbed) tributary joins Pennypack Creek at the point where it turns from flowing in a southwest direction to flow in an east-northeast direction and how the southwest oriented Pennypack Creek segment appears to line up with the southwest oriented Tookany (Tacony) Creek segment. This evidence suggests the southwest oriented Tookany (Tacony) and Pennypack Creek valley segments were eroded headward along the same southwest oriented flow channel.

The point where Pennypack Creek turns from flowing in a southwest direction to flow in an east-northeast direction and the barbed tributary provide evidence the Pennypack Creek valley eroded headward across a southwest oriented complex of diverging and converging flow channels. Headward erosion of the south oriented Pennypack Creek valley first beheaded and reversed flow on a west-southwest oriented channel that converged further downstream (on the beheaded channel) with a southwest oriented channel and the combined channel then continued along the present day southwest oriented Tookany (Tacony) Creek segment alignment. The barbed Pennypack Creek tributary is evidence of reversed flow on the beheaded southwest oriented flow channel. The reversed flow formed an east-northeast oriented channel that then captured flow in the converging southwest oriented flow channel, which beheaded flow moving across the present day drainage divide to the southwest-oriented Tookany (Tacony) Creek segment.

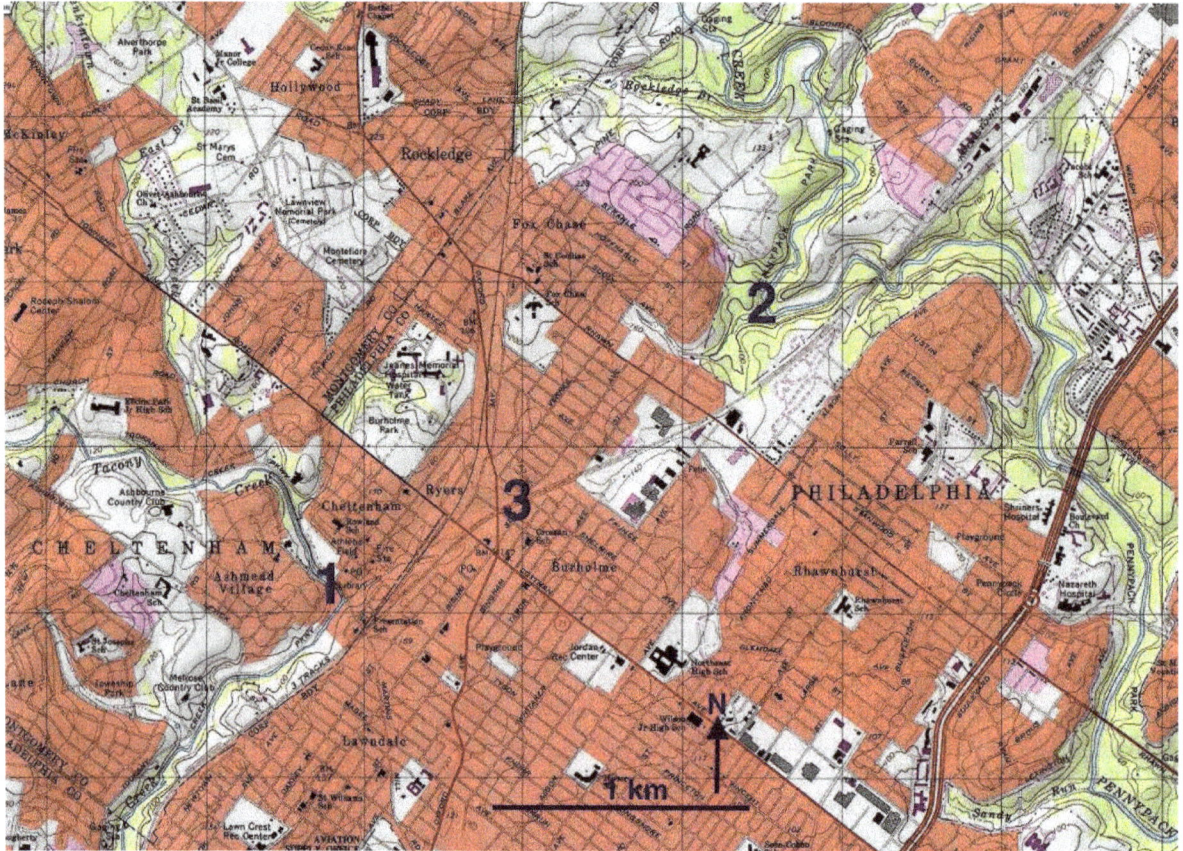

Figure 6. USGS 1:24,000 topographic map (with a 20-foot contour interval) taken from Pennsylvania DCNR Interactive Map Resources website showing the Tookany (Tacony) Creek abrupt direction change at Cheltenham (1), Pennypack Creek southwest-east-northeast direction change (2), and railroad in the shallow through valley crossing the Pennypack-Tookany Creek drainage divide (3)

3.6 Barbed and other Tributaries to the Tookany Creek Valley Downstream from Jenkintown

W.M. Davis probably was most familiar with the Tookany Creek valley downstream from Jenkintown (see figure 7) simply because it was closest to the Davis family home and would have been an attraction not only for recreation, but also for any young person developing an interest in the natural sciences. In addition young Davis probably had excellent access to much of the land in this area as his father, in addition to his coal company interests, also headed the Chelten Hills Land Association, which in 1854 purchased 1000 acres of farmland between the railroad and the Philadelphia city line, including lands adjoining the Tookany Creek valley (Rothschild, 1976, p. 68).

Between Jenkintown and the Philadelphia city line the Tookany Creek valley is eroded into the Wissahickon Formation schist with the deepest valley segment being between Jenkintown and Elkins Park. Today at several locations along that valley segment Tookany Creek flows over patches of exposed bedrock while bedrock outcrops can be seen along the creek banks and on valley walls and in the numerous railroad cuts. This valley segment in places has a V-shaped profile and depending on where and how measurements are made is from 100 to 200 feet deep. Tookany Creek down cutting is today proceeding extremely slowly as is retreat of the steep valley walls. Anyone making observations of present day erosion rates only in this Tookany Creek valley segment could reasonably conclude that similar conditions operating over extremely long periods of time developed the Tookany Creek valley as it exists today.

However, several features seen in figure 7 suggest the Tookany Creek valley had a very different history than the modern day erosion rates suggest. First is Mill Run, a northeast oriented Tookany Creek tributary originating south of the former Davis home and joining the southeast oriented Tookany Creek segment downstream from Elkins Park. Water in Mill Run makes a U-turn as it first flows in a northeast direction to join southeast oriented Tookany Creek, which then turns abruptly to flow in a southwest direction. The Mill Run evidence suggests headward erosion of a deep southeast oriented Tookany Creek valley segment beheaded and reversed a southwest oriented flow channel while the southwest oriented Tookany Creek valley segment suggests it had been eroded headward along a southwest oriented flow channel.

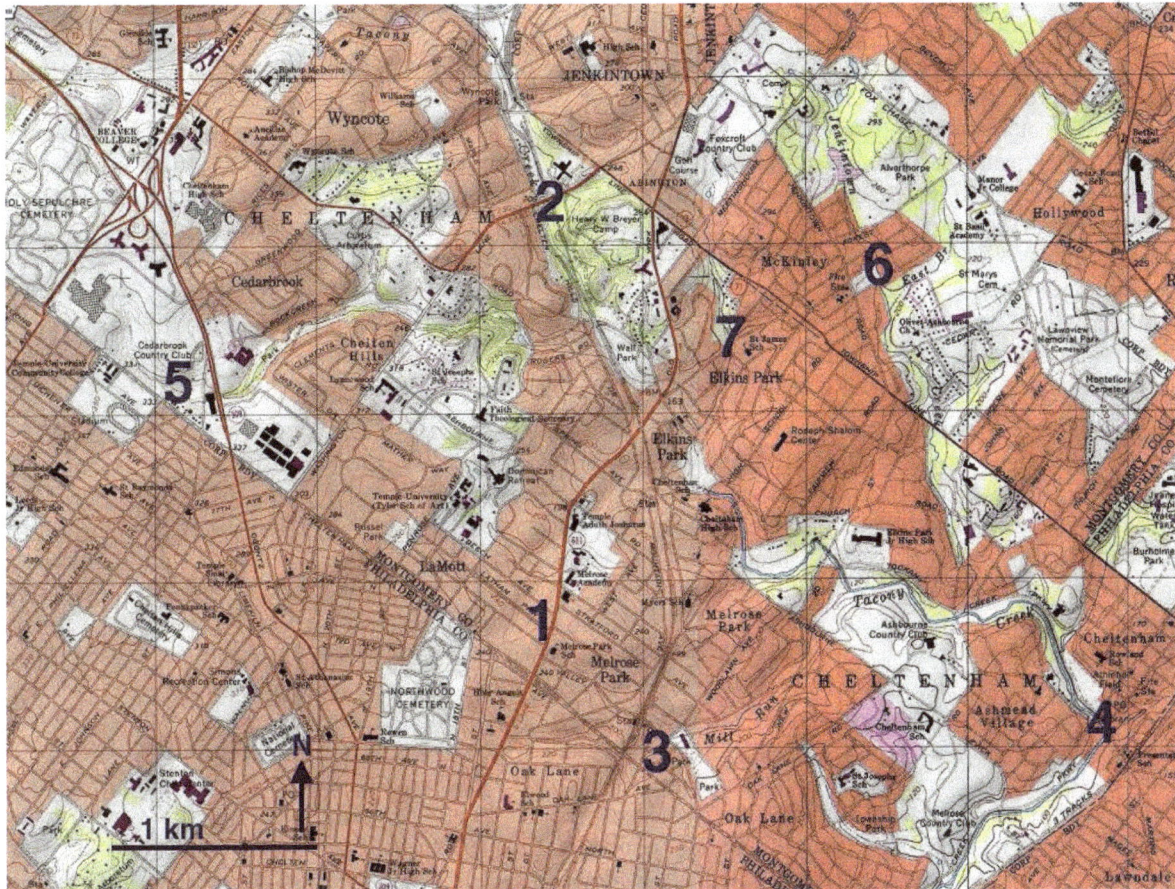

Figure 7. USGS 1:24,000 topographic map (with 10-foot contour interval in west and 20-foot contour interval in the east) taken from Pennsylvania DCNR Interactive Map Resources website showing Davis home location (1), Tookany (Tacony) Creek valley downstream from Jenkintown (2), Mill Run headwaters (3), Tookany (Tacony) Creek Cheltenham direction change location (4), Cedar Brook headwaters (5), Jenkintown Creek (6), and an unnamed south oriented Tookany (Tacony) Creek tributary (7)

A close look at other Tookany Creek tributaries shows a deep northeast oriented valley extending through Wyncote and draining to Tookany Creek near the Jenkintown railroad station (the creek in that valley is shown in figure 1 but has since been incorporated into the Wyncote sewer system). This creek is another barbed tributary and provides further evidence the deep south oriented Tookany Creek valley eroded headward across southwest oriented flow. A tributary (Cedar Brook) originates at Cedarbrook Country Club and flows across lands the Chelten Hills Land Association once owned in a south and then northeast direction between Cedarbrook and Chelten Hills before continuing in an east direction to join Tookany Creek across from Wall Park. The east and northeast oriented Cedar Brook segments probably were formed when headward erosion of the deep south-oriented Tookany Creek valley beheaded and reversed west and southwest oriented flow channels and the south oriented headwaters valley was eroded by captured southwest oriented flow from what at that time was the yet to be beheaded and reversed southwest oriented channel that was later reversed to create the northeast oriented stream through Wyncote to the north.

East of the Tookany Creek valley short southwest oriented valleys suggest erosion by southwest oriented flow, but the two longest tributaries including Jenkintown Creek flow in south directions. The unnamed westernmost of these two south oriented tributaries turns in a southwest direction before joining Tookany Creek while Jenkintown Creek has southwest oriented headwaters and at least two southwest oriented tributaries. This evidence suggests the south oriented tributary valleys eroded headward across southwest oriented flow moving water into the newly eroded Tookany Creek valley with the unnamed south oriented tributary capturing the flow first and headward erosion of the south oriented Jenkintown Creek valley capturing the southwest oriented flow next. While not seen in figure 7, Pennypack Creek valley headward erosion east of figure 7 beheaded all southwest oriented flow to the Jenkintown Creek valley.

4. Discussion

The erosion history that emerges when wind gaps, through valleys, elbows of capture, and tributary and Tookany Creek valley orientations are explained is one of massive southwest oriented floods flowing across the entire region on what was probably a low gradient topographic surface equivalent in elevation to or higher than the highest ridges seen today. Floodwaters flowed in diverging and converging channels and were first captured by headward erosion of the deep south-oriented Wissahickon Creek valley. Headward erosion of the deep south-oriented Tookany Creek valley next captured the southwest oriented flow, but due to headward erosion of west oriented tributary valleys from the newly eroded Wissahickon Creek valley (in the Chester Valley area) was unable to erode headward across the Edge Hill quartzite ridge. Finally headward erosion of the deep south-oriented Pennypack Creek valley beheaded all southwest oriented flow routes moving to the Tookany Creek valley and also to the Chester Valley eastern end. This erosion history describes erosion events in a logical sequence and explains all observed landform features.

The origin of the massive southwest oriented floods described here cannot be determined from Tookany Creek or adjacent drainage basin landform evidence. All that can be conclusively determined is the floodwaters came from north and east of the region and were of great enough volume and duration to erode deep valleys into the preexisting topographic surface (and perhaps lowered that preexisting surface significantly). Melting of an Antarctic sized continental ice sheet, which may have contained 26,384,368 cubic kilometers of water according to a NASA distributed middle school exercise (Parkinson, 1999) could have produced floods capable of overwhelming all existing drainage systems and may have been responsible for the massive floods documented here. If so the floods crossed what are today south oriented Neshaminy Creek and Delaware River valleys located to the east and north of the Tookany Creek drainage basin. Evidence from other regions suggests the south oriented Delaware valley segment was eroded late in geologic time, perhaps between the Miocene and early Pleistocene (Braun, D. D., Pazzaglia, F. J., and Potter, N., 2003 p. 220). If so the previously established sequence of Wissahickon, Tookany, and Pennypack Creek valley headward erosion can be extended to include the Neshaminy Creek valley and the south and southeast oriented Delaware River valley segment (with headward erosion of the south oriented valleys beginning in the southwest and continuing to the northeast). This sequence may be evidence the south oriented valleys eroded headward from the head of the southwest oriented Delaware River valley segment (to which Tookany, Pennypack, and Neshaminy Creek now flow) as it eroded headward in a northeast direction along what was at that time a major southwest oriented flood flow channel.

The Tookany Creek drainage basin erosion history determined here is radically different from anything W. M. Davis ever described or that his erosion cycle or uniformitarianism paradigms even permit. Davis did not consider his geographical cycle or uniformitarianism ideas to be based on any specific geographic region (e.g. Davis, 1899), although in his mind he must have compared them many times with observations made while living in the Tookany Creek drainage basin. Davis did recognize his erosion cycle and uniformitarianism paradigms did not explain all landform evidence, but he knew of no satisfactory alternative paradigm that could. In his 1889(b) "Rivers and valleys of Pennsylvania" *National Geographic Magazine* paper, he stated, "If this theory of the history of our rivers is correct, it follows that any one river as it now exists is of so complicated an origin that its development cannot become a matter of general study and must unhappily remain only a subject for special investigation for some time to come" suggesting he knew his erosion cycle and uniformitarianism concepts could not explain much of the observed evidence.

The Tookany Creek drainage basin erosion history described here demonstrates that with a paradigm defined by headward erosion of deep valleys into a low gradient topographic surface over which immense floods moved water in shallower diverging and converging channel complexes it is possible to explain all observed wind gaps, through valleys, barbed tributaries, elbows of capture, and many valley segment orientations. The Davis uniformitarianism paradigm by not permitting catastrophic floods severely underestimated the amount of melt water released during continental ice sheet melting and unfortunately may be causing modern day geomorphologists to overlook wind gaps, through valleys, barbed tributaries, valley orientations, elbows of capture, and similar erosional landform features as important evidence that needs to be explained. Those erosional landform features exist and landform evolution studies of regions containing them must develop explanations that can explain all wind gap, through valley, barbed tributary, elbow of capture, and similar types of erosional landform evidence that may be present.

References

Baker, V. R (Ed.) (1981). *Catastrophic flooding, the origin of the channeled scabland*: Benchmark Papers in Geology (55), Dowden, Hutchinson & Ross, 360p.

Bascom, F., Clark, W. B., Darton, N. H., Knapp, G. N., Kuemmel, H. B., Miller, B. L., & Salisbury, R. D. (1909). Philadelphia folio: Norristown, Germantown, Chester, and Philadelphia, Pennsylvania-New Jersey-Delaware. Folios of the Geologic Atlas: 162.

Bischke, R, E. (1980). The Abington-Cheltenham, PA. Earthquake Sequence of March-May, 1980: *Pennsylvania Geology, 11*(5), 10-13.

Bishop, P. (2007). Long-term landscape evolution: linking tectonics and surface processes. *Earth Surface Processes and Landforms,* (32), 329-365. https://doi.org/10.1002/esp.1493

Braun, D. D., Pazzaglia, F. J., & Potter, N. (2003), Margin of Laurentide Ice to the Atlantic Coastal Plain: Miocene-Pleistocene landscape evolution in the Central Appalachians (p. 220). In Easterbrook, D. (Ed.), *Quaternary Geology of the United States: INQUA 2003 Field Guide Volume*: Desert Research Institute, Reno, NV.

Bretz, J. H. (1923). The Channeled Scabland of the Columbia Plateau. *Journal of Geology*, (31), 617-649. https://doi.org/10.1086/623053

Camburn, R. S. (1977). *The Story of Greater Glenside including the towns of Ardsley, Edge Hill, North Hills, and Weldon*: revised edition (p.120), Glenside Free Library, Glenside, PA.

Chorley, R. J., Beckinsale, R. P., & Dunn, A. J. (1973). *The History of the Study of Landforms or the Development of Geomorphology: Volume 2: The life and work of William Morris Davis*. Routledge, New York. 874p.

Clausen, E. (2015). Exploring the Geography of William Morris Davis' Pennsylvania roots. *Pennsylvania Geographer*, 53(1), 44-65.

Davis, W. M. (1889a). A river pirate. *Science* 13, pp. 108-9

Davis, W. M. (1889b). The rivers and valleys of Pennsylvania. *National Geographic Magazine* (1), pp. 183-253.

Davis, W. M. (1895). Bearing of physiography on uniformitarianism. *Bulletin Geological Society of America*, (7), 8-11.

Davis, W. M. (1899). The Geographical Cycle: *Geographical Journal*, 14, (pp. 481-504). Reprinted in: Davis, W. M., 1909, *Geographical Essays*: Ginn and Company, New York, pp. 249-278

Hack, J. T. (1957). *Studies of longitudinal stream profiles in Virginia and Maryland.* United States Geological Survey Professional Paper 294-B, 45-97.

Horton, R. E. (1945). Erosional development of streams and their drainage basins: hydrophysical approach to quantitative morphology. *Bulletin of the Geological Society of America*, (56), 275-370. https://doi.org/10.1130/0016-7606(1945)56[275:EDOSAT]2.0.CO;2

Orme, A. R. (2007) The rise and fall of the Davisian cycle of erosion: prelude, fugue, coda, and sequel: *Physical Geography*, (28), 474-506. https://doi.org/10.2747/0272-3646.28.6.474

Parkinson, C. L. (1999) Ice Sheets and sea level rise: *from* The PUMAS Collection. Retrieved 10/20/2016 from http://pumas.jpl.nasa.gov/examples/index.php

Pennsylvania Department of Conservation and Natural Resources (2016). DCNR Interactive Map Resources, Retrieved 10/20/2016 from http://www.dcnr.state.pa.us/learn/interactivemapresources/index.htm

Potter, N. (1999). Part V: Physiography: In C. H. Schultz (Ed), *The Geology of Pennsylvania* (pp. 342-344), Pennsylvania Geological Survey and Pittsburgh Geological Society, Harrisburg and Pittsburgh, PA.

Rothschild, E. W. (1976). *A History of Cheltenham Township: Montgomery County, PA*. Cheltenham Township Historical Commission, Cheltenham Township, PA. 81p.

Strahler, A. N. (1952). Dynamic basis of geomorphology. *Geological Society of America Bulletin*, (63), 923-938. https://doi.org/10.1130/0016-7606(1952)63[923:DBOG]2.0.CO;2

United States Geological Survey Historical Map Collection (1896). Germantown, PA 1:62,500 scale topographic map. Retrieved 10/20/2016 from http://ngmdb.usgs.gov/maps/TopoView/viewer/#

United States Geological Survey Historical Map Collection (1997). Frankford, PA 1:24,00 scale topographic map. Retrieved 10/24/2016 from http://ngmdb.usgs.gov/maps/TopoView/viewer/#

United States Geological Survey Historical Map Collection (1997). Germantown, PA 1:24,000 scale topographic map. Retrieved 10/20/2016 from http://ngmdb.usgs.gov/maps/TopoView/viewer/#

United States Geological Survey Historical Map Collection (1999). Ambler, PA 1:24,00 scale topographic map. Retrieved 10/20/2016 from http://ngmdb.usgs.gov/maps/TopoView/viewer/#

United States Geological Survey Historical Map Collection (1999). Hatboro, PA 1:24,00 scale topographic map. Retrieved 10/20/2016 from http://ngmdb.usgs.gov/maps/TopoView/viewer/#

8

Mapped Fractures and Sinkholes in the Coastal Plain of Florida and Georgia to Infer Environmental Impacts from Aquifer Storage and Recovery (ASR) and Supply Wells in the Regional Karst Floridan Aquifer System

Wenjing Xu[1], Sergio Bernardes[1], Sydney T. Bacchus[1] & Marguerite Madden[1]

[1] Center for Geospatial Research, Department of Geography, University of Georgia, Athens, Georgia 30602-2502, USA

Correspondence: Marguerite Madden, Center for Geospatial Research, Department of Geography, University of Georgia, Athens, Georgia 30602-2502, USA. E-mail: mmadden@uga.edu

Abstract

The regional Floridan aquifer system (FAS) extends from the submerged carbonate platform of the Atlantic Ocean, Gulf of Mexico, and Straits of Florida in the southeastern United States (US), throughout Florida and the coastal plain of Alabama, Georgia, and South Carolina. This carbonate aquifer system is characterized by bedding planes, fractures, dissolution cavities, and other karst features that result in preferential flow of ground water, particularly in response to anthropogenic perturbations such as groundwater withdrawals and aquifer injections. The FAS was divided into six sub-regions for groundwater-modeling purposes in 1989, with results concluding that breaches of those groundwater divides had occurred and those breaches were attributed to large withdrawals of ground water in the US southeastern coastal plain. Those results suggest the model did not elucidate preferential flow conditions through fractures and other karst conduits. We hypothesized that incorporating fractures and sinkholes into groundwater models could improve results and predict adverse impacts to environmentally sensitive areas. We analyzed extensive fracture networks and sinkholes previously mapped throughout Florida and in Dougherty County, Georgia. Some of those fractures extend from one sub-region into an adjacent sub-region of the FAS and may be facilitating the breaching of groundwater divides described in the 1989 groundwater model for this regional aquifer system. The greater total fractures and fracture density in Dougherty County (1,225 and 141.3/100 km^2, respectively) compared to 21 north-Florida counties (10-91 fractures per county and 0.6-3.8/100 km^2, respectively) presumably is due to the scale of fracture mapping and shorter mean lengths of mapped fractures in Dougherty County (1.2 km), compared to north Florida counties (26-118 km), rather than to orders of magnitude increases in fracture densities in that part of the FAS. The number of sinkholes identified in Dougherty County in a recent, unrelated project using 2011 Advanced Spaceborne Thermal Emission and Reflection Radiometer (ASTER) images, was approximately an order of magnitude greater than the number of sinkholes mapped in analog form in that county and published in 1986. Extension of the dense network of those fractures that occurred within the boundaries of a Priority Amphibian and Reptile Conservation Area (PARCA) that encompassed Dougherty County covered the Elmodel Wildlife Management Area (WMA) and ASR demonstration well in Baker County, Georgia. Those extensions also passed through numerous agricultural areas with center-pivot irrigation wells in southwest Georgia; intersected other Georgia PARCAs near the Florida-Georgia state line; and clumped in two areas of dense sinkhole clusters in northwest Florida. No determination has been made regarding the contributions of pirated water from the Apalachicola-Chattahoochee-Flint (ACF) River Basins and Wakulla Springshed from the magnitude and extent of agricultural, municipal, and industrial groundwater withdrawals in Georgia's coastal plain, that exceed groundwater withdrawals in Florida for that area of the FAS, to the increase in sinkholes in Dougherty County and the dense clusters of sinkholes in northwest Florida, via preferential flow through fractures. Similarly, the survival and recovery of at least 24 animal species in Georgia that are either federally listed or high-priority state species may be jeopardized by adverse direct, indirect, and cumulative impacts from preferential flow through fractures, sinkholes, and other karst conduits in response to aquifer injections and withdrawals that have not been evaluated. Currently no regional groundwater model has been constructed to evaluate such preferential groundwater flow in the FAS. A model incorporating preferential flow via mapped fractures and sinkholes is essential to determine the magnitude and extent of environmental impacts from ASR wells and other supply and disposal wells in this regional aquifer system, such as pirated water from the ACF and other river basins, alterations in submarine groundwater discharge to Apalachicola Bay and other coastal areas, saltwater intrusion, upconing of saline ground water and resulting impacts to federally endangered and threatened species and high-priority state species.

Keywords: Apalachicola-Chattahoochee-Flint (ACF) River Basins, arsenic, breached groundwater divides, carbonate aquifer system, geographic information system (GIS), pirated water, saltwater intrusion, upconing

1. Introduction

1.1 Regional Floridan Aquifer System

The regional Floridan aquifer system (FAS) extends throughout Florida, the coastal plain of Alabama, Georgia, and South Carolina (Johnston & Bush, 1988) and offshore from those states, to the limits of the submerged carbonate platform of the Atlantic Ocean, Gulf of Mexico, and Straits of Florida in the southeastern US. This aquifer system is the primary source of municipal water for those states. It also is a critical source of water for environmentally sensitive public and private lands, federally endangered and threatened species, and high-priority state species, including coastal species, but groundwater quantity and quality are declining without acknowledging the role of preferential groundwater flow in those declines. Carbonate aquifer systems, including the FAS, are characterized by bedding planes, fractures, dissolution cavities, piping, and other karst features. These result in preferential flow of ground water, particularly in response to anthropogenic perturbations such as groundwater withdrawals and aquifer injections (Bacchus, 2002; 2007; Bacchus, Bernardes, Jordan, & Madden, 2014; Bacchus, Bernardes, Xu, & Madden, 2015a; Bush & Johnston, 1988; Ford, Palmer, & White, 1988; Ford & Williams; 1989; Kohout, 1967; Krause & Randolph, 1989; Meyer, 1989; Neuendorf, 2005; Popenoe, Kohout, & Manheim, 1984; Yassin, Muhammad, Taib, & Al-Kouri, 2014).

Krause and Randolph (1989) divided the FAS into six sub-regions for groundwater modeling purposes. They concluded that breaches of those groundwater divides had occurred, attributing the breaches to large withdrawals of groundwater in the US southeastern coastal plain. Breaching probably is the result of extensive networks of fractures that have been reported by the US Army Corps of Engineers (ACOE, 2004), and mapped by Florida Department of Transportation (FDOT, 1973), and Vernon (1951) throughout Florida, facilitated by remote sensing. In the northwestern extent of the FAS, Brook and Allison (1986) mapped similar extensive fracture networks in Dougherty County, Georgia, located in the northwest portion of the FAS. Remote sensing continues to be a useful tool for identifying fractures, including in non-karst and Cretaceous formations (Oden, Okpamu, & Amah, 2012).

1.2 Background

1.2.1 Terminology Related to Aquifer Storage and Recovery (ASR)

Bacchus et al. (2015a) provided definitions for terminology related to "aquifer storage and recovery" (ASR) that is used by regulatory agencies, municipalities, and representatives of the ASR industry. That terminology, created primarily by the ASR industry and regulatory agencies, generally does not conform to standard or scientific definitions or concepts and is misleading. Examples include "aquifer storage and recovery," "bubble," "confining," "excess water," "lost to tide," "recharge," "reservoir," "restoration," "target storage volume" (TSV), and "water banking" (Bacchus et al., 2015a). Neither definitions nor descriptions of some of these terms, as applied to ASR, have been included in scientific peer-reviewed publications with dictionaries or glossaries. Therefore, in some cases (e.g., "aquifer storage and recovery," "target storage volume," and "water banking"), the default sources for definitions and descriptions of terms are documents produced by representatives of the ASR industry. Because of those constraints, terms that are used in an unorthodox manner in regulatory and other documents referenced in this paper are provided in quotations marks, to avoid confusion and misrepresentation.

McNeill (2000) concluded that municipal sewage effluent injected into deep saline aquifers for decades at Miami-Dade, Florida's South District Wastewater Treatment Plant (SDWWTP) was flowing vertically upward. More than a decade latter, Walsh and Price (2010) and Walsh (2012) concluded that "injectate remained chemically distinct as it migrated upwards through rapid vertical pathways via density-driven buoyancy" at that site. In fact, numerous hydraulic and geochemical effects of the waste injections into the FAS already had been documented by the early 1970s. Examples included: (1) reversal of natural hydraulic gradients, with establishment of potential for up-dip migration of fluids; (2) increased localized aquifer permeability and transmissivity; (3) dissolution of the carbonate aquifer and localized cavern development beneath the injection sites; (4) upward movement of wastes; and (5) evolution of high concentrations of hydrogen sulfide, nitrogen, methane, and other gases (Kaufman, 1973). Additional evidence that more buoyant, low-salinity water injected into saline aquifer zones discharges vertically into overlying surface waters was published by numerous scientists in the 1980s based on independent research, as described by Bacchus (2001; 2002). That extensive evidence spanning decades, that low-salinity water injected into saline aquifer zones flows vertically upward, was not addressed by Walsh and Price (2010) and Walsh (2012) in their conclusion that "only a one-time pulse of injectate into the overlying aquifers" occurred at Miami-Dade's North District Wastewater Treatment Plant (NDWWTP) "due to improper well construction" or the unsupported statement that "[D]eep well injection into non-potable saline aquifers of treated domestic wastewater has been used in Florida for decades as a safe and effective alternative to ocean outfall disposal (Walsh, 2010).

Walsh and Price (2010) and Walsh (2012) also concluded that no fracturing had been reported at either the SDWWTP or NDWWTP injection-well sites. Based on the locations of those injection wells in the FDEP UIC database, the SDWWTP site is located immediately northeast of a fracture mapped by FDOT (1973) and north of the intersection of fracture extensions from the FDOT (1973) fracture data set. The NDWWTP site is located immediately southwest of the extension of a fracture mapped by FDOT (1973), but that NDWWTP location does not appear to coincide with the location of the NDWWTP shown in Figure 1 of Walsh and Price (2010). It is important to note that the SDWWTP site is west of submarine groundwater discharge (SGD) sites "b" and "a," south of SGD site "a," and north of SGD "d," all of which are in Biscayne Bay, and both SDWWTP and NDWWTP sites are associated with extensions of a fracture mapped by FDOT (1973) that extends to a coral reef in the Florida Keys subjected to eutrophication (Bacchus et al., 2014). The data and samples analyzed by Walsh and Price (2010) and Walsh (2012) were from the existing UIC monitoring wells, rather than samples and data from fractures associated with those two aquifer-injection sites. It also is important to note that Walsh's (2012) conclusion regarding fractures was that "seismic data acquisition is recommended for any future injection sites, as it may be able to optimize location of future injection sites in areas where these subsurface features are not found." Since 2012, FDEP has issued 2,216 UIC permits, but none of those permits required seismic data acquisition (Joe Haberfeld, FDEP Geological Survey, UIC Program Geologist, pers. comm., 04/25/16 and 04/27/16). This deficiency does not appear to be confined to Florida. There was no evidence that seismic data acquisition was required prior to the permitting and construction of the recent ASR demonstration well in Georgia, the recent ASR wells in Hilton Head, South Carolina (Pyne, 2015) or any of the other ASR wells permitted in South Carolina. The analysis of ASR for Hilton Head Island does not even address the presence of fractures or the ability of ASR to increase saltwater intrusion, rather than provide a solution to that serious problem (Pyne, 2015).

1.2.2 The Theory of ASR

The theory of ASR is temporary artificial aquifer "recharge" that consists of three components: (1) aquifer injections of fluids ("recharge"); (2) withdrawals of the injected fluids ("recovery"); and (3) a period of time ("cycle") between the injections and withdrawals (Bacchus et al., 2015a). Bacchus, Bernardes, Xu, and Madden (2015b) discussed the difference between natural and artificial aquifer recharge and how natural recharge exceeds actual "recovery" documented from Florida ASR wells.

The period of time between aquifer injections and aquifer withdrawals from an ASR well is considered to be "storage" of the injected fluids. The injected water generally is termed "excess water." The concepts of "excess water" and water "lost to tide" during the wet season ignore the beneficial/essential roles of natural pulses of uncontaminated, nonsaline surface and ground water to coastal ecosystems. The source of this temporary artificial aquifer "recharge" includes: (1) stormwater runoff (containing agricultural, industrial and/or municipal contaminants) pumped out of canals, mine pits or other areas; (2) treated sewage effluent (also known as reclaimed water, reuse water, bright water) previously mined from the aquifer system and resulting in induced (forced) recharge; (3) surface water diverted or extracted from natural streams, lakes, and other surfacewater ecosystems during the wet season; and (4) ground water withdrawn from one layer or zone of the aquifer system and injected into another. "Augmentation" and "blending" are examples of terms applied to withdrawing ground water from one aquifer zone for injecting into another zone.

The US Environmental Protection Agency (USEPA) regulates these types of underground injections pursuant to the federal Safe Drinking Water Act (SDWA) Underground Injection Control (UIC) Program. Although the USEPA also is responsible for enforcing the federal Clean Water Act and complying with the federal Endangered Species Act, direct, indirect, and cumulative environmental impacts have not been evaluated for: (1) diverting natural surfacewater discharges during wet seasons; (2) aquifer-injections of water with different chemical composition than existing ground water; and (3) discharges of injected water into surface waters via fractures and other karst conduits (Bacchus, et al., 2014; Bacchus et al., 2015a; Bacchus et al., 2015b).

1.2.3 "Disposal" vs. "Storage" and Reported "Recovery" vs. Actual "Recovery"

Aquifer injections into the FAS for "disposal" of various contaminated fluids occur via wells at similar depths as the ASR "storage" and "recovery" wells. These fluids include sewage effluent treated to various degrees; contaminated stormwater runoff; and agricultural, industrial, and other municipal wastes (e.g., brine from reverse osmosis production of municipal water). Most "disposal" of contaminated fluids through aquifer injections are reported as exceeding "1 million gallons per day (MGD)." The presumption for these aquifer "disposals" of contaminated fluids through injection wells is that the contaminated fluids constantly flow away from the injection wells as additional injections of contaminated fluids occur. These "disposal" wells also are regulated under the SDWA by the USEPA's UIC program. The governing presumption of this regulatory program is that contaminated fluids injected into the aquifer do not flow vertically upward and into an overlying aquifer zone that is used or potentially available for potable water. Horizontal flow of those injected contaminants, which include discharges to inland and coastal surface waters, is ignored.

Despite the established fact that fluids injected into the FAS flow away from the point of aquifer injection, federal and state regulatory agencies, and the ASR industry continue to claim that aquifer injections of water into ASR wells remain at the base of the well in a "bubble" waiting for withdrawal as "recovery" of the injected water (Bacchus et al., 2014; 2015a; 2015b). The reported "recovery" in the US Geological Survey (USGS) review of ASR wells and well data in Florida (Reese, 2002, Table 5) was based on the chloride concentration established under the SDWA for potable water (250 mg/L), rather than the much lower chloride concentration of the water injected into those ASR wells. Actual "recovery" was determined by adjusting the results of cycle testing conducted in ASR test wells constructed in south Florida to compensate for differences between chloride concentration of water injected into ASR wells and water withdrawn from those wells (i.e., reported chloride concentration/(recovered/injected concentration)), as summarized by Bacchus et al. (2014; 2015a; 2015b). Actual "recovery," based solely on chloride content of the withdrawn water adjusted to match the chloride content of the injected water ranged from 0% to 16.8% for "storage" periods of 15 and 8 days, respectively for those ASR wells (Table 3 of Bacchus et al., 2015a). That evaluation by Reese (2002) conducted at selected ASR wells appears to be the most extensive compilation of the chloride content of ground water at the ASR location and the chloride content of "recovered" water, which exceeded the chloride concentration of the ground water routinely, while still being considered as "recovered" water.

Those shocking results illustrate the lack of evidence supporting the presumption of "storage" or "recovery" of water injected into ASR wells and that ASR wells respond differently than aquifer disposal wells, despite agency and industry terminology that attempts to support distinct differences. After the evaluation of 18 ASR sites by Reese (2002, Table 5), in the USGS review of Florida's ASR wells, 13 of those sites were reported as abandoned by Bloetscher, Sham, Danko III, and Ratick (2014). That suggests percent "recovery" was misrepresented because water continued to be withdrawn after the chloride content of the withdrawn water was considerably greater than the chloride content of the water injected into ASR wells, making claims of the volume and duration of injected water "stored" questionable. Another ASR well site evaluated by Reese (2002, i.e., San Carlos) later was reported in the Florida Department of Environmental Protection (FDEP) database as inactive. The names of those 14 ASR sites are shown in bold in Table 1 (Bacchus et al., 2015b). Examples of other abandoned ASR sites in Collier, Lee, Miami-Dade, and Okeechobee Counties are shown in Table 1 of Bloetscher et al. (2014).

An additional process, gravity flow in shallow ground water of surficial aquifer zones in the FAS, is well documented as a transport mechanism for contaminants at locations such as non-human concentrated animal feeding operations (CAFOs) and similar human operations, such as the City of Tallahassee municipal sewage effluent sprayfield (Bacchus & Barile (2005); Kincaid, Davies, Hazlett, Loper, Dehan, & McKinlay (2004); and Kincaid, Davies, Werner, & DeHan (2012)). The established fact that ground water is not static but flows raises the question of how the same water that is injected into ASR wells can be "recovered." Similarly, the question arises regarding whether aquifer injections can be considered as artificial aquifer recharge, as ASR proponents claim, or if the injected water simply flows rapidly from the point of injection via karst conduits and then discharges to surface waters (Bacchus et al., 2015a; 2015b). We found no published literature or reports or permits for ASR wells in the FAS describing the chemical characterization or "fingerprinting" of the native ground water and water injected for "storage" and "recovery" that would support the claims of "storage" and "recovery."

1.3 Objectives

This investigation reports on the dynamics of potential pathways from the subsurface to the surface in Florida and the coastal plain of Georgia and South Carolina, all of which are underlain by the FAS. The first objective was to compare locations, lengths, and densities of the three referenced sets of fracture networks in Florida (ACOE, 2004; FDOT, 1973; Vernon, 1951) to the fracture networks in Dougherty County, mapped by Brook and Allison (1986) in the northwestern Georgia portion of the FAS. Those fracture networks, originally mapped in analog format, were georeferenced and converted to digital formats to update those previous studies and facilitate our comparisons. The second objective was to compare the proximity of those fractures and fracture extensions to permitted ASR wells in Florida, Georgia, and South Carolina, and relict and modern sinkholes mapped in Florida and Dougherty County, Georgia.   The third objective was to compare the locations of those mapped fractures and extensions in Florida and Georgia to areas of federally listed or high-priority state animal species, selected federal and state lands, watersheds for the Apalachicola-Chattahoochee-Flint (ACF) River Basins (adapted from Seaber, Kapinos, & Knapp, 1987) and watersheds throughout the coastal plain designated by the USGS.

2. Study Area

Our study area covers the extent of the regional FAS from the submerged carbonate platform of the Atlantic Ocean, Gulf of Mexico, and Straits of Florida, throughout Florida and the coastal plain of Alabama, Georgia, and South Carolina in the southeastern US. Krause and Randolph (1989) divided the FAS into six sub-regions for groundwater modeling purposes. They identified those sub-regions as D, E, F, G, H, and an unnamed sub-region between D and H, where available data were insufficient for groundwater modeling. Figure 1A illustrates the approximate submerged extent of the Floridan aquifer system and the location of these six sub-regions, with the previously unnamed sub-region labeled U.

More detailed focus in the study area (Figure 1B) included the ACF River Basins, Dougherty County, and adjacent counties in Georgia where fractures and sinkholes have been mapped and preferential flow documented through fractures (Brook, 1985; Brook & Allison, 1986; Brook, Carver, & Sun, 1986; Brook & Sun, 1982; Brook, Sun, & Carver, 1988). The headwaters of the Chattahoochee and Flint Rivers originate in Alabama and Georgia, north of the northern extent of the Florida aquifer system, and were not considered in our study. Those rivers, however, flow into the Apalachicola River and eventually discharge into Apalachicola Bay and the Gulf of Mexico. The counties in Georgia and South Carolina included in the study also were based on the location of an ASR demonstration well recently constructed in Baker County, Georgia and counties in South Carolina where ASR wells have been installed.

A

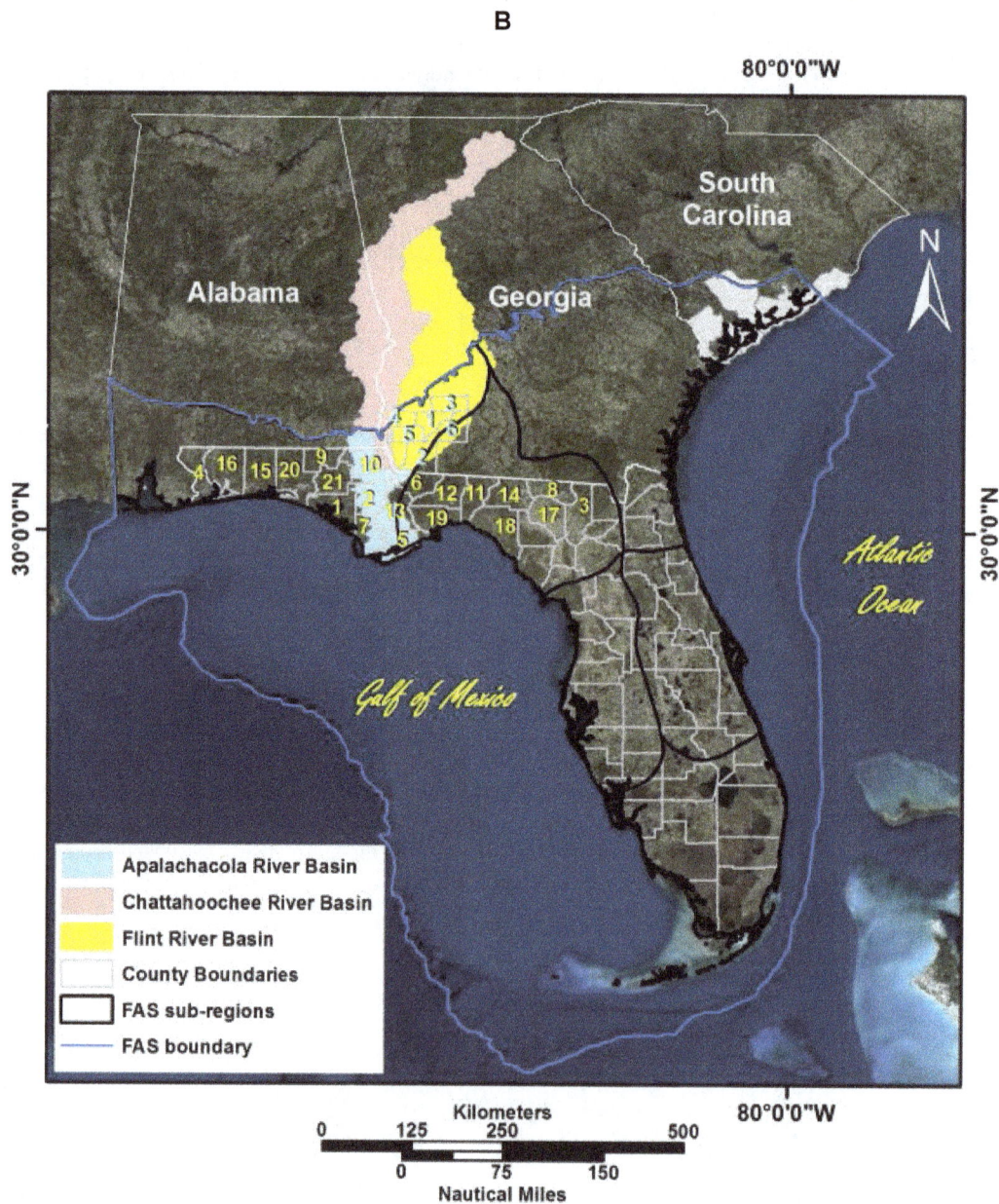

Figure 1. Approximate submarine extent of the Floridan aquifer system in the Atlantic Ocean and Gulf of Mexico, and throughout Florida and southeastern coastal plain of Alabama, Georgia, and South Carolina, with:

A. the six sub-regions (D, E, F, G, H, and unnamed (U)) designated by Krause & Randolph (1989), and

B. detailed view showing the Apalachicola-Chattahoochee-Flint (ACF) River Basin, and counties in north Florida (numbered), Georgia, and South Carolina with mapped fractures, relict sinkholes or ASR wells included in the study

Having such a large data set allowed us to evaluate the relationship of fractures and sinkholes to key watersheds of concern, as well as sub-regions of the regional aquifer system designated by the USGS. Our large data set also facilitated evaluating the relationship of fractures and sinkholes to examples of federal and state lands in Florida, Georgia, and South Carolina, and selected Priority Amphibian and Reptile Conservation Areas (PARCAs) as part of our study. Table 1 lists the counties, and federal and state lands in north Florida, Georgia, and South Carolina that were evaluated in more detail in our study area.

Table 1. Counties in north Florida, Georgia, and South Carolina in the Floridan aquifer system with more detailed analysis of mapped fracture data, permitted ASR wells, and federal and state lands, and towns in the vicinity of those fractures or ASR wells.

ID	North Florida Counties	Federal and State Lands and Towns
1	Bay	Panama City
2	Calhoun	Blountstown
3	Columbia	Lake City
4	Escambia	Pensacola
5	Franklin	Apalachicola NF[1], Tate's Hell State Forest[2]
6	Gadsen	Quincy
7	Gulf	Port St. Joe
8	Hamilton	Jasper
9	Holmes	Bonifay
10	Jackson	Marianna
11	Jefferson	St. Marks National Wildlife Refuge (NWR)[3]
12	Leon	Apalachicola NF[1]
13	Liberty	Apalachicola NF[1], Tate's Hell State Forest[2]
14	Madison	Madison
15	Okaloosa	Crestview
16	Santa Rosa	Milton
17	Suwannee	Live Oak
18	Taylor	St. Marks NWR[3]
19	Wakulla	Apalachicola NF[1], St. Marks NWR[3], Wakulla Springs SP
20	Walton	DeFuniak Springs
21	Washington	Chipley
	Georgia Counties	
1	Baker	Elmodel Wildlife Management Area, ASR demonstration well
2	Decatur	Bainbridge
3	Dougherty	Albany, Pretoria, Putney
4	Early	Damascus
5	Miller	Cooktown
6	Mitchell	Camilla
	South Carolina Counties	
	Beaufort	Pinckney Island NWR
	Charleston	Charleston, Kiawah, Mt. Pleasant
	Colleton	Edisto Beach[4]
	Jasper	Savannah NWR, Tybee NWR

[1]Apalachicola National Forest is the largest U.S. National Forest in Florida at 2,561.2 km^2 (988.9 mi^2) and includes Bradwell Bay Wilderness Area, Mud Swamp/New River Wilderness Area and Leon Sinks Geological Area.
[2]Tate's Hell SF encompasses 819 km^2 (316.2 mi^2).
[3]St. Marks National Wildlife Refuge includes the estuaries of seven north Florida streams and encompasses 280 km^2 (108.1 mi^2).
[4]Proposed ASR well and RO facility.

3. Methods

3.1 Shapefiles and Data from Other Sources

3.1.1 Shapefiles and Data for Florida

The locations of the Florida ASR wells permitted by the FDEP were created from location information included in the FDEP database and provided by that agency. Locations of all aquifer injection wells permitted by the FDEP in Florida are available online (http://ca.dep.state.fl.us/mapdirect/?focus=uic). The locations for modern sinkholes (used synonymously with subsidence features) reported in Florida were created from the FDEP Florida Geological Survey (FGS) shapefile for subsidence features mapped in Florida through October 2014. Shapefiles not available from the Army Corps of Engineers (ACOE) for the lineaments representing extensive networks of fractures shown as the "Lineament map of south Florida" (Figure 3-7 in ACOE (2014), originally from ACOE (2004)), were created by converting the analog file to a digital file as described in Bacchus et al. (2015a). The acquisition methods for the FDOT (1973) and Vernon (1951) lineaments in Florida were described in Bacchus et al. (2014) and Lines et al. (2012).

3.1.2 Shapefiles and Data for Georgia and South Carolina

Locations for fractures and relict sinkholes in Dougherty County were based on analog versions identified by Brook and Allison (1986), who used topographic maps and 1:24,000 scale, color infrared images to identify sinkholes, based on surfacewater features, vegetation and soil moisture patterns, and topographic expression. Mapped sinkhole distribution and color infrared images then were used to map fractures and compare the mapped fracture orientations with regional trends for that area. Modern sinkholes in Dougherty County were identified independently, by the NASA DEVELOP National Program for Georgia Disasters and Water Resources for the period of 2002 to 2011, identifying as many as 18,557 sinkholes in 2011 for that county

(Cahalan, Amin, Berry, Xu, Hu, & Milewski, 2015). The methods used for that NASA DEVELOP project were the same methods used to evaluate sinkholes in a smaller area in southeast Dougherty County (Cahalan, Milewski & Durham, in press). Relict sinkholes mapped by Brook and Allison (1986) were digitized to convert from analog to digital format.

We located the ASR demonstration well in the Elmodel Wildlife Management Area (WMA), in Baker County, Georgia in the approximate center of the boundaries of the Elmodel WMA, because specific location information (i.e., latitude and longitude or shapefile) for that ASR well was not provided by the Georgia Department of Natural Resources (GDNR) that permitted that well (Jim Kennedy, GDNR, State Geologist, pers. comm., 12/15) or the Georgia Environmental Finance Authority (GEFA) Program that funded that well (Shane Hix, Director of Public Affairs, pers. comm., 12/15). Specific location information also was not available from the public and private utility companies in South Carolina for the 68 ASR wells permitted in that state, but a spreadsheet for those ASR wells, including the latitude and longitude coordinates for each well, was provided by the South Carolina Department of Environment and Health Control (SCDEHC, Christopher Wargo, pers. comm., 3/16). We converted the coordinates for those ASR wells into point-files. Funding of a proposed ASR well and associated reverse osmosis (RO) system in Edisto Beach, South Carolina is pending approval by the Edisto Beach council and mayor. Therefore, that pending ASR well was not included in the spreadsheet provided by the SCDEHC, but Colleton County, where the pending ASR well would be located, was included in the figures we created for this paper.

A shapefile for the locations of the Georgia PARCAs from Figure 1 of Jenson (2015) was provided by J.J. Apodaca (Professor of Conservation Biology at Warren Wilson College, in Asheville, North Carolina) to facilitate our determination of the proximity of those PARCAs to mapped fractures and fracture extensions. Shapefiles for the USGS watershed basin dataset (WBD) hydrologic unit codes (HUC) that intersect the state boundaries of Georgia are those used by the River Basin Center (RBC), University of Georgia, Odum School of Ecology in Athens, Georgia, as extracted from http://nhd.usgs.gov/wbd.html, and were provided by Duncan Elkin (RBC Postdoctoral Research Associate). He also provided the shapefile for the ACF River Basins (adapted from Seaber et al., 1987) that extend from Georgia through northwest Florida.

3.1.3 Shapefiles for Federal and State Lands, and Basemaps for Figures

Shapefiles for federal and state lands (e.g., National Wildlife Refuges and state Wildlife Management Areas) were obtained online from http://gapanalysis.usgs.gov/padus/data/download/. The basemap source for Figures 1A and 6A was provided as "Esri, HERE, DeLorme, TomTom, Intermap, GEBCO, USGS, FOA, NPS, NRCAN, GeoBase, IGN, Kadaster NL, Ordnance, Survey, Esri Japan, METI, Esri China." The basemap source for Figures 1B, 2A-B, 3A-B, and 6B-C was provided as, "Esri, Digital Globe, GeoEye, i-cubed, Earthstar, Geographics, CNES/Airbus DS, USDA, USGS, AEX, Getmapping, Aerogrid, IGN, IGP, swisstopo, and the GIS User Community."

3.2 Analog to Digital Conversion of Mapped Sinkholes and Lineaments Representing Fractures and Data Analysis

Geospatial analysis to assess fracture lineaments, sinkholes, and wells was conducted in ArcGIS Version 10.2, as described in Bacchus et al. (2015a). Sub-regions of the FAS were delineated by following widely-accepted map vectorization procedures, including the control-point based georeferencing of the sub-regions map in Krause and Randolph (1989) and the heads up digitizing of the resulting georeferenced map image. Political boundaries from the US Census Bureau, including county limits, were used as reference during georeferencing. Spatial frequency analysis of previously mapped linear features indicative of fractures also was conducted as described in Bacchus et al. (2015a).

Brook and Allison (1986) used topographic maps and 1:24,000 scale, color infrared images to create analog maps of sinkholes and fractures, based on the presence of surface water features, vegetation and soil moisture patterns, and topographic expression, and to compare the mapped fracture orientations with regional trends. We used ArcGIS to georeference their analog maps for Dougherty County. During georeferencing, 25 evenly distributed control points were identified over a digital copy of the fractures map and the county boundary reference layer. The resulting georeferenced map was then used to manually trace the fractures using the "Create Features" tool in ArcMap.

4. Results

4.1 Previously Mapped Fractures, Sinkholes, and ASR Wells in the FAS Sub-Regions

Figure 2A shows the locations of fractures mapped by Vernon (1951), the Remote Sensing Section of FDOT (1973), and unidentified source(s) originating in an unauthored draft report by ACOE (2004) and included in the ACOE Final TDR (2014) in southern Florida. Also shown in Figure 2A are the modern sinkholes in the Florida database (FDEP) and the ASR wells permitted throughout Florida, in Baker County, Georgia, and throughout South Carolina, in addition to the boundaries of the six sub-regions of the FAS. The FDOT (1973) fractures were mapped throughout Florida, in all six sub-regions of the FAS. The Vernon (1951) fractures were mapped only in sub-regions E and F and the Florida portion of sub-regions D and U. The ACOE (2004; 2014) fractures were mapped primarily in sub-region G, but also extended into sub-regions E and F. Dense clusters of modern sinkholes reported to the FDEP in Florida have been mapped in sub-regions E, F and U.

Fractures mapped in Florida that occurred in proximity to Florida ASR wells were extended to the submerged and landward FAS boundaries. Figure 2B shows the locations of these fracture extensions. The northern extension of one of the fractures mapped by FDOT (1973), that is in the immediate vicinity of a Florida ASR well in sub-region E, also is associated with the operational ASR wells in Hilton Head, Long Cove, and Palmetto Bay in South Carolina The northern portion of other fracture extensions from the Florida boundary of sub-region U continues through the Georgia counties surrounding the Baker County ASR demonstration well (Figure 2B). The absence of lineaments representing fractures in the center of south Florida coincides with the location of Lake Okeechobee and results from the inability to detect linear features within water bodies remotely, with aerial photography and satellite images used for those data sets. Therefore, the absence of mapped lineaments in that area of Figures 2A and B does not imply that fractures do not underly Lake Okeechobee.

A

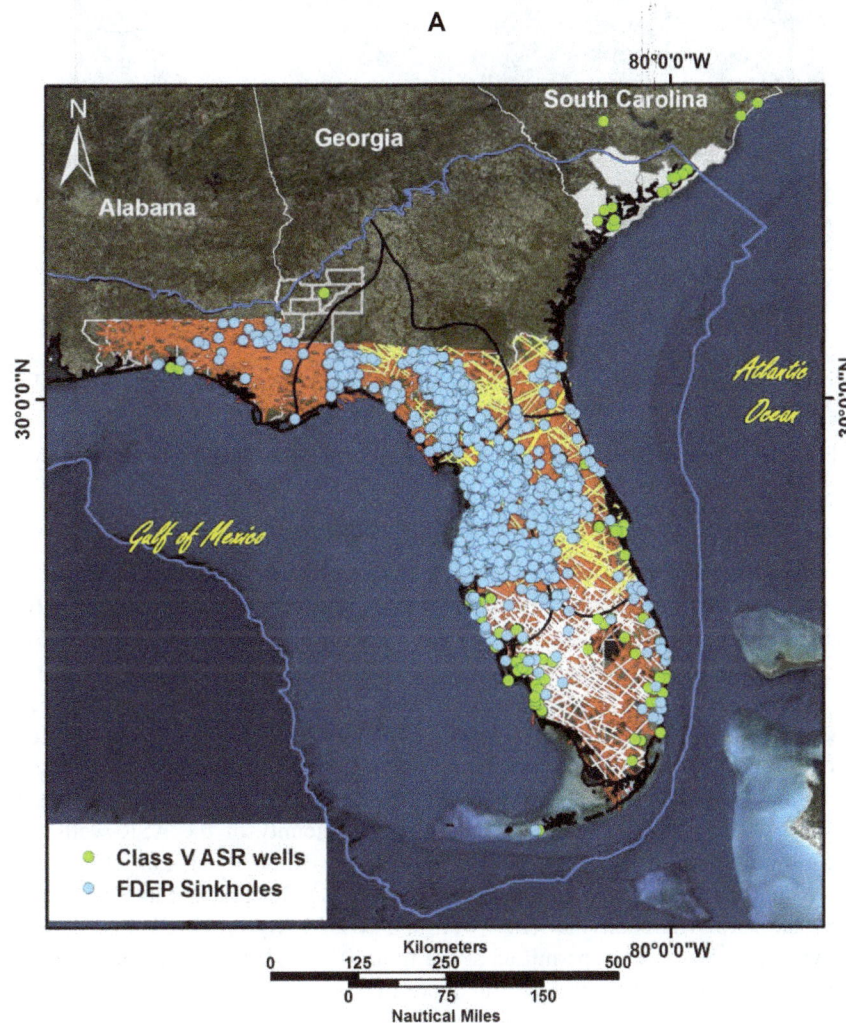

B

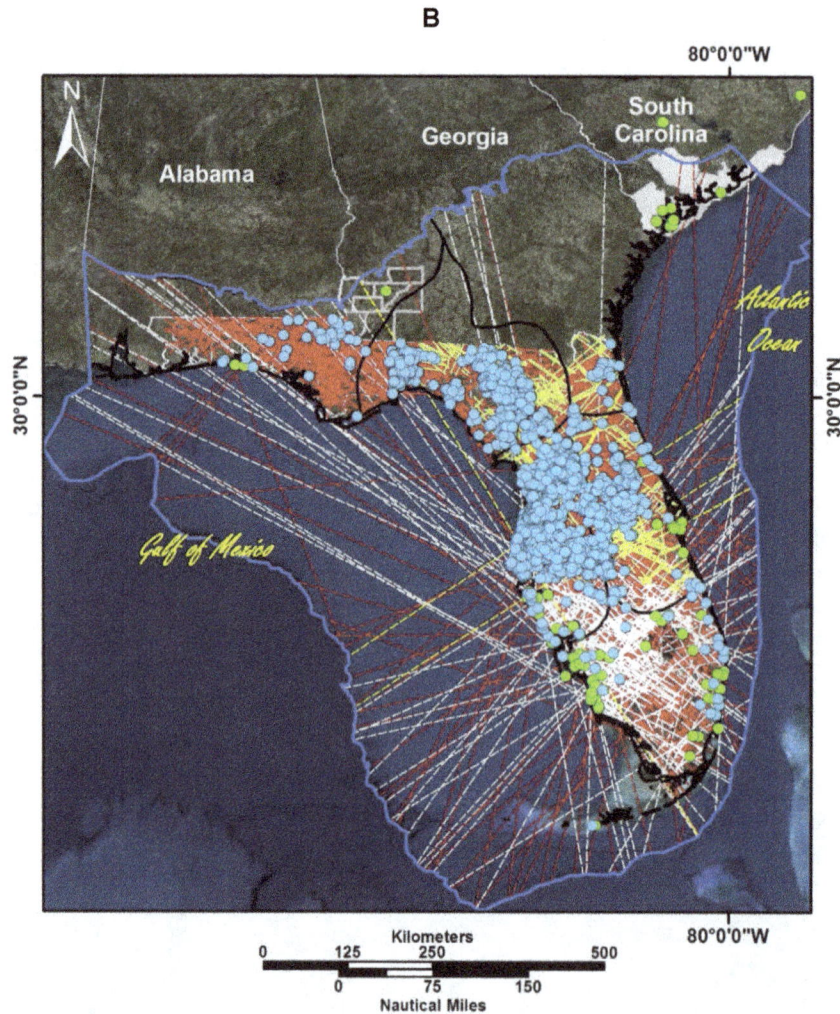

Figure 2. Proximity of modern sinkholes in Florida (blue circles) and permitted ASR wells in Florida, Georgia, and South Carolina (green circles) to:

A. fractures in Florida, as reported by the ACOE (2004, diagonal white lines) and mapped by FDOT (1973, diagonal red lines) and Vernon (1951, diagonal yellow-green lines), and

B. extensions of those fractures in proximity to ASR wells (dashed diagonal lines of same colors)

Enlarging the vicinity of the permitted Elmodel WMA ASR demonstration well in Baker County, Georgia (Figure 3A), shows the proximity to the dense network of fractures and sinkholes mapped by Brook and Allison (1986) in Dougherty County, north of Baker County (1). The dark green areas in Figure 3A for Dougherty County, where no fractures were mapped, are streams and riparian wetlands that obscure features indicative of fractures, similar to the Lake Okeechobee open-water area in south Florida. The extensive circular and rectangular light areas in Figure 3A for Baker County (1) and Mitchell County (6) are center-pivot irrigation areas and other agricultural fields that also are irrigated. The yellow and red diagonal dashed lines southwest and northeast of the Elmodel WMA ASR demonstration well in Figure 3A are extensions of fractures mapped in Florida in proximity to ASR wells in that state. Enlarging the vicinity of the ASR wells in South Carolina (Figure 3B) shows the proximity of the operational ASR wells in Hilton Head, Long Cove, and Palmetto Bay (Beaufort County) to the red diagonal dashed line extensions of fractures mapped in Florida, that are associated with ASR wells in that state. The additional ASR well and RO facility proposed for Edisto Beach (Colleton County) is not shown because it was not permitted at the time of acceptance of this paper for publication. That ASR well and facility would be located on the north shore of the mouth of the Edisto River, between the ASR wells in Beaufort County (southwest) and the ASR wells in Charleston County (north east), in proximity to the two intersecting fracture extensions also associated with ASR wells in Florida.

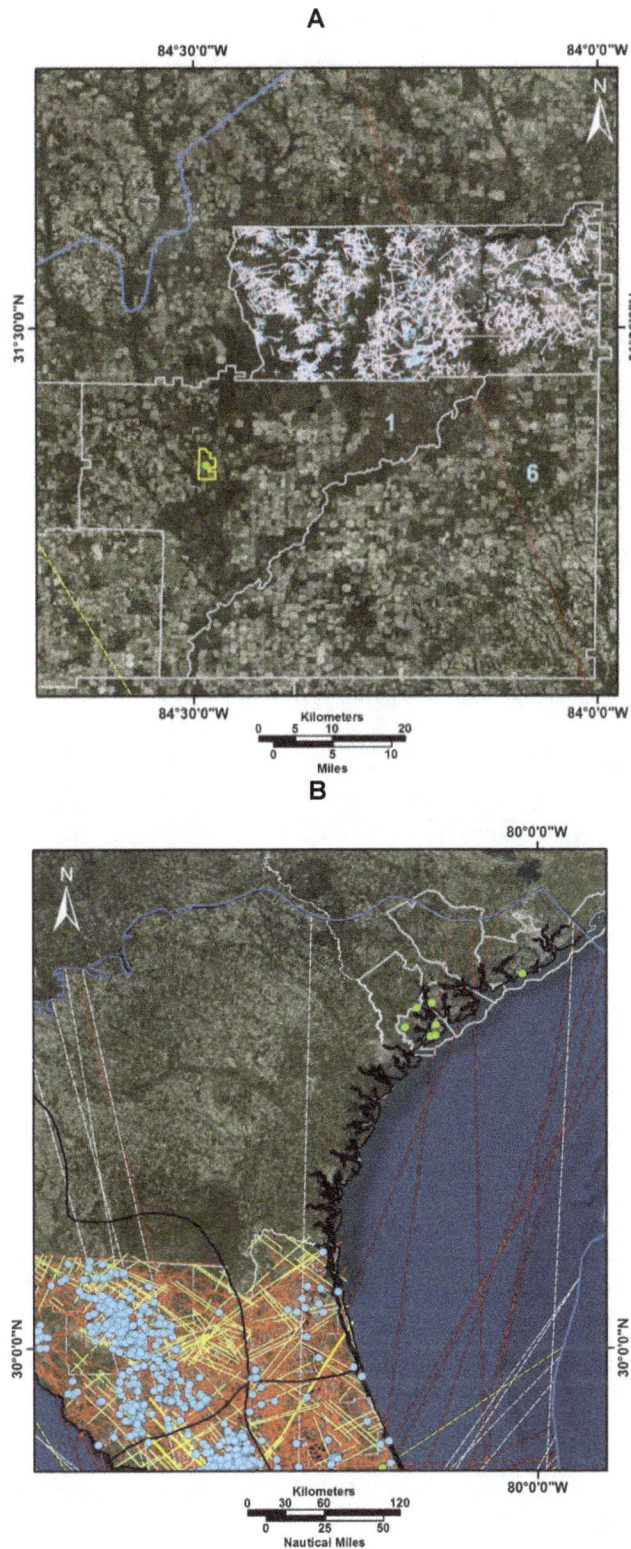

Figure 3. Enlargement of fracture extensions (dashed lines, as described in Figure 2), showing proximity to:
A. fractures and relict sinkholes mapped in Dougherty County, Georgia by Brook and Allison (1986, pink lines and blue polygons, respectively) and the permitted Elmodel Wildlife Management Area ASR demonstration well in Baker County, Georgia and

B. permitted ASR wells in South Carolina within the FAS boundaries (in Beaufort, Charleston, Colleton, and Jasper Counties) and beyond the FAS boundaries (in Columbus, Georgetown, Horry, Marion, and Orangeburg Counties)

4.2 Length, Frequency, and Density of Fractures Mapped in Florida and Georgia

Table 2 summarizes the density of fractures, total number of fractures and the shortest, longest, and mean fracture lengths mapped in Dougherty County by Brook and Allison (1986) and mapped by Vernon (1951) and by FDOT (1973) in the 21 counties in north Florida selected for comparison in our study. Based on our georeferenced and digitized data 1,225 fractures were mapped in Dougherty County by Brook and Allison (1986). The lengths of the shortest and longest fractures were 0.1 km (0.1 mi) and 8.6 km (5.3 mi), respectively, and the mean fracture length was 1.2 km (0.8 mi). By comparison, the longest and most numerous fractures in the 21 north Florida counties in our study were mapped by FDOT (1973), with the shortest and longest fractures 17 km (10.6 mi) and 373 km (231.8 mi), respectively. Mean fracture lengths from the FDOT (1973) data set ranged from 68.4 km (42.5 mi) to 117.5 km (73 mi). The most numerous fractures mapped in those north Florida counties by FDOT (1973) occurred in Washington County, which included 60 fractures (Table 2). Only six of the 21 counties in north Florida included fractures mapped by Vernon (1951). The most numerous fractures mapped in north Florida counties for that data set occurred in Columbia County, which included 36 fractures. The length of the shortest and longest fractures in the Vernon (1951) data set was 5 km (3.1 mi) and 65 km (40.4 mi) and mean fracture lengths ranged from 25.6 km (15.9 mi) to 38.4 km (23.9 mi). The mapped fractures reported in Florida by the ACOE (2004) did not include north Florida counties (Table 2 and Figure 2). Total density of fractures and fracture density per 100 km^2 for all three data sets, by county, also is provided in Table 2.

Table 2. Density and frequency of fractures, lengths of shortest and longest fractures, and mean fracture lengths for Dougherty County, Georgia[1] and 21 counties in north Florida[2]

County	Number of Fractures/ 100 km^2			Total Number of Fractures			Total Combined Fractures	Shortest-Longest Fractures (km)			Mean Fracture Length (km)		
	B&A	FDOT	Vernon	B&A	FDOT	Vernon		B&A	FDOT	Vernon	B&A	FDOT	Vernon
Dougherty	141.3	-	-	1,225	-	-	1,225	0.1-8.6	-	-	1.2	-	-
Bay	-	2.8	-	-	54	-	54	-	30-355	-	-	93.4	-
Calhoun	-	3.6	-	-	54	-	54	-	20-373	-	-	100.6	-
Columbia	-	2.7	1.7	-	55	36	91	-	32-373	5-65	-	111.5	28.0
Escambia	-	0.6	-	-	10	-	10	-	20-163	-	-	76.6	-
Franklin	-	1.3	-	-	19	-	19	-	36-222	-	-	100.8	-
Gadsden	-	2.6	-	-	36	-	36	-	26-373	-	-	113.8	-
Gulf	-	1.9	-	-	29	-	29	-	31-159	-	-	81.3	-
Hamilton	-	2.2	2.2	-	30	30	60	-	31-373	8-52	-	111.3	26.7
Holmes	-	2.6	-	-	32	-	32	-	35-236	-	-	79.4	-
Jackson	-	2.2	-	-	53	-	53	-	20-373	-	-	85.4	-
Jefferson	-	2.5	1.3	-	39	20	59	-	21-373	10-51	-	117.5	28.6
Leon	-	2.9	0.6	-	53	10	63	-	23-373	30-51	-	105.3	38.4
Liberty	-	2.6	-	-	57	-	57	-	26-373	-	-	106.2	-
Madison	-	2.0	-	-	37	-	37	-	31-373	-	-	103.3	-
Okaloosa	-	1.4	-	-	34	-	34	-	35-222	-	-	80.9	-
Santa Rosa	-	1.1	-	-	28	-	28	-	20-163	-	-	68.4	-
Suwannee	-	2.5	1.6	-	44	29	73	-	21-355	7-49	-	96.5	25.6
Taylor	-	1.4	0.1	-	37	4	41	-	27-182	14-48	-	87.1	32.0
Wakulla	-	2.7	-	-	43	-	43	-	32-163	-	-	91.1	-
Walton	-	1.8	-	-	50	-	50	-	17-355	-	-	92.3	-
Washington	-	3.8	-	-	60	-	60	-	30-373	-	-	96.7	-
								0.1-8.6	17-373	5-65			

[1] from Brook and Allison (1986), georeferenced/digitized for our study
[2] from FDOT (1973) and Vernon (1951)

Figure 4 provides a comparison of the length and frequency of fractures mapped in the 21 counties in north Florida (from FDOT, 1973 and Vernon, 1951), to length and frequency of fractures mapped in Dougherty County by Brook and Allison (1986). The scale of the X and Y axes for the north-Florida counties are the same, with a maximum of 60 fractures and 400 km, respectively. The scale of the X and Y axes for Dougherty County was adjusted for a maximum of 1,200 fractures and 10 km, respectively, because of the shorter, more numerous fractures mapped in that county.

4.3 Distances Between Previously Mapped Fractures and Sinkholes in Georgia and Florida

A histogram showing the distances between the 1,225 fractures and 934 relict sinkholes mapped in Dougherty County by Brook and Allison (1986) is provided in Figure 5A. All of those sinkholes were within 1.75 km (1.09 mi) of a fracture. Only six sinkholes occurred more than 0.75 km (0.47 mi) from the nearest fracture mapped by Brook and Allison (1986). The majority of those sinkholes, 627, were less than 0.25 km (0.78 mi) from a fracture, and 281 and 19 sinkholes were located between 0.25 and 0.75 km (0.16 mi and 0.47 mi), respectively, from the nearest fracture.

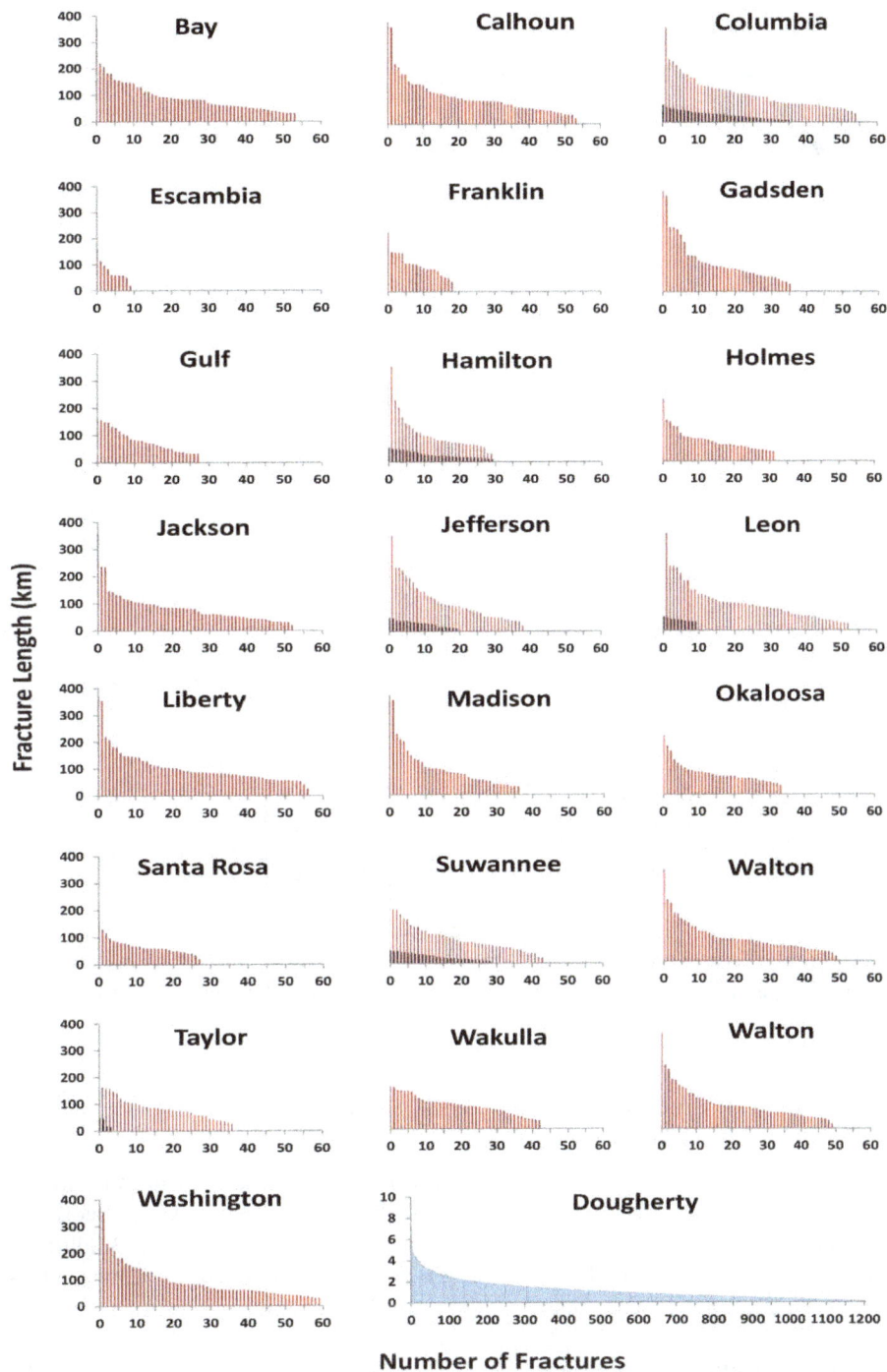

Figure 4. Length and frequency of fractures mapped in 21 counties in north Florida (red from FDOT (1973) and black from Vernon (1951)) and Dougherty County, Georgia (blue from Brook & Allison (1986))

A similar histogram is provided in Figure 5B, showing the distances of sinkholes within 5 km (3.11 mi) from each of the three sources of mapped fractures in Florida (ACOE, 2004; FDOT, 1973; and Vernon, 1951) and those combined fractures to the 2,814 modern sinkholes reported in Florida, based on the FDEP database. Only the FDOT (1973) fracture data set covered the entire state of Florida, while the ACOE (2004) data set was located in the southern part of Florida and the Vernon (1951) data set was located further north in the state. Half of the sinkholes reported in Florida (1,407) occurred within 1.25 km from the nearest mapped fracture and more than 75% of the sinkholes reported in Florida (2,115) occurred within 2.75 km (1.71 mi) from the nearest mapped fracture. All but 12 of the remaining sinkholes reported in Florida occurred between 2.75 km and 5 km.

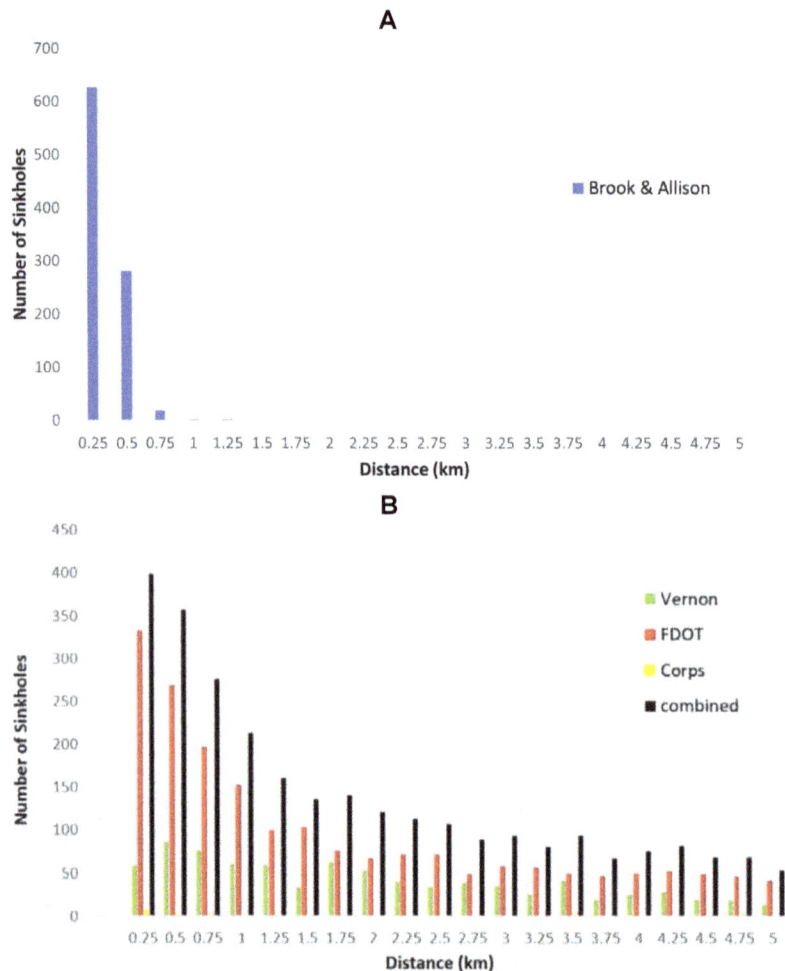

Figure 5. Distances from fractures mapped in: A. Dougherty County, Georgia to relict sinkholes in the same area (Brook & Allison, 1986) and B. Florida (ACOE, 2004; FDOT, 1973; and Vernon, 1951) to modern sinkholes in the state database

4.4 Proximity of Previously Mapped Fractures in Georgia and Florida and Extensions to ASR Wells, Sinkhole Clusters, and Federal, State, and Other Environmentally Sensitive Areas

Figure 6A illustrates the proximity of ASR wells (as described in Figure 2) and fractures mapped in Florida and Dougherty County (as described in Figure 5) to examples of federal lands (i.e., Apalachicola National Forest, Florida; Okefenokee National Wildlife Refuge, Georgia; Pinckney Island National Wildlife Refuge, South Carolina; and St. Marks National Wildlife Refuge, Florida). Figure 6A also shows proximity of those ASR wells and mapped fractures to examples of state lands (i.e., Elmodel Wildlife Management Area, Georgia; Okefenokee State Park, Georgia; Tate's Hell State Forest, Florida; and Wakulla Springs State Park, Florida); and Georgia's Priority Amphibian and Reptile Conservation Areas (PARCAs), in addition to the ACF River Basins (adapted from Seaber et al., 1987).

Figure 6B shows the proximity of modern sinkholes mapped in Florida (FDEP database) to the extensions (dashed white lines) of only fractures mapped in Dougherty County by Brook and Allison (1986) that are within the boundaries of the Chickasawhatchee Swamp/Ichauway Plantation PARCA (#13). Figure 6B also shows clusters of those fracture extensions that obscure PARCA 13, the Elmodel WMA/ASR demonstration well, and the Apalachicola and lower Flint River Basins identified in Figure 6A. Those clusters of fracture extensions also intersect PARCAs 15, 16, and 18; the Apalachicola National Forest and Tate's Hell State Forest; and they coincide with the clusters of Florida sinkholes in sub-regions H and U. An enlargement of Figure 6B, in the vicinity of the ACF River Basins, the six Georgia counties in our study (outlined in red), and the clusters of fracture extensions in sub-regions D, H, and U, is provided in Figure 6C.

A

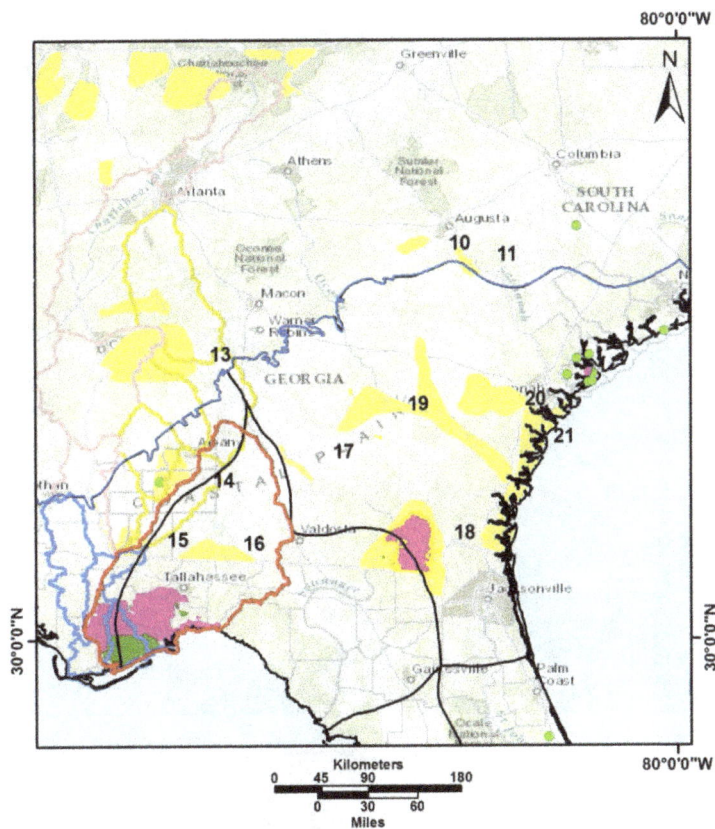

B

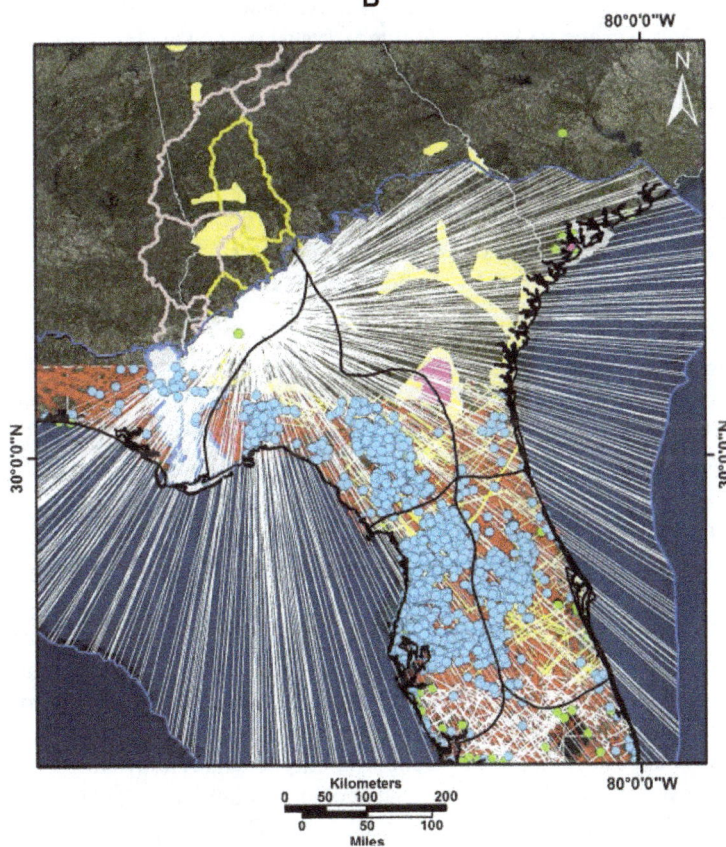

C

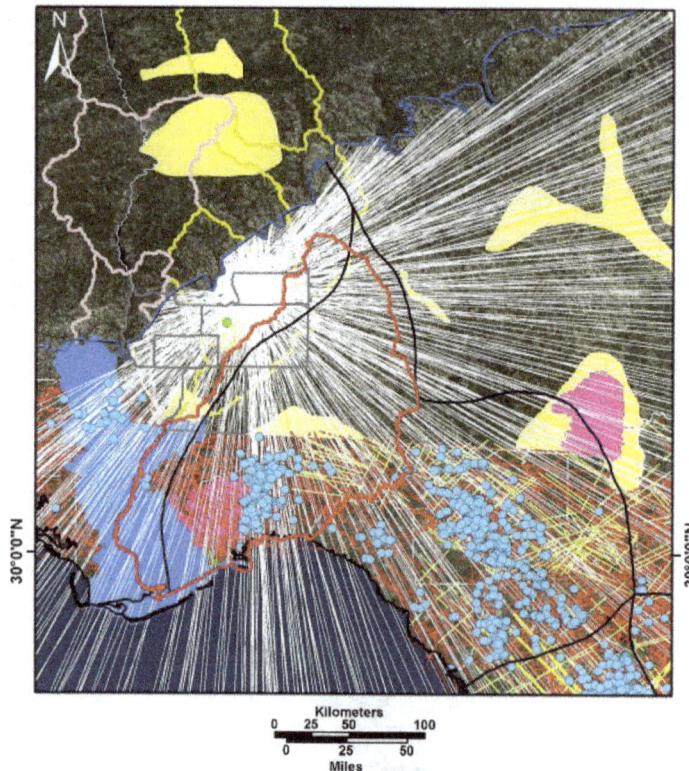

Figure 6. Fractures in Florida and Dougherty County, Georgia and ASR wells (as described in Figures 2 and 3) associated with the model boundaries for the Wakulla Springshed (from Kincaid et al., 2012 - outlined in red): A. federal lands (Apalachicola National Forest, Okefenokee National Wildlife Refuge, Pinckney Island National Wildlife Refuge, and St. Marks National Wildlife Refuge - fuchsia); state lands (Elmodel Wildlife Management Area, Okefenokee State Park, Tate's Hell State Forest, and Wakulla Springs State Park – dark green); Georgia's Priority Amphibian and Reptile Conservation Areas (yellow); the Apalachicola-Chattahoochee-Flint River Basins (as shown in Figure 1B);

B. extensions of fractures mapped in Dougherty County and within the boundaries of PARCA 13 (dashed white lines), modern sinkholes mapped in Florida, federal and state lands, and Georgia PARCAs;

C. Enlargement of clustered fracture extensions obscuring PARCA 13 and the Elmodel WMA/ASR demonstration well, and intersecting PARCAs 16 and 18, and sinkhole clusters (blue) in sub-regions H and U

The PARCAs addressed in our study are listed in Table 3, which identifies the sub-regions where the PARCAs occur. Those PARCAs are identified in Table 3 and Figure 6 by the numbers assigned to those PARCAs in Figure 1 of Jenson (2015). The Ft. Gordon PARCA (10) is north of the FAS boundary. All of the remaining PARCAs listed in Table 3 occur, at least in part, in Georgia's coastal plain, which coincides with the FAS in Georgia.

For the federal lands, the Apalachicola National Forest is split between sub-regions H and U; the Okefenokee Swamp Wilderness Area is split between sub-regions D and U; the St. Marks National Wildlife Refuge is in sub-region U; and the Pinckney Island National Wildlife Refuge is in the northeast portion of sub-region D. For the selected state lands, the Elmodel WMA occurs in sub-region H; Okefenokee State Park occurs in sub-region U; Tate's Hell State Forest occurs in sub-regions H and U; and Wakulla Springs State Park occurs in sub-region U. The lower extent of the ACF River Basins occurs within sub-region H, with the exception of a small portion of the southeastern Flint River Basin, which extends into sub-region U. The same environmentally sensitive areas included in Figure 6A are shown in Figure 6B, in proximity to sinkholes (as described in Figures 2 and 3A), mapped fractures and extensions of fractures mapped in Dougherty County that coincide with the Chickasawhatchee Swamp/Ichauway Plantation PARCA (13). Those fractures were extended to the FAS boundaries.

Table 3. Georgia's Priority Amphibian and Reptile Conservation Areas (PARCAs) and locations in FAS sub-regions.

ID	Name	FAS Sub-regions
10	Ft. Gordon[2]	-
11	Yuchi WMA/Plant Vogtle	D/-
13	Ft. Benning/Western Fall Line Hills[2]	-
14	Chickasawhatchee Swamp/Ichauway Plantation	H
15	Lake Seminole Region	H
16	Georgia Red Hills	U
17	Alapaha River and Sandhills	D
18	Okefenokee Swamp[3]	D/U
19	Altamaha-Ocmulgee-Ohoopee River Corridors	D
20	Ft. Stewart	D
21	Barrier Islands and Salt Marshes	D

[1] Priority Amphibian and Reptile Conservation Areas (PARCAs) from Jenson (2015).
[2] Located north of the FAS boundary.
[3] Located partially in sub-region D and partially in sub-region U.

5. Discussion and Conclusions About Influence of Fractures and Sinkholes on Hydrologic and Environmental Impacts

5.1 River Basins, Groundwater Basins, Breached Groundwater Divides, and Pirated Water

5.1.1 Pirated Water

Ground-surface elevations generally are used to determine the boundaries of river basins, such as the ACF River Basin. The boundaries of aquifers, also known as groundwater basins or groundwater reservoirs, rarely coincide with the boundaries of overlying river basins, as illustrated by the ACF River Basins (Figures 1B and 6A). The American Geological Institute's (AGI) Glossary of Geology (Neuendorf, 2005) defines groundwater basin and groundwater reservoir as follows:

Groundwater basin

(a)   A subsurface structure having the character of a basin with respect to the collection, retention, and outflow of water.

(b)   An aquifer or system of aquifers, whether basin-shaped or not, that has reasonably well defined boundaries and more or less definite areas of recharge and discharge.

Groundwater reservoir

(a)   *Aquifer.*

(b)   A term used to refer to all the rocks in the saturated zone, including those containing permanent or temporary bodies of groundwater

Krause and Randolph (1989) determined that breaches of the groundwater divides in the six sub-regions of the FAS had occurred in response to large withdrawals of ground water in the US southeastern coastal plain. The diversion of ground water also is known as captured or pirated water, as defined in the AGI Glossary of Geology (Neuendorf, 2005). The piracy can involve the capture of entire streams or portions of streams, including water from another valley, which is known as "beheading" when it involves the "cutting-off of the upper part of a stream and the diversion of its headwaters into another drainage system by capture." The terminology also is applied to the area within an aquifer in which ground water flows to a well under the influence of pumping conditions (Neuendorf, 2005). Although streamflow depletion from piracy can be determined by deducting the amount of water that flows out of an area from the amount of water that flows into the area (Neuendorf, 2005), a comparable determination cannot be made for water captured from depressional wetlands in a watershed or throughout a regional aquifer like the Floridan aquifer system. By the late 1980s in west-central Florida (sub-region F), pirated water from unsustainable FAS groundwater withdrawals, addressed as unintended subsurface interbasin flow, had resulted in environmental damage so severe that "Water Use Caution Areas" were established (Bacchus, 2000). Despite that designation, groundwater withdrawals in that area were estimated at approximately 246 MGD by 1993 and were causing both adverse economic and environmental impacts, including premature decline and death of longleaf pine in uplands and pond-cypress in wetlands, catastrophic destructive wildfires, and declines in native wildlife (Bacchus, 2000).

An important contribution of the Woodville Karst Plain Hydrologic Research Program was the groundwater model created by Kincaid, Meyer, and Day (2012), which identified pirating of ground water from the Wakulla Springshed from agricultural and nonagricultural groundwater withdrawals in southwest Georgia. A springshed identifies the area that provides water to the spring system, similar to a watershed or river basin that identifies the surface area that contributes surface water to a river system. The northern extent of the groundwater basin for the Wakulla Springshed is located in southwest Georgia, with the southern extent at the Gulf Coast of Florida. The Wakulla Springshed is encompassed by the boundaries of the groundwater model (GWM) domain, which encompass a larger area for determining the boundaries of the Wakulla Springshed for the GWM (Figures 10 and 11 of Kincaid et al., 2012). Consolidated locations of agricultural and nonagricultural groundwater pumping were identified in their research in and surrounding the GWM domain. The distribution and magnitude of groundwater pumping for agricultural irrigation and municipal use in Florida and Georgia counties intersecting the GWM domain were provided in millions of gallons per day (MGD) in Tables 12 and 13, respectively, in Kincaid et al. (2012), based on 2005 data they obtained. Consistent with other groundwater models created for areas of the FAS, that GWM was not designed to include preferential flow through fractures, but we have incorporated those model boundaries into Figure 6A to show the proximity of this area of pirated ground water to mapped fractures and sinkholes included in our study.

The volume of groundwater extractions within that GWM domain (Kincaid et al., 2012) included the Florida and Georgia vicinities of our study. The total agricultural pumping for the 15 Georgia counties and eight Florida counties within that GWM domain was 182.88 MGD and 29.03 MGD, respectively. The largest groundwater withdrawals per county occurred in Baker, Decatur and Mitchell Counties, Georgia and were 30.75 MGD, 34.95 MGD, and 28.88 MGD, respectively.   The number of extraction points (wells) for those three counties (Figure 1B and Table 1) were 373, 601, and 718, respectively. The withdrawals from each of those counties exceeded the total for all agricultural groundwater withdrawals in Florida counties within the GWM domain. The total municipal groundwater withdrawals from Georgia and Florida counties within the GWM domain were 53.92 MGD and 48.70 MGD, respectively. The largest volume of municipal groundwater withdrawals from the Georgia counties in the GWM domain was 13.35 MGD from the two suppliers in Dougherty County. The largest volume of municipal groundwater withdrawals from the Florida counties in the GWM domain was 39.10 MGD from the 11 suppliers in Leon County (Kincaid et al., 2012). The next step in GWM development for this area should be to incorporate preferential flow of these withdrawals through mapped fractures.

The results of the research and GWM by Kincaid et al. (2012) led to understanding the link between groundwater extractions in Georgia and the water budget of north Florida, without knowledge of the role fracture systems play in the pirating of ground water. The presumption prior to their study was that pumping in Florida represented the largest anthropogenic impact to the springs and coastal freshwater discharge. Based on the data compiled during that study, however, agricultural and municipal pumping in Georgia was identified as having a much larger magnitude than all groundwater pumping in the north-Florida model area. Therefore, it was clear that the data for Georgia groundwater withdrawals must be included in any modeling analysis designed to predict the impacts of pumping or sea-level rise on spring flows. Specifically, groundwater withdrawals in that area of Georgia have affected and will continue affecting the boundaries of the springsheds by reducing upland storage. Reduced storage, in turn, reduces hydrologic heads and flows, particularly in times of drought, rendering management actions that focus exclusively on Florida ineffectual. Additionally, arbitrary delineations of permeability designed to match heads (e.g., elevation, hydraulic, and pressure) are not likely to result in accurate or defensible delineations of springshed boundaries or groundwater travel-times (Kincaid et al., 2012).

The 182.88 MGD of groundwater withdrawals identified by Kincaid et al. (2012), in an area that includes southeastern portions of the ACF River Basins, are similar to the magnitude of groundwater withdrawals that resulted in interbasin transfer of ground water in west-central Florida (Bacchus, 2000).   That strongly suggests ground water also is being pirated from the Flint and Apalachicola River Basins. One could conclude that the damming and surfacewater withdrawals of the ACF near the Florida/Georgia state line should not be the primary focus of the tri-state water wars that have existed between Alabama, Florida and Georgia for decades, but that preferential flow through fractures, in response to groundwater withdrawals should be of equal concern.

5.1.2 Geographically Isolated vs Hydrologically Connected Wetlands

Whether natural wetlands are geographically isolated, but hydrologically connected depends on characteristics of the underlying aquifer system. Research has shown that natural depressional wetlands that are characteristic of the FAS (e.g., pond-cypress wetlands and wet prairies) have historic surfacewater connections to navigable waters and occur in relict sinkholes that are aligned along fractures and connected to the underlying regional karst aquifer system, as summarized by Bacchus (1998; 2000; 2006; 2007) and Bacchus et al. (2003). Therefore, those wetlands are not hydrologically isolated from waters of the US or exempt from federal regulation under the Clean Water Act (CWA).

The 2001 US Supreme Court ruling in Solid Waste Agency of Northern Cook County (SWANCC) v. US Army Corps of Engineer (531 US 159) resulted in misconceptions about the application of that ruling by agencies charged with regulating wetlands and other waters of the US in the FAS, as well as with consultants and scientists who work with those wetlands. That Supreme Court case involved the ACOE's attempt to assert jurisdiction under the CWA over an abandoned sand and gravel pit in northern Illinois, a man-made wetland. The ruling simply clarified that CWA jurisdiction did not include that man-made mine pit, simply because the mine pit included wetlands and migratory birds were utilizing that area. That mine pit, with wetlands, had no scientific or legal relevance to the natural depressional wetlands characterizing the regional karst FAS and the southeastern coastal plain of the US.

For example, Wise, Annable, Walser, Switt, and Shaw (2000) acknowledged the hydraulic connectivity (also known as hydrologic connection) of the aquifer and natural depressional wetlands in Florida, but then referred to those wetlands as "isolated wetlands." The Georgia State Wildlife Action Plan (SWAP) is another example of the misapplication of the term "isolated wetlands" to natural depressional wetlands in the FAS (Albanese, Wiseniewskiki, & GaschoLandis, 2015; GDNR, 2015; Jenson, 2015; Schneider, & Keyes, 2015).

Recognizing the hydrological connection of natural depressional wetlands is important because those wetlands cannot be recreated at random locations.   As emphasized by Wise et al. (2000), knowledge of the underlying groundwater interactions is required to assess the ecological impacts from anthropogenic changes in groundwater levels. More importantly, they described the misapplication of the federal "no net loss" policy adopted during the administration of former President George Bush (1989–1993) to depressional wetlands. That policy, largely in response to the National Wetlands Forum of 1988, presumed that those types of wetlands destroyed by developments or other means could be "replaced" elsewhere by man-made wetlands similar to the very wetlands that the SWANCC ruling concluded were not regulated. The reason those natural depressional wetlands cannot be replaced by man-made wetlands at random locations, such as excavated stormwater and mine pits, is because the natural depressional wetlands were established in relict sinkholes aligned along fractures, in response to groundwater conditions that cannot be recreated at random locations that are convenient for develop, industry and agriculture (Bacchus, 2007). Solutions to the incorrect application of terminology would be to reference natural depressional wetlands in the FAS that are hydrologically connected via ground water either as geographically isolated or simply as natural depressional wetlands.

The knowledge that natural depressional wetlands in the FAS are aligned along fractures also is an important consideration for interpreting the results of our study. For example, comparing the proximity of the relict sinkholes to fractures mapped in Dougherty County by Brook and Allison (1986) and shown in Figures 3A and 5A to the greater distances between sinkholes and the three sets of mapped fractures in Florida (Figure 5B) suggests that there are additional fractures in Florida that have not been mapped. Further support that additional, unmapped fractures occur in Florida is provided by the greater density of fractures mapped in the Dougherty County study area.

5.1.3 Adverse Environmental Impacts Associated with Pirated Water from ASR and Other Aquifer Withdrawals and Injections

Removing water from and injections into aquifers can create adverse environmental impacts. Examples of adverse environmental impacts throughout the extent of the FAS associated with pirated water from aquifer withdrawals (mechanical and nonmechanical) and injections, including ASR, have been described by Bacchus (1998; 2000; 2001; 2002; 2006; 2007); Bacchus and Barile (2005); Bacchus and Brook (1996); Bacchus et al. (2000; 2003; 2005; 2011; 2014; 2015a); Bernardes, He, Bacchus, Madden, and Jordan (2014); Cunningham, Renken, Wacker, and Zygnerski, (2003); Fitterman and Deszcz-Pan (1999); Hofstetter and Sonenshein (1990); Lines et al. (2012); Maslia and Prowell (1990); McNeill (2000); Price and Pichler (2004); Renken, Shapiro, Cunningham, Harvey, Zygnerski, Metge, and Wacker (2004); Sonenshein and Hofstetter (1990); and Wilcox, Solo-Gabriele, and Sternberg (2004). This large body of literature has identified many categories of adverse impacts. Examples include: (1) depletion of groundwater reserves; (2) intrusion of water of undesirable quality (e.g., lateral saltwater intrusion, upconing of brackish and saline water through fractures and other karst conduits, and contaminants such as arsenic and benzene); (3) contravention of existing water rights; (4) excessive depletion of streamflow by induced infiltration/recharge; (5) land subsidence (e.g., "reactivating" relict sinkholes by increasing flow through infilled sediments and debris); (6) reductions of levels and/or extent of lakes and wetlands, invasion of alien/nuisance species and premature decline and death of trees from insects and pathogens, with consequent loss of valued habitat; (7) reductions in extent of areas where water is available to plants that use the capillary fringe, followed by catastrophic destructive wildfires and loss of habitat; and (8) reductions of groundwater outflow to coastal waters, with consequent impacts to coastal wetlands and/or nearshore benthic marine habitats.

The co-location of groundwater-extraction wells at an area established as a wilderness park in west-central Florida, renamed the J.P. Starkey Wellfield and Wilderness Park (Starkey Wilderness Park), is similar to the co-location of the proposed ASR demonstration well for groundwater extraction in the Elmodel Wildlife Management Area (Elmodel WMA) and Chickasawhatchee Swamp/Ichauway Plantation PARCA (13) in Georgia. A permit authorized 15 MGD of groundwater withdrawals from Starkey Wilderness Park, but irreversible environmental damage occurred when pumping of only 12 MGD was reached. The environmental damage included catastrophic, destructive wildfires that spread during a routine prescribed burn, killing longleaf pines in the uplands and pond-cypress in the wetlands designated for protection in the Starkey Wilderness Park, and igniting and consuming the organic sediments in wetlands, destroying the roots. Similar catastrophic, destructive wildfires are typical throughout the FAS in other areas of unsustainable groundwater withdrawals, also known as groundwater mining (Bacchus, 2000). Photographs of the irreversible damage to depressional wetlands characteristic of wildfires caused by unsustainable groundwater withdrawals are included in Bacchus (2007).

Adverse environmental impacts associated with pirated water from ASR and other aquifer withdrawals and injections that are characteristic and diagnostic of unsustainable groundwater withdrawals in the FAS include the premature decline and death of native upland and wetland trees, particularly oaks, longleaf pine and pond-cypress, and accompanying infestations of insects (e.g., beetles) and pathogens. Such adverse environmental impacts routinely and erroneously are attributed to "drought" rather than chronic water stress from unsustainable groundwater withdrawals. Unsustainable groundwater withdrawals and pirated water are not restricted to mechanical dewatering (e.g., pumping from groundwater supply wells, including ASR wells), but also result from nonmechanical dewatering (e.g., evaporative loss from excavated pits). One approach for estimating the volume of water being diverted (pirated) from wetlands in the Everglades National Park (ENP) by municipal wellfields, directly and indirectly, used isotopic analysis of the water (Wilcox et al., 2004), but did not assess the damage to ENP wetlands from the pirated water. The irreversible degradation of habitat leads to declines in native wildlife (Bacchus, 2000). Examples of these and other adverse environmental impacts in the FAS are described in Bacchus (1998; 2000; 2001; 2006; 2007) and Bacchus et al. (2000; 2003; 2005; 2011; 2015a).

Assessing impacts on wildlife, Georgia's 2005 and 2115 State Wildlife Action Plans (SWAP) included a list of 25 Standardized Threat Descriptions for the state's wildlife. Table 4 includes the headings of the original 2005 rankings and revised 2015 statewide, Atlantic and Gulf rankings of those threats. The headings, descriptions and rankings of 17 standardized threats from the 2005 SWAP that result from or result in pirated water from mechanical and nonmechanical aquifer withdrawals and injections, including ASR, are provided in Appendix A. Headings in red in Appendix A and Table 5 were identified in the 2015 SWAP as particularly relevant to aquatic species.

At least the following three threats (preceded by the original ranking number) included in Table 4 result from groundwater withdrawals and injections: 3. Altered Fire Regimes; 10. Disease; and 21. Invasive/Alien Species. Additionally, the following examples of standardized threats (preceded by the original ranking number) included in Table 4 result in alterations of hydrology and water quality: 2. Incompatible Agricultural Practices; 6. Commercial/Industrial Development; 7. Conversion to Agriculture; 8. Dam and Impoundment Construction; 9. Development of Roads or Utilities; 14. Incompatible Forestry Practices; and 15. Global Warming/Climate Change. It could be useful for public understanding to revise the standardized threats to include actions resulting from and in alterations of hydrology and water quality that can be intensified and spread far beyond the surface footprint of those threats by the presence of fractures.

The types of hydrologic impacts that can spread beyond the surface boundaries of projects, river basins, and groundwater basins via preferential flow through factures can result in irreversible degradation to habitat for high priority state and federally listed species. Tables 5A-C list examples of Georgia's high priority state and federally listed species in the ACF River Basins and coastal plain with jeopardized survival and recovery by adverse direct, indirect, and cumulative impacts from the threats listed in Appendix A in Florida, Georgia, and South Carolina.

Table 4. Rankings of Standardized Threats to High Priority Fish and Aquatic Invertebrate Species in Georgia in 2005 and 2015.

Statewide Rankings[2]	Atlantic Rankings[2]	Gulf Rankings[2]
#1 – 5. Altered Water Quality	#1 – 4. Altered Hydrology	#1 – 4. Altered Hydrology
#2 – 2. Incompatible Agricultural Practices	#2 – 5. Altered Water Quality	**#2 – 11. Ground/Surface Water Withdrawal**
#3 – 4. Altered Hydrology	#3 – 14. Incompatible Forestry Practices	#3 – 5. Altered Water Quality
#4 – 23. Residential Development	#4 - 8. Dam & Impoundment Construction	#4 – 2. Incompatible Agricultural Practices
#5 – 8. Dam & Impoundment Construction	#5 – 2. Incompatible Agricultural Practices	#5 - 8. Dam & Impoundment Construction
#6 – 11. Ground/Surface Water Withdrawal	#6 – 6. Commercial/Industrial Development	#6 – 21. Invasive/Alien Species
#7 – 6. Commercial/Industrial Development	#7 – 15. Climate Change	#7 - 23. Residential Development
#8 – 14. Incompatible Forestry Practices	#8 – 23. Residential Development	#8 - 9. Develop. of Roads/Utilities
#9 – 20. Industrial/Municipal Pollution	#9 – 21. Invasive/Alien Species	#9 - 14. Incompatible Forestry Practices
#10 – 21. Invasive/Alien Species	#10 – 18. Incompat. Mining/Mineral Extraction	**#10 - 19. Incompat. Road/Utilities Mgmt**
#11 – 9. Develop. of Roads/Utilities	**#11 – 7. Conversion to Agriculture**	#11 – 20. Industrial/Municipal Pollution
#12 – 15. Climate Change	**#12 – 11. Ground/Surface Water Withdrawal**	#12 - 6. Commercial/Industrial Development
#13 – 7. Conversion to Agriculture	**#13 –19. Incompatible Road/Utilities Mgmt**	#13 – 15. Climate Change
#14 – 19. Incompatible Road/Utilities Mgmt.	#14 – 20. Industrial/Municipal Pollution	
#15 – 18. Incompat. Mining/Mineral Extraction	**#15 –25. Vehicle-Induced Mortality**	
#16 – 24. Unmanaged Recreation		
#17 – 17. Incompatible Fisheries Practices		
#18 – 25. Vehicle-Induced Mortality		

from Albanese et al. (2015); threats particularly relevant to aquatic species in red font
2015 rankings Statewide and for Atlantic and Gulf areas (#) followed by 2005 rankings for each standardized threat

Table 5A. Examples of Georgia's high priority state and federally listed bird species in the ACF River Basins and coastal plain with jeopardized survival and recovery by adverse direct, indirect, and cumulative impacts from Floridan-aquifer injections and withdrawals in Florida, Georgia, and South Carolina.

Bird Species	Status	Examples of Habitat & Adverse Impacts
Little blue heron *Egretta caerulea*	**Special Concern**	**Coastal plain– dewatering/altering natural hydroperiods & water quality of marshes, flood plains & swamps required for nesting/feeding**
Tricolored heron *Egretta tricolor*	**Special Concern**	**Coastal zone– dewatering/altering natural hydroperiods & water quality of wetlands required for nesting/feeding**
Florida Sandhill Crane *Grus canadensis pratensis*	**Special Concern**	**Coastal plain– dewatering/altering natural hydroperiods & water quality of depressional wetlands required for nesting/feeding; Okefenokee NWR is only breeding site in Georgia**
Bald Eagle *Haliaeetus leucocephalus*	**High Priority**	**Lower Chattahoochee River (CR) – edges of lakes & large rivers; seacoasts**
Wood Stork *Mycteria americana*	**Federally Threatened**	**Coastal plain– dewatering/altering natural hydroperiods & water quality of depressional wetlands required for nesting/feeding**
Red-cockaded Woodpecker *Picoides borealis*	**Federally Threatened**	**Pine savannas, 13,16,20– destruction of critical nest trees from chronic water stress related to dewatering/altering natural hydroperiods**

[1]**Information from Schneider and Keyes (2015) and two-digit numbers indicating Priority Amphibian and Reptile Conservation Areas (PARCAs) described in Jenson (2015)**

Table 5B. Examples of Georgia's high priority state and federally listed amphibian and reptile species in the ACF River Basins and coastal plain with jeopardized survival and recovery by adverse direct, indirect, and cumulative impacts from Floridan-aquifer injections and withdrawals in Florida, Georgia, and South Carolina.

Species	Status	Examples of Habitat & Adverse Impacts
Amphibians		
Reticulated Flatwoods Salamander *Ambystoma bishopi*	High Priority	14–dewatering/altering hydroperiods & water quality of depressional wetlands required for reproduction
Eastern Tiger Salamander *Ambystoma tigrinum*	High Priority	13,14,16,17,19,20–dewatering/altering hydroperiods & water quality of depressional wetlands required for reproduction
Chamberlain's Dwarf Salamander *Eurycea chamberlaini*	Petitioned Species	Lower Chattahoochee River (CR) – elimination of seepage in ravines
Georgia Blind Salamander *Eurycea wallacei*	Petitioned Species	Floridan aquifer system – dewatering & water quality alterations
Gopher Frog *Lithobates capito*	Petitioned Species	13,14,16,17,19,20–dewatering/altering hydroperiods & water quality of depressional wetlands required for reproduction
Striped Newt *Notophthalmus perstriatus*	Federal Candidate	13,14,17-20 – dewatering/altering hydroperiods & water quality of depressional wetlands required for reproduction
Reptiles		
Eastern Indigo Snake *Drymarchon couperi*	Federally Threatened	15,17-19 – destruction of micro-climate conditions related to altered hydroperiods
Gopher Tortoise *Gopherus polyphemus*	Federal Candidate	10,11,13-20 - destruction of micro-climate conditions in burrows related to altered hydroperiods
Barbour's Map Turtle *Graptemys barbouri*	Federal Candidate	Lower CR, 12-14 – reduced flow of rivers & large creeks of Apalachicola River drainage
Alligator Snapping Turtle *Macrochelys temminckii*	Petitioned/ Threatened	Lower CR, 14 – reduced flow of large streams & rivers; impoundments; river swamps

[1]Information and two-digit numbers indicating Priority Amphibian and Reptile Conservation Areas (PARCAs) from Jenson (2015)

Table 5C. Examples of Georgia's high priority state and federally listed fish and invertebrate species in the ACF River Basins and coastal plain with jeopardized survival and recovery by adverse direct, indirect, and cumulative impacts from Floridan-aquifer injections and withdrawals in Florida, Georgia, and South Carolina.

Species	Status	Examples of Habitat & Adverse Impacts
Fish		
Gulf Sturgeon *Acipenser oxyrinchus desotoi*	Federally Threatened	Lower Chattahootchee River (CR) – reduced fresh water in estuaries; deep pools at lower end of large rivers
Spotted Bullhead *Ameiurus serracanthus*	Rare	Lower CR – reduction in moderate currents of large streams with rock-sand substrate
River Redhorse *Moxostoma carinatum*	Rare	Etowah River[2] – reduced flow in swift waters of medium to large rivers
Robust Redhorse *Moxostoma robustum*	Endangered	Broad, Ocmulgee, Oconee & Ogeechee Rivers – reduced flow in medium to large rivers, shallow riffles to deep flowing water with moderately swift current
Broadstripe Shiner *Pteronotropis euryzonus*	Rare	Lower CR – reduced flow of medium sized streams associated with sandy substrate and woody debris or vegetation
Invertebrates		
Southern Elktoe *Alasmidonta triangulata*	Endangered	Lower CR – reduced flow in large to small streams
Rayed Creekshell *Anodontoides radiatus*	Threatened	Lower CR – reduction in flow of large to small streams on Gulf Coastal Plain
Delicate Spike *Elliptio arctata*	Endangered	Lower CR – reduction in flow of large to medium-sized streams

[1]Information from Albanese, Wiseniewskiki & GaschoLandis (2015)
[2]Not in the ACF River Basin or coastal plain, but includes a proposed ASR "water bank"

5.2 Strategic Location of ASR and Other Injection and Withdrawal Wells in Florida, Georgia, and South Carolina

Years before ASR wells were permitted in Florida, under the guise of aquifer "storage" and "recovery," aquifer injections were permitted and constructed at similar depths for "disposal" of agricultural, industrial and municipal wastes, such as minimally treated sewage effluent. Evidence that these injected wastes rapidly flow through preferential pathways in the karst aquifer system, then discharge into nearshore coastal water, was summarized by Bacchus et al. (2014). Bacchus et al. (2015a) illustrated the proximity and frequencies of permitted ASR wells, other injection and withdrawal wells, to reported modern sinkholes, mapped fractures and fracture intersections in Florida.

It has been documented at least since the 1990s that the upper Floridan aquifer is connected to underlying aquifer zones by fractures that resulted in upconing of saline water from those underlying zones in response to unsustainable pumping from the upper Floridan (Bacchus, 2000; Odum et al., 1998; Spechler, 1994; Spechler & Phelps, 1997). The logical presumption is that pumping from those underlying aquifer zones results in induced recharge from the Floridan and surficial aquifer zones through fracture networks, as shown in Figure 3 of Odum et al. (1998) and Figure 37 of Spechler (1994). Brook and Sun (1982) and Brook et al. (1986; 1988) also documented the dependence of well yield on proximity of wells to fractures.

We found no peer-reviewed publications with results from scientific investigations in the FAS using groundwater tracers to determine where the fluids injected into ASR wells and disposal wells are going or isotopic analysis to identify the "fingerprint" of the native ground water prior to any injections and withdrawals associated with a new ASR well, to compare to the "fingerprint" of water that is "recovered" from ASR wells, similar to the isotopic analysis of water by Wilcox et al. (2004). Without such evidence, no assumptions can be made that injections into and withdrawals from ASR wells in the FAS function differently than wells used solely for water supply or for aquifer "disposal."

The published literature we have referenced suggests that the proximity to fractures of ASR wells and other injection and withdrawal wells throughout the FAS plays a significant role in the performance of those wells, including ASR wells functioning as typical water-supply wells and disposal wells. For example, the strategic location of ASR wells near inland rivers, lakes, and depressional wetlands associated with fractures could increase the productivity of those ASR wells considerably as withdrawals divert water from those natural features. Conversely, other ASR wells in proximity to fractures may increase the time of occurrence, magnitude and public costs of those wells from water-quality deterioration from lateral saltwater intrusion, upconing of brackish and saline water and contaminants, such as arsenic and benzene via fractures and other karst conduits. Because fractures in the FAS also have been shown to extend under coastal waters, coastal ASR wells may be more vulnerable than inland wells for saltwater contamination of the aquifer system.

5.2.1 Abandoned, Inactive and Active ASR Wells in Florida

Bloetscher et al. (2014) identified 32 ASR well sites in Florida that were no longer active because of problems including arsenic contamination, clogging, "recovery" problems or water quality deterioration, with 13 ASR well sites in Florida reported as a abandoned in that study. Their study incorporated data from Herczeg, Rattray, Dillon, Pavelic, and Barry (2004), that previously described arsenic contamination as a common problem associated with ASR wells, and included data collected through July 1, 2013. Neither study evaluated the hydrological influence that fractures might have on the long-term operation of ASR wells in Florida and other areas with karst aquifer systems, such as the proximity of ASR wells to fractures. Bacchus et al. (2015b) described the 13 ASR well sites in Florida that were evaluated by Reese (2002), and later identified by Bloetscher et al. (2014) as abandoned. Some of those 13 abandoned ASR well sites had multiple ASR wells (Bacchus et al., 2015b). The consulting firm CH2M Hill was involved in the construction, operation, or testing associated with the ASR cycle-test data for the Boynton Beach, Broward County, City of Delray, Lake Okeechobee, Marathon, Miami-Dade W Well Field, San Carlos Estates, and West Palm Beach ASR wells. Ironically, all but one of those ASR sites, including the West Palm Beach ASR sites, was reported as abandoned by Bloetscher et al. (2014). The West Palm Beach, Florida ASR well was described as the largest capacity ASR well in the world (Pyne, 2004). That ASR well was located near the coast, but included no record of the chloride levels for injected or "recovered" water (Reese, 2002; Bacchus et al., 2015a). The proximity of those ASR wells to fractures and fracture extensions was shown in Bacchus et al. (2015a; 2015b). That suggests the proximity of abandoned and inactive ASR wells to fractures may have resulted in favorable "recovery" initially, followed by increasing chloride levels that resulted in comparable increases in operating costs.

Although our study did not attempt to determine if additional ASR wells in Florida had been abandoned after the 2013 data-collection date used for Bloetscher et al. (2014), Bacchus et al. (2015b) identified another ASR site evaluated in Florida by Reese (2002, San Carlos) that was listed by FDEP as inactive, but not identified as an abandoned or inactive ASR well by Bloetscher et al. (2014). It is unclear how many of those 33 ASR well sites in Florida were abandoned or were inactive because of preferential flow through fractures or other karst conduits of the aquifer injections and withdrawals associated with those ASR wells. Also unclear is if attempts are being made to strategically locate ASR wells in proximity to fractures in Florida, Georgia, and South Carolina to increase the productivity of those ASR wells, particularly in cases where the capacity of the aquifer has been depleted and new supply wells are receiving more scrutiny.

The ASR wells in the vicinity of Lake Okeechobee and the Peace River also are associated with fracture networks (Bacchus et al., 2015a; 2015b) and appear to be additional examples of strategically located ASR wells capable of tapping water from those natural resources. The Peace River ASR site, located along the west bank of the Peace River, reportedly is the oldest ASR project in Florida, with initial investigations beginning in 1983 and full-time operations initiated in 1988 (Brown, 2005). Pyne (2004) described the Peace River ASR site as the largest ASR wellfield in the eastern US. There are no investigations assessing the role of preferential flow through fractures and other karst conduits in response to ASR operations at the Peace River site in the formation of sinkholes described by Patten and Klein (1989). Aquifer injections into the 22 ASR wells associated with fracture networks at the Peace River ASR site are of particular concern, considering that the largest groundwater withdrawals in that area are from phosphate mining located in proximity to those ASR wells.

Mining is widespread in Florida and results in mechanical and nonmechanical extractions of large volumes of water from the FAS, depleting ground water from natural resources and availability for municipal use. For example, the 25th modification of the Water Use Permit (WUP) for current phosphate mining in south-central Florida allows groundwater pumping that ranges from "69,600,000 gpd" to "87,000,000 gpd" through 2032, in a five-county area (Southwest Florida Water Management District WUP 20011400.025 issued 2/1/12). That does not include the significant evaporative loss of water from the large open mine pits that remain in perpetuity after the mining is completed or additional ground water used for the mining, but permitted as separate agricultural permits. The cost to the mining industry for this water use is confined to the cost of the permits and fuel for the pumps to extract ground water from supply wells. Although this water use depletes ground water that is the primary source of municipal water supply, the mining industry does not pay the high costs of constructing ASR wells, treating the water that is injected into those wells, or treating the water that is withdrawn from ASR wells, such as the cost of constructing, operating and maintaining the 22 ASR wells at the Peace River site. Those costs are paid by the public for a "supplemental" source of municipal water perceived to be "recovery" rather than ground water. Yet no data have been produced to show that the water injected into those 22 ASR wells is not being withdrawn by the neighboring phosphate mining supply wells via the fracture network.

5.2.2 Recent ASR Wells in Georgia

Unlike Florida and South Carolina, Georgia only recently permitted and constructed the first ASR well in the state, following expiration of a state-imposed moratorium on ASR wells. This well was permitted by GDNR as an ASR "demonstration" well. On December 21, 2012, a statement of qualifications for the ASR "Demonstration Program" was submitted to the Southwest Georgia Regional Commission by David Pyne, President of ASR Systems, LLC. At that time, the moratorium on ASR wells was still in effect in Georgia. The location selected for the ASR "Demonstration Program" was the Elmodel WMA and the purpose was to augment the dry season flow of the Flint River and its tributaries. Pyne referenced 30 years of engineering services provided to CH2M Hill as evidence of his ASR expertise. His letter also referenced Ted Belser, another CH2M Hill employee (in the capacity of lead design engineer, department manager, and senior project manager), as an additional member of the "team" for the proposed Elmodel ASR "Demonstration Program."

Belser's expertise listed in the statement of qualifications, indicated "ASR systems for which he has provided design services in recent years include projects for: ... Orangeburg, SC; Hilton Head (North Island), SC; Hilton Head (South Island), SC; City of Bradenton, FL; Beaufort-Jasper Water Supply Authority, SC; Orange County, FL; ... the City of Rockledge, FL; and Charlotte County, FL." The FDEP ASR database dated December 20, 2014: (1) does not include an ASR well identified as the Orange County ASR well(s); (2) indicates that the Rockledge, Florida ASR well is inactive; and (3) does not include an ASR well identified as the Charlotte County ASR well. Confirmation was obtained from FDEP staff that: (1) the Orange County ASR well has been cycle-tested from 2010 to the present, with "storage" ranging from 14 to 120 days; (2) neither the Rockledge ASR well nor the Rockledge ASR facility is active, although construction occurred between 2008 and 2010; and (3) there is no record of an ASR well referenced as the Charlotte County ASR well (Joe Haberfeld, FDEP

Geological Survey, UIC Program Geologist, pers. comm., 02/15/16).

The referenced statement of qualifications dated December 21, 2012, also stated that the Elmodel Project ASR Systems Team members have "represented Georgia in the joint studies and the negotiations with Alabama and Florida on the water resource conflicts in the Apalachicola-Chattahoochee-Flint (ACF) river Basin" and "secured an agreement with EPD that the agency will provide a copy of the State's numerical ground water flow model that will cover all of southwest Georgia including the relevant aquifers underlying the Elmodel Site."

That statement of qualifications also stated that a final report would be prepared, "including a recommended ASR wellfield expansion plan for the Claiborne and Clayton Aquifers of the lower Flint River Basin, with an ASR recovery capacity of up to 250 MGD. The wellfield expansion plan will consider alternate use of ground water during winter months as the source of supply for recharge, and also treated surface water during times when river and tributary flows are high." It is important to note that at the time ASR Systems, LLC was proposing to create a groundwater model for the vicinity of the Elmodel ASR demonstration well, "deemed most suitable to meet flow augmentation goals, utilizing approximately 30,000 acres of state-owned lands," a groundwater model for that area already had been created and released for that vicinity showing that groundwater withdrawals in that area were dewatering the Wakulla Springshed that occurs in southwest Georgia and northwest Florida (Kinkaid et al., 2012). That 250 MGD production capacity proposed by ASR Systems, LLC exceeds the 182.88 MGD total agricultural groundwater withdrawals for 15 Georgia counties in that vicinity that have contributed to the dewatering of the Wakulla Springshed (Kinkaid et al., 2012). The 250 MGD production capacity proposed by ASR Systems, LLC also exceeds the approximately 246 MGD occurring in the west-central Water Use Caution Area by 1993, that caused both adverse economic and wide-spread environmental impacts, including premature decline and death of longleaf pine in uplands and pond-cypress in wetlands, catastrophic destructive wildfires, and declines in native wildlife (Bacchus, 2000).

On July 8, 2015, the GEFA issued a media release announcing that dual-aquifer ASR well and two monitoring wells, that were designed, constructed, administered, and analyzed by CH2M Hill, had not performed as predicted. Specifically, the media release stated that the dual-aquifer ASR well at the Elmodel WMA is "unlikely to produce enough water for stream-flow augmentation in Chickasawhatchee Creek." That media release further announced that instead of moving forward with the dual-aquifer ASR well at the Elmodel WMA, GEFA and the Georgia EPD plan to have a Claiborne aquifer production well designed and installed next to the monitoring wells. The media release also stated "the information will support the state's evaluation and management of these aquifers as a potential alternative water source for the region."

The research by Brook and Sun (1982) and Brook et al. (1986; 1988) documented the dependence of well yield on proximity to fractures, specifically in the vicinity of the areas referenced in the GEFA press release. In fact, the following statement from the GEFA media release dated July 8, 2015 and the extensive research by USGS (Odum et al., 1998; Spechler, 1994; 1997) support the conclusion that the site-specific higher yields from the underlying Claiborne and Clayton aquifer zones result from induced recharge from the overlying Floridan and surficial aquifer zones:

The yield from the Claiborne and Clayton aquifers appears to be site-specific, which suggests more yield is possible at other locations in the basin. For example, a Claiborne production well drilled this spring to support irrigation-related research at the Stripling Irrigation Research Park, which is roughly 9 miles from the Elmodel WMA, showed significantly more yield.

The GEFA media release also indicated that a technical peer review of the Elmodel WMA ASR demonstration project was conducted by Cardno. Based on information on the Cardno website (12/09/15), the company is based in Australia, with US merger partners. The website stated that Cardno "provided groundwater modeling, well field design, construction oversight and water supply planning using our state-of-the-art aquifer storage and recovery (ASR) pretreatment system design" to "help supply fresh drinking water to Marco Island" (Florida). That ASR well was not included in the FDEP ASR database dated December 20, 2014. The FDEP staff confirmed that the Marco Island ASR well was not included in the FDEP database because that project was never constructed (Joe Haberfeld, FDEP Geological Survey, UIC Program Geologist, pers. comm., 12/14/15).

The process for the initial ASR demonstration well and program in Georgia suggests that the FDEP and GEFA are not aware of the body of peer-reviewed scientific publications related to adverse impacts associated with aquifer injections and withdrawals, including those associated with ASR wells. Rather, the consulting companies benefitting financially from designing, constructing, operating and monitoring/reviewing these types of wells appear to have been the sources of information for this ASR demonstration project. Additionally, the location of the Georgia ASR demonstration well in the Elmodel WMA to augment the dry season flow of the

Flint River and its tributaries in Georgia appears to be related to two water-depletion problems. The first and most widely known problem is the on-going "tri-state water wars" that began decades ago, based on allegations by Florida and Alabama that Georgia is depleting water in those downstream states because of excessive water withdrawals from and dams on the ACF. The second problem, most recently documented by Kincaid et al. (2012), is the depletion of ground water essential for maintaining Wakulla Springs in Florida that is resulting from the unsustainable groundwater withdrawals in the vicinity of the Elmodel WMA. Based on the body of scientific literature we have described, the continued and proposed increases in groundwater withdrawals in the Elmodel WMA area will increase the diversion of ground water from Florida and result in irreversible adverse environmental impacts to Georgia's PARCAs within the FAS, particularly those associated with the fractures and fracture extensions shown in Figure 6.

5.2.3 ASR Wells in South Carolina

The SCDEHC has issued permits to 11 water supply facilities for a total of 68 UIC ASR wells in South Carolina. No UIC "disposal" wells have been permitted in South Carolina. Facilities in Columbus, Georgetown, Horry, Marion, and Orangeburg Counties are beyond the FAS boundaries and account for 54 of the permitted ASR wells in South Carolina. A total of 34 of those 54 wells originally were conventional municipal supply wells operated by the Grand Strand Water and Sewer Authority, but resulted in a large cone of depression from extensive withdrawals and have been converted to ASR wells. Permit dates for those ASR wells range from April 15, 1994 to May 15, 2015. Nine of those ASR wells are operated by the Georgetown County Water and Sewer District, five are operated by Mount Pleasant Water Works (permitted from May 28, 2003 to November 27, 2012), four are operated by the Orangeburg Department of Public Utilities (permitted from January 10, 2008 to March 24, 2011), and one is operated by the Little River Water and Sewerage Company, Inc. (permitted on May 20, 2015). One of the ASR wells in Mount Pleasant and the only ASR well permitted in Myrtle Beach on May 16, 1995 and September 14, 1990, respectively, are reported as permanently abandoned. No reason was included in the SCDEHC data for either abandonment (SCDEHC, Christopher Wargo, pers. comm., 3/16). The figures showing locations for the South Carolina ASR wells include only one location dot for each facility rather than for each well.

The ASR wells permitted or pending in South Carolina within the FAS boundaries are located along the coast in Beaufort, Charleston, Colleton, and Jasper Counties. The pending ASR well and RO facility in Colleton County would be located in Edisto Beach, south of the ACE Basin National Wildlife Refuge. Neither the proposed well nor the ACE Basin NWR are shown in Figures 2 or 3, but the intersecting fracture extensions shown in Figures 2B and 3B are in the vicinity of the proposed Edisto Beach ASR well and RO facility. Permit dates for the four ASR wells operated by the Beaufort Jasper Water and Sewer Authority range from June 1, 1999 to January 7, 2011. The two ASR wells operated by the Charleston Water System were permitted on May 24, 1994 and May 26, 1999. The two ASR wells operated by the Hilton Head Public Service District (HHPSD) were permitted on November 8, 2010 and December 20, 2010 and both of those ASR wells are associated with a fracture extension (Figures 2B and 3B). Both of the ASR wells operated by the Kiawah Island Utility, Inc. were permitted on May 30, 2001. The four remaining ASR wells in South Carolina's coastal counties of the FAS are operated by South Island Public Service District and were permitted from July 11, 2012 to October 16, 2013. They include the Long Cove and Palmetto Bay ASR wells that are associated with the same fracture extension as the two ASR wells operated by the HHPSD. The only one of those wells reported as "permanently abandoned" was the ASR well permitted for the Charleston Water System on May 26, 1999.

Extensive mapping of fracture networks, similar to the mapping conducted throughout Florida and Dougherty County, Georgia, was not available for South Carolina. Therefore, the inferences in our study about fractures in the portion of the FAS that extends into South Carolina were confined to extensions of fractures mapped in Florida and Georgia's coastal plain.

Although South Carolina's total number of ASR wells is sizeable, the paucity of information available about those wells is similar to the paucity of information about the sole ASR demonstration well constructed in Georgia. We found no publications for South Carolina comparable to the USGS review of ASR wells and well data in Florida (Reese, 2002). The most extensive information available about the ASR wells in South Carolina was included in the statement of qualifications by ASR Systems, LLC to secure the contract for Georgia's Elmodel WMA ASR demonstration well that failed. That statement of qualifications indicated "[P]rior to five years ago ASR Systems staff provided ASR consultant services to ... the Kiawah Island Utility, SC." The following quotes from that statement of qualifications regarding ASR wells in South Carolina have been included because similar information isn't available from sources such as peer-reviewed journal publications:

Hilton Head Public Service District, SC (HHPSD)

ASR feasibility assessment report, design, permitting, construction management, cycle testing and placing an ASR well in operation with a recovery capacity of 2.5 MGD. Facilities included one ASR well, two monitor wells, pump, motor, wellhead piping, wellhouse and telemetry. Freshwater wells on Hilton Head Island are being lost to saltwater intrusion in the Upper Floridan aquifer. Imported water from Beaufort Jasper Water and Sewer Authority is available at half cost during winter months. This water is blended with offpeak water produced from the HHPSD Reverse Osmosis (RO) plant, and also water from remaining fresh waterwells. The blended water is used for ASR recharge. The storage zone is the Middle Floridan aquifer, which is brackish. Water is recovered during summer months to help meet peak demands. The project was completed in a record short time of 23 months and came in under budget. Work was completed during 2012.

South Island Public Service District, SC (SIPSD)

Following the loss during 2010 of one of their ten production wells in the Upper Floridan Aquifer due to salt water intrusion, SIPSD embarked on a comprehensive water facilities upgrade and expansion program, based upon the expectation that all of their freshwater supply wells will be lost during the next 40 years. ASR Systems completed an ASR feasibility study, then proceeded with design of two ASR wells in the Middle Floridan Aquifer; five monitor wells in the Upper, Middle and Lower Floridan aquifers; and one Lower Floridan Aquifer production well to provide a backup water supply source in the event that the SIPSD Cretaceous Aquifer water supply well fails. This well is 3,800 ft deep and supplies 3.5 MGD to a reverse osmosis (RO) Water Treatment Plant (WTP) during peak demand periods and is the backbone of the SIPSD water supply system. Wellhead facilities design was completed and all facilities were permitted for construction, which is currently underway. Well construction is expected to be completed by April 2013, to be followed by interim recharge into each ASR well. Wellhead facilities construction will then begin and should be completed by the end of 2013. This will be followed by cycle testing. The source of water supply for recharge is fresh water from the remaining SIPSD wells in the Upper Floridan aquifer, supplemented by water available from the RO WTP.

Town of Edisto Beach, Edisto Island, SC

ASR Systems was part of a URS Team that prepared a water supply master plan for the Town, completed during 2012. Salt water intrusion is causing a deterioration of drinking water quality, shortened life of household appliances, and an increase in the salt concentration of reclaimed water that is utilized for golf course irrigation. The final plan included recommended construction of a reverse osmosis (RO) water treatment plant and brine disposal outfall structure. It also included an ASR wellfield to store drinking water during winter months when water demand is very low, for recovery during summer months with peak demands. ASR wells will provide water supply reliability since ground elevations are quite low, well below the hurricane surge elevation. Drinking water stored underground can provide a water supply after a major hurricane until such time as other infrastructure has been repaired and placed back into service.

5.3 Laws, Regulations and Costs without Scientific Basis

5.3.1 Construction Costs of ASR without Scientifically Designed and Executed Studies, Monitoring and Modeling

Considerable public funds are being spent for ASR wells in the FAS without any scientific basis to support the claims that injected water can be "stored" and "recovered" or to evaluate the role of hydrological connections between ASR wells and environmentally sensitive areas via fractures. For example, the cost of the unsuccessful Elmodel WMA ASR demonstration well in Georgia was $1,395,712.16. That cost included: (1) $625,455.26 for construction of the Claiborne and Clayton monitoring wells and the Floridan supply well; (2) $133,150 for data analysis and groundwater modeling related to the Claiborne and Clayton monitoring wells, which included analysis of the geophysical logging, performance of a geochemical assessment, and geologic core sample analysis; and (3) $637,106.90 for planning, surveying, designing, permitting, and other administrative and legal expenses (GEFA media statement, 07/08/15). There was no documentation or explanation why or how that ASR well failed.

The cost of the Hilton Head North Island ASR well, that was operational under HHPSD #1 in South Carolina at the time of our study, was $3.9 million and was funded by general obligation (GO) bonds. Construction of this ASR well and associated equipment began in approximately 2010. Aquifer injections of "240 million gallons" began in October 2012 and continued to February 2013. Those aquifer injections were known as "building the bubble" for a "buffer" that was not to be withdrawn. The second series of aquifer injections began in October 2013, continued until February 2014 and totaled "241.631 Million Gallons." Withdrawals from that ASR well totaling "242.951 Million Gallons" and "237.761" began in May 2014 and continued through September 2014 and May 2015 through September 2015, respectively. The approximate chloride content of the injected water was 30 mg/L and increased to 164 mg/L in water withdrawn in September. The fact that "recovered" water has a chloride content that is 5.5 to 8 times (or more) greater than the chloride content of injected water is proof that the injected water is not remaining at the end of the ASR well as a "bubble" and may be increasing chloride contamination of the aquifer system rapidly via fractures. The HHPSD #1 reverse osmosis (RO) plant only produces "3 MGD," but it also buys water from the Beaufort Jasper Water and Sewer Authority (BJWSA), on the mainland. The Savannah River is the source of water for the BJWSA. This ASR well is dependent on the treated river water from BJWSA, sold to HHPSD #1 at off-peak rates in the winter months (Bill Davis, HHPSD #1, pers. comm., 12/22/15). Daily "recovery" from this ASR well for 2014 and 2015 was approximately 1.59 MGD and 1.55 MGD, respectively, based solely on the initial volumes withdrawn from this ASR well during these first two years and disregarding increases in chloride content to approximately 5.5 times the chloride content of the injected water.

An example of another costly ASR venture occurred in north and central Florida when the St. John's River Water Management District (SJRWMD) designated $47 million for ASR systems in its district from 2001-2006. In south Florida, $45 million originally was proposed for the ASR pilot projects, with $1.7 billion for more than 330 ASR wells originally proposed for construction throughout the Everglades. The ACOE's Final ASR Report and groundwater model (ACOE, 2014) revised the recommendation to 232 ASR wells for the Greater Everglades Basin (Bacchus et al., 2015b). Using the recent funding figure of approximately $1.4 million for the ASR well from the unsuccessful Elmodel ASR demonstration project in Georgia, that would result in a cost of approximately $324.8 million tax dollars for the proposed Everglades ASR wells. Based on the funding figure of $3.9 million for the Hilton Head North Island ASR well, those 232 proposed ASR wells would cost approximately $904.8 million tax dollars.

All of these publically funded ASR wells have the potential to fail, due to lateral saltwater intrusion, upconing of brackish and saline water and contaminants (e.g., arsenic and benzene) via preferential flow through fractures and other karst conduits. The injections and withdrawals associated with these ASR wells also have the ability to result in adverse impacts to environmentally sensitive state and federal lands. Additionally, the costs for ASR projects exclude long-term operation and maintenance costs, which are energy intensive and also exclude the costs of extensive evaluations of adverse environmental impacts (e.g., hydroperiod alterations and contamination of surface waters with arsenic) and the reversal of those impacts. Natural recharge in the FAS requires no operation or maintenance and has none of the harmful consequences of ASR injection and withdrawal wells (Bacchus et al., 2015b; Fernald, Purdum, Anderson, & Krafft, 1998). Our study supports the conclusion that mapped fracture networks are so extensive in the FAS that it would be difficult to find any sites in the carbonate platform of Florida, Georgia, and South Carolina that would not be influenced by the fracture network and associated sinkholes.

On October 7, 2015, the US Office of Management and Budget, Council on Environmental Quality, and Office of Science and Technology Policy within the Executive Office of the President directed federal agencies to incorporate ecosystems services into the planning and decision making for those agencies. Clearly ASR wells permitted as UICs under the federal Safe Drinking Water Act, without consideration of adverse impacts to the federal Endangered Species Act and the Clean Water Act meets the intent of this federal directive. A minimum requirement for future reporting and permitting for all states with fractured aquifers should be the identification of fractures in the vicinity of permitted or proposed UIC wells and testing that involves tracer and isotopic analyses.

5.3.2 Ground Water Is Not an "Alternative" Source for Ground Water

As described above, decades of research results and other evidence have documented preferential flow laterally and vertically between every zone of the FAS induced in response to groundwater withdrawals from every FAS zone, beginning with the surficial aquifer and extending to the deepest zone, although the permitting of ASR wells presumes no hydrologic connections between the lower, middle, and upper Floridan and the surficial aquifer. Therefore, it is difficult to understand why millions of tax dollars currently are being proposed for theoretical artificial aquifer "storage" and "recovery" using ASR wells in Florida and Georgia. In fact, the following excerpt from the Florida Statutes implies that ground water can be used as an "alternative" source of for ground water, which the scientific data clearly shows is not possible for the FAS:

> Alternative water supply (AWS) "Salt water; brackish surface water and groundwater; surface water captured predominately during wet-weather flows; sources made available through the addition of new storage capacity for surface water or groundwater, water that has been reclaimed after one or more public supply, municipal, industrial, commercial, or agricultural uses; the downstream augmentation of water bodies with reclaimed water; stormwater; and, any other water supply source that is designated as nontraditional for a water supply planning region in the applicable regional water supply plan" (Section 373.019, Florida Statutes).

This example of state law reflects the apparent lack of lawmakers' scientific understanding of groundwater responses in a regional karst aquifer system and the interconnections between all zones of the aquifer system and surface waters. This lack of understanding also appears to extend to the attorneys with practices that involve environmental litigation, based on the recent legal analysis of Florida's new water law termed a "marvel of compromise," that will take effect July 1, 2016 (if signed into law by the governor), and which acknowledges the following (Alderman, 2016):

> ...significant portions of Florida do not have enough water reserves from traditional groundwater sources to sustain continued growth. This dilemma has sparked the need to promote or even require development of alternative water supplies and to adopt additional limitations on withdrawals from traditional groundwater sources.

> Finally, the Bill addresses the multiple existing programs for protection of the South Florida natural environment...

> In addition to the threat of diminishing water supplies, continued concern for Florida's premier springs brought about the creation of a new regulatory category to afford them special protection, together with associated development limitations and remediation plans. Additional protections have also been afforded to help remediate impaired water bodies throughout the state, but particularly the ecosystems in south Florida.

> A new protected class of waters is created: the Outstanding Florida Spring (OFS) OFSs include all historic first magnitude springs and their associated spring runs and the following: De Leon, Peacock, Poe, Rock, Wekiwa and Gemini Springs, and their spring runs.

> Development of alternative water supplies for water-starved areas is encouraged through provision for pilot projects to be undertaken by the three largest water management districts: St. Johns, South Florida (SFWMD) and Southwest Florida.

The decades of scientific research and data addressed in our study clearly show that the state's environmentally sensitive lands and waters, particularly the springs, cannot be protected under this proposed new law with the proposed continuation of groundwater withdrawals, injections, and depletions from large excavations referenced as "reservoirs" and "water farming." Therefore, the environment has gained nothing, but lost additional protection under this law. Hopefully those economically and environmentally costly proposed approaches will be replaced with closed loop systems that are capable of minimizing adverse environmental impacts, to earn the claim of true compromise.

6. Summary and Conclusions

Fractures and sinkholes in the FAS have been mapped throughout Florida and at least one coastal-plain county of Georgia, although this regional karst aquifer also extends throughout the coastal plain of Alabama and South Carolina, from the submerged carbonate platform of the Atlantic Ocean, Gulf of Mexico, and Straits of Florida. This carbonate aquifer system is characterized by bedding planes, fractures, dissolution cavities, and other karst features that result in preferential flow of ground water, particularly in response to anthropogenic perturbations such as groundwater withdrawals and aquifer injections. Extensive networks of fractures have been reported by the US Army Corps of Engineers (ACOE, 2004), and mapped by Florida Department of Transportation (FDOT, 1973), and Vernon (1951) throughout Florida and mapped in Dougherty County by Brook and Allison (1986). Krause and Randolph (1989) divided the FAS into six sub-regions for groundwater modeling and concluded breaches of those groundwater divides had occurred, attributing those breaches to large withdrawals of groundwater in the US southeastern coastal plain. Some of those mapped fractures extended from one sub-region into an adjacent sub-region in the FAS and may be facilitating the breaching of groundwater divides described by Krause and Randolph (1989) in that regional aquifer system.

The greater total fractures and fracture density in Dougherty County (1,225 and 141.3/100 km^2, respectively) compared to 21 north-Florida counties (10-91 fractures per county and 0.6-3.8/100 km^2, respectively) presumably is due to the scale of fracture mapping and shorter mean lengths of mapped fractures in Dougherty County (1.2 km), compared to north Florida counties (26-118 km), rather than to orders of magnitude increases in fracture densities in that part of the FAS. Extensions of the Dougherty County fractures, however, did not suggest that those short fractures were fragmented segments of longer fractures. The number of sinkholes (934) mapped in Dougherty County by Brook and Allison (1986) in analog form and converted to georeferenced digital format in our study increased by more than an order of magnitude, based on the number of sinkholes identified in a NASA DEVELOP project (18,557), using 2011 Advanced Spaceborne Thermal Emission and Reflection Radiometer (ASTER) images. When we extended the dense network of those fractures that occurred within the boundaries of Chickasawhatchee Swamp/Ichauway Plantation PARCA, that encompassed Dougherty County, those extensions covered the Elmodel WMA and ASR demonstration well in Baker County, Georgia. Those extensions also passed through numerous agricultural areas with center-pivot irrigation wells in southwest Georgia; intersected other Georgia PARCAs near the Florida/Georgia state line; and clumped in two areas of dense sinkhole clusters in northwest Florida (sub-regions H and U). The magnitude and extent of agricultural, municipal, and industrial groundwater withdrawals in Georgia's coastal plain exceed groundwater withdrawals in Florida for that area of the FAS. No determination has been made regarding the contributions of those groundwater withdrawals in Georgia to water pirated, via preferential flow through fractures, from the Apalachicola-Chattahoochee-Flint (ACF) River Basins and Wakulla Springshed or to the increase in sinkholes in Dougherty County and the dense clusters of sinkholes in northwest Florida. Similarly, the survival and recovery of at least 24 animal species in Georgia that are either federally listed or high-priority state species may be jeopardized by adverse direct, indirect, and cumulative impacts from preferential flow through fractures, sinkholes, and other karst conduits in response to aquifer injections and withdrawals that have not been evaluated. Currently no regional groundwater model has been constructed to evaluate preferential groundwater flow through the mapped fractures, sinkholes, and other karst conduits in the FAS. Such a model is essential to determine the magnitude and extent of environmental impacts from ASR wells and other supply and disposal wells in this regional aquifer system, such as pirated water from the ACF and other river basins, alterations in submarine groundwater discharge to Apalachicola Bay and other coastal areas, saltwater intrusion, upconing of saline ground water, and resulting impacts to federally listed and high-priority state species.

Acknowledgments

We are grateful for the comments of anonymous reviewers that resulted in significant contributions to our final manuscript. We also thank J.J. Apodaca for providing shapefiles for the the locations of Georgia's PARCAs; Bill Davis for providing the address of the Hilton Head ASR; Duncan Elkin for providing shapefiles for the ACF River Basins and for hydrologic units intersecting the boundaries of Georgia; Todd Kincaid for providing a copy of the 2012 GeoHydros, LLC Report to FDEP and KMZ file for the groundwater model boundaries surrounding the Wakulla Springshed; and Christopher Wargo for providing a comprehensive list of the UIC/ASR wells permitted in South Carolina and the latitude and longitude for those wells.

References

Albanese, B., Wiseniewskiki, J. M., & GaschoLandis, A. (2015). *Appendix E. Fishes and Aquatic Invertebrates Technical Team Report*. Social Circle, GA: Georgia Department of Natural Resources.

Alderman, S. (2016). New Water Law Will Affect Everyone Who Uses Water in Florida. Retrieved April 1, 2016, from http://www.jdsupra.com/legalnews/new-water-law-will-affect-everyone-who-67097

Bacchus, S. T. (1998). *Determining sustainable yield in the Southeastern Coastal Plain: a need for new approaches*. In J. Borchers & C. D. Elifrits (Eds.), Proceedings of the Joseph F. Poland Symposium on Land Subsidence, Belmont, CA.

Bacchus, S. T. (2000). Uncalculated impacts of unsustainable aquifer yield including evidence of subsurface interbasin flow. *Journal of the American Water Resources Association, 36*(3), 457-481. http://dx.doi.org/10.1111/j.1752-1688.2000.tb04279.x

Bacchus, S. T. (2001). Knowledge of groundwater responses - A critical factor in saving Florida's threatened and endangered species. Part I: Marine ecological disturbances. *Endangered Species Update, 18*(3), 79-90.

Bacchus, S. T. (2002). The "ostrich" component of the multiple stressor model: Undermining Florida. In J. W. Porter & K. G. Porter (Eds.), *The Everglades, Florida Bay, and Coral Reefs of the Florida Keys - An Ecosystem Sourcebook* (pp. 669-740): CRC Press.

Bacchus, S. T. (2006). Nonmechanical dewatering of the regional Floridan aquifer system. *Perspectives on Karst Geomorphology, Hydrology, and Geochemistry, 404,* 219-234. http://dx.doi.org/10.1130/2006.2404(18)

Bacchus, S. T. (2007). More inconvenient truths: Wildfires and wetlands, SWANCC and Rapanos. *National Wetlands Newsletter, 29*(11), 15-21.

Bacchus, S. T., & Barile, P. J. (2005). Discriminating sources and flowpaths of anthropogenic nitrogen discharges to Florida springs, streams and lakes. *Environmental and Engineering Geoscience, 11*(4), 347-369. http://dx.doi.org/10.2113/11.4.347

Bacchus, S. T., & Brook, G. A. (1996). Geophysical characterization of depressional wetlands: a first step for determining sustainable yield of groundwater resources in Georgia's Coastal Plain *Technical Completion Report* (pp. 36 + appendices). Atlanta, GA.

Bacchus, S. T., Archibald, D. D., Britton, K. O., & Haines, B. L. (2005). Near infrared model development for pond-cypress subjected to chronic water stress and *Botryosphaeria rhodina. Acta Phytopathologica et Entomologica Hungarica, 40*(2-3), 251-265.

Bacchus, S. T., Archibald, D. D., Brook, G. A., Britton, K. O., Haines, B. L., Rathbun, S. L., & Madden, M. (2003). Near infrared spectroscopy of a hydroecological indicator: New tool for determining sustainable yield for Floridan aquifer system. *Hydrological Processes, 17,* 1785-1809. http://dx.doi.org/10.1002/hyp.1213

Bacchus, S. T., Bernardes, S., Jordan, T., & Madden, M. (2014). Benthic macroalgal blooms as indicators of nutrient loading from aquifer-injected sewage effluent in environmentally sensitive near-shore waters associated with the south Florida Keys. *Journal of Geography and Geology, 6*(4), p. 164. http://dx.doi.org/10.5539/jgg.v6n4p16

Bacchus, S. T., Bernardes, S., Xu, W., & Madden, M. (2015a). Fractures as preferential flowpaths for aquifer storage and recovery (ASR) injections and withdrawals: implications for environmentally sensitive near-shore waters, wetlands of the Greater Everglades Basin and the regional karst aquifer system. *Journal of Geography and Geology, 7*(2), 117-155.

Bacchus, S. T., Bernardes, S., Xu, W., & Madden, M. (2015b). What Georgia Can Learn from Aquifer Storage and Recovery (ASR) in Florida. In McDowell R. J., Pruitt, C. A., Bahn, R. A. (Eds.) *Proceedings of the 2015 Georgia Water Resources Conference, held April 28-29, 2015, at The University of Georgia, Athens, GA.*

Bacchus, S. T., Hamazaki, T., Britton, K. O., & Haines, B. L. (2000). Soluble sugar composition of pond-cypress: a potential hydroecological indicator of groundwater perturbations. *Journal of the American Water Resources Association, 36*(1), 55-65. http://dx.doi.org/10.1111/j.1752-1688.2000.tb04248.x

Bacchus, S. T., Masour, J., Madden, M., Jordan, T., & Meng, Q. (2011). Geospatial analysis of depressional wetlands near Peace River watershed phosphate mines, Florida, USA. *Environmental and Engineering Geoscience, 17*(4), 391-415. http://dx.doi.org/10.2113/gseegeosci.17.4.391

Bernardes, S., He, J., Bacchus, S. T., Madden, M., & Jordan, T. (2014). Mitigation banks and other conservation lands at risk from preferential groundwater flow and hydroperiod alterations by existing and proposed northeast Florida mines. *Journal of Sustainable Development, 7*(4), 37. http://dx.doi.org/10.5539/jsd.v7n4p225

Bloetscher, F., Sham, C., Danko, III. J., & Ratick, S. (2014). Lessons Learned from Aquifer Storage and Recovery (ASR) Systems in the United States. *Journal of Water Resource and Protection, 6,* 1603-1629. doi: 10.4236/jwarp.2014.617146

Brook, G. A. (1985). Geological factors influencing well productivity in the Dougherty Plain covered karst region of Georgia. In *Proceedings of the Ankara - Antalya Symposium* (pp. 87-99). IAHS Publication no.161.

Brook, G. A., & Allison, T. L. (1986). Fracture mapping and ground subsidence susceptibility modeling in covered karst terrain - the example of Dougherty County, Georgia. In Dougherty, P. H. (Ed.) *Environmental Karst* (pp. 91-108). GeoSpeleo Publications, Cincinnati, OH.

Brook, G. A., & Sun, C. H. (1982). Predicting the specific capacities of wells penetrating the Ocala Aquifer beneath the Dougherty Plain, Southwest Georgia. Environmental Resources Center, Atlanta, GA.

Brook, G. A., Carver, R. E., & Sun, C. H. (1986). Predicting well productivity using principal components analysis. *Professional Geographer, 38*(4), 324-331.

Brook, G. A., Sun, C. H., & Carver, R. E. (1988). Predicting water well productivity in the Dougherty Plain, Georgia. *Georgia Journal of Science, 46,* 190-203.

Brown, C. J. (2005). *Planning decision framework for brackish water aquifer, storage and recovery (ASR) projects.* (PhD Dissertation), University of Florida, Gainesville, FL.

Bush, P. W., & Johnson, R. (1988). Ground-water hydraulics, regional flow, and ground-water development of the Floridan aquifer system in Florida and in parts of Georgia, South Carolina. *US Geological Survey Professional Paper 1403-C*, 80 + 17 plates.

Cahalan, M., Amin, M., Berry, K., Xu, W., Hu, T., & Milewski, A. (2015). *Georgia Disasters & Water Resources: Utilizing NASA Earth Observations to Monitor Sinkhole Development and Identify Risk Areas in Dougherty County, GA (Unpublished)*. University of Georgia, United States. Retrieved from https://drive.google.com/file/d/0B68PDhnQDnhxZ1pmLVZOdjdWclU/view?usp=sharing

Cahalan, M., Milewski, A. M., & Durham, M. C. (in press). Sinkhole formation dynamics and geostatistical-based prediction analysis in a mantled karst terrain.

Cunningham, K. J., Renken, R., Wacker, M., & Zygnerski, M. (2003). *Application of carbonate cycle stratigraphy to delineate porosity and preferential flow and to assess advective transport in the karst limestone of the Biscayne Aquifer*. Paper presented at the Proceedings of GSA Annual Meeting and Exposition.

Fernald, E. A., Purdum, E. Anderson, J. R., & Krafft, P. A. (1998). Water resources atlas of Florida. Institute of Science and Public Affairs, Florida State University, Tallahassee, FL.

Fitterman, D. V., & Deszcz-Pan, M. (1999). *Geophysical mapping of saltwater intrusion in Everglades National Park. US Geological Survey Open-File Report 99-181*. Boca Raton, FL: US Geological Survey.

Florida Department of Transportation. (1973). *Map of lineaments in the state of Florida*. Tallahassee, FL.

Ford, D. C., Palmer, A. N., & White, W. B. (1988). Landform development; karst. In W. Back, J. S. Rosenshein, & P. R. Seaber (Eds.), *Hydrogeology: The Geology of North America* (pp. 401-412). Boulder, Colorado: Geological Society of America.

Ford, D., & Williams, P. W. (1989). *Karst geomorphology and hydrology*. London ; Boston: Unwin Hyman.

Georgia Department of Natural Resources. (2015). *Georgia State Wildlife Action Plan*. Social Circle, GA.

Herczeg, A. L., Rattray, K. J., Dillon, P. J., Pavelic, P., & Barry, K. E. (2004). Geochemical processes during five years of aquifer storage recovery. *Ground Water, 42*(3), 438-445. http://dx.doi.org/10.1111/j.1745-6584.2004.tb02691.x

Hofstetter, R. H., & Sonenshein, R. S. (1990). *Vegetative Changes in a Wetland in the Vicinity of a Well Field, Dade County, Florida. US Geological Survey Water-Resources Investigations Report 89-4155*. US Geological Survey.

Jenson, J. B. (2015). *Appendix D. Reptiles and Amphibians Technical Team Report*. Social Circle, GA: Georgia Department of Natural Resources.

Johnston, R. H., & Bush, P. W. (1988). *Summary of the Hydrology of the Floridan Aquifer System in Florida and in Parts of Georgia, South Carolina, and Alabama*. US Geology Survey.

Kaufman, M. I. (1973). Subsurface wastewater injection, Florida. American Civil Engineers Proceedings. Paper 9598. *Journal of Irrigation and Drainage Div., 99*(53-70).

Kincaid, T. R., Davies, G. J., Hazlett, T. J., Loper, D., Dehan, R., & McKinlay, C. (2004). *Groundbreaking characterization of the karstfield Floridan aquifer in the Woodville Karst Plain of North Florida*. In Geoscience in a Changing World: Annual Meeting and Exposition, Denver, CO.

Kincaid, T., Davies, G., Werner, C., & DeHan, R. (2012). *Demonstrating interconnection between a wastewater application facility and a first magnitude spring in a karstic watershed: Tracer study of the Southeast Farm Wastewater Reuse Facility, Tallahassee, Florida*. Tallahassee, FL: Florida Geological Survey.

Kincaid, T., Meyer, M. S., & Day, K. (2012). Woodville Karst Plain Hydrologic Research Program – Report on Tasks Performed in 2011 & 2012 Under FDEP Contract RM100 Amendment #3 (pp. 562).

Kohout, F. A. (1967). Ground-water flow and the geothermal regime of the Floridian Plateau. Trans.: *Gulf Coast Assoc. Geol. Soc., 27*, 339-354.

Krause, R. E., & Randolph, R. B. (1989). *Hydrology of the Floridan aquifer system in southeast Georgia and adjacent parts of Florida and South Carolina*. US Geological Survey.

Lines, J. P., Bernardes, S., He, J., Zhang, S., Bacchus, S. T., Madden, M., & Jordan, T. (2012). Preferential groundwater flow pathways and hydroperiod alterations indicated by georectified lineaments and sinkholes at proposed karst nuclear power plant and mine sites. *Journal of Sustainable Development, 5*(12), 78-116. http://dx.doi.org/10.5539/jsd.v5n12p78

Maslia, M. L., & Prowell, D. C. (1990). Effect of faults on fluid-flow and chloride contamination in a carbonate aquifer system. *Journal of Hydrology, 115*(1-4), 1-49. http://dx.doi.org/10.1016/0022-1694(90)90196-5

McNeill, D. (2000). *A review of upward migration of effluent related to subsurface injection at Miami-Dade Water and Sewer South District Plant*: Prepared for the Sierra Club - Miami Group.

Meyer, F. W. (1989). *Hydrogeology, Ground-Water Movement, and Subsurface Storage in the Floridan Aquifer System in Southern Florida.* US Geology Survey Professional Paper 1403-G, 59.

Neuendorf, K. K. E. (2005). *Glossary of geology* (5th ed.). Alexandria, VA: American Geological Institute.

Oden, M. I., Okpamu, T. A., & Amah, E. A. (2012). Comparative Analysis of Fracture Lineaments in Oban and Obudu Areas, SE Nigeria. *Journal of Geography and Geology, 4*(2), 36-47. http://dx.doi.org/10.5539/jgg.v4n2p36

Odum, J. K., Stephenson, W. J., Williams, R. A., Worley, D. M., Toth, D. J., Spechler, R. M., & Pratt, T. L. (1998). *Land-based high-resolution seismic reflection image of a karst sinkhole and solution pipe on Fort George Island, Duval County, northeastern Florida.* In R. S. Bell, M. H. Powers & T. Larson (Eds.), Symposium on the Application of Geophysics to Environmental and Engineering Problems at the Annual Meeting of the Environmental and Engineering Geophysical Society, Chicago, IL.

Patten, T. H., & Klein, J. G. (1989). *Sinkhole formation and its effect on Peace River hydrology.* In B. F. Beck (Ed.), Proceedings of the Third Multidisciplinary Conference on Sinkholes and the Engineering and Environmental Impacts of Karst, St. Petersburg Beach, FL.

Popenoe, P., Kohout, F., & Manheim, F. (1984). *Seismic-reflection studies of sinkholes and limestone dissolution features on the northeastern Florida shelf.* Paper presented at the Proceedings of First Multidisciplinary Conference on Sinkholes, Orlando, FL.

Price, R. E., & Pichler, T. (2004). *Arsenic and aquifer storage and recovery in southwest Florida: Source, abundance, and mobilization mechanism, Suwannee Limestone, upper Floridan aquifer.* In Aquifer Storage Recovery IV: Science, Technology, Management and Policy Conference, Tampa, FL.

Pyne, R. D. G. (2004). ASR dynamics, issues and solutions. Paper presented at Aquifer Storage Recovery IV: A two-day information-exchange forum, Tampa, FL.

Pyne, R. D. G. (2015). Aquifer storage recovery: An ASR solution to saltwater intrusion at Hilton Head Island, South Carolina, USA. *Environmental Earth Science, 73*(12), 7851-7859. http://dx.doi.org/10.1007/s12665-014-3985-z

Reese, R. S. (2002). *Inventory and Review of Aquifer Storage and Recovery in Southern Florida. US Geological Survey Water Resources Investigation Report 02-4036.* US Geological Survey.

Renken, R. A., Shapiro, A. M., Cunningham, K. J., Harvey, R. W., Zygnerski, M. R., Metge, D. W., & Wacker, M. (2004), *Pathogen transport in a sole source karst aquifer near a public well field–will everglades limestone mine expansion pose a clear and present danger to public health?* In Geoscience in a Changing World: Annual Meeting and Exposition, Denver, Colorado.

Schneider, T., & Keyes, T. (2015). *Appendix B. Birds Technical Team Report.* Social Circle, GA: Georgia Department of Natural Resources.

Seaber, P. R., Kapinos, F. P., & Knapp, G. L. (1987). Hydrologic Unit Maps: *US Geological Survey Water-Supply Paper 2294,* 63 p.

Sonenshein, R. S., & Hofstetter, R. H. (1990). *Hydrologic Effects of Well-Field Operations in a Wetland, Dade County, Florida. US Geological Survey Water-Resources Investigations Report 90-4143.* US Geological Survey.

Spechler, R. M. (1994). *Saltwater intrusion and the quality of water in the Floridan aquifer system, northeastern Florida. US Geological Survey Water-Resources Investigations Report 92-4174.* US Geological Survey.

Spechler, R. M., & Phelps, G. G. (1997). *Saltwater intrusion in the Floridan aquifer system, northeastern Florida.* In Georgia Water Resources Conference, Athens, GA.

United States Army Corps of Engineers. (2004). *Lineament analysis, South Florida region. Draft technical memorandum prepared by the USACE-SAJ.* Jacksonville, FL: US Army Corps of Engineers.

United States Army Corps of Engineers. (2014). *Central and Southern Florida Project, Comprehensive Everglades Restoration Plan, Final Technical Data Report, Aquifer Storage and Recovery Regional Study, October 2014.* Jacksonville, FL: US Army Corps of Engineers.

Vernon, R. O. (1951). Geology of Citrus and Levy Counties, Florida. *Florida Geological Society, Geological Bulletin, 35,* 256.

Walsh, V. (2012). Geochemical Determination of the Fate and Transport of Injected Fresh Wastewater to a Deep Saline Aquifer. *FIU Electronic Theses and Dissertations, 692.*

Walsh, V., & Price, R. M. (2010). Determination of vertical and horizontal pathways of injected fresh wastewater into a deep saline aquifer (Florida, USA) using natural chemical tracers. *Hydrogeology Journal, 18,* 1027-1042. http://dx.doi.org/10.1007/s10040-009-0570-8

Wilcox, W. M., Solo-Gabriele, H. M., & Sternberg, L. O. R. (2004). Use of stable isotopes to quantify flows between the Everglades and urban areas in Miami-Dade County Florida. *Journal of Hydrology, 293*(1-4), 1-19. http://dx.doi.org/10.1016/j.jhyrol.2003.12.041

Wise, W. R., Annable, M. D., Walser, J. A. E., Switt, R. S., & Shaw, D. T. (2000). A wetland-aquifer interaction test. *Journal of Hydrology, 227*(1-4), 257-272. http://dx.doi.org/10.1016/S0022-1694(99)00188-2

Yassin, R. R., Muhammad, R. F., Taib, S. H., & Al-Kouri, O. (2014). Application of ERT and aerial photographs techniques to identify the consequences of sinkholes hazards in constructing housing complexes sites over karstic carbonate bedrock in Perak, peninsular Malaysia. *Journal of Geography and Geology, 6*(3), 55. http://dx.doi.org/10.5539/jgg.v6n3p55

Appendix A.

2005 Rankings/Descriptions of Standardized Threats to High Priority Fish and Aquatic Invertebrate Species in Georgia related to Groundwater Withdrawals and Injections in the FAS. (* from Albanese et al. (2015); threats from 2005 SWAP Plan particularly relevant to aquatic species in bold red font.)

2. Incompatible Agricultural Practices

Includes agricultural practices that impact the environment well outside the actual agricultural operation through releases of excess nutrients, toxins, or sediments. Includes practices that degrade stream or wetland habitat quality.

3. Altered Fire Regimes:

Includes fire exclusion, fire suppression, alteration of habitats through unnatural timing, Frequency, or intensity of prescribed burns, and other incompatible fire management practices. Fire regimes are affected by altered community composition (e.g., increase of non-pyric species such as oak) and habitat fragmentation. Fire is an important ecological process that drives many of the terrestrial habitats in Georgia.

4. Altered Hydrology

Includes construction and use of ditches, levees, dikes, and drainage tiles, flow diversion,

dredging, channelization, filling of wetlands and headwater streams, destabilization of stream banks or channels, head-cutting, and other alterations to stream morphology or hydrologic regimes. Results in degradation or destruction of aquatic and wetland habitats.

5. Altered Water Quality

Includes various forms of point and non-point source pollution, such as herbicides, pesticides, sediments, nutrient loading, and thermal modifications that directly impact water quality. Sources are quite varied and include waste water discharges, excessive soil disturbance near streams, increased impermeable surface area resulting from development, and loss of vegetation in riparian buffers.

6. Commercial/Industrial Development

Includes development of structures and infrastructure (buildings, utilities, driveways and roads) for commercial or industrial purposes, usually in an urban setting. Impacts may include direct habitat destruction, fragmentation, altered thermal regimes, and indirect pollution sources that alter water quality.

7. Conversion to Agriculture

Includes the conversion of natural habitats to anthropogenic habitats managed for agricultural crops, pasture, horticulture, or silviculture. Usually involves removal of native vegetation, site preparation, and planting of off-site or non-native species. Results in habitat destruction or fragmentation and may impact water quality.

8. Dam and Impoundment Construction

Includes the construction of dams and impoundments (from agricultural ponds to large reservoirs) that directly affect stream flows and fragment aquatic habitat. Results in impacts to the impounded portion of the stream as well as habitats above and below the dam.

9. Development of Roads or Utilities

Includes construction of new roads (interstate highways, state highways, and county roads) and utility right-of-ways (e.g., electrical transmission lines, water/sewer, gas pipelines) that result in habitat destruction or fragmentation and creation of new avenues for invasion by exotic species.

10. Disease

Includes fatal or debilitating disorders resulting from infections, poisons, pathogenic microorganisms, or parasites. The most serious impacts generally result from introduced vectors or pathogens (e.g., sudden oak death, hemlock wooly adelgid, chestnut blight). Impacts can be devastating to the species directly attacked as well as natural communities.

11. Excessive Groundwater and Surface Water Withdrawal

Includes direct groundwater and surface water withdrawals for agricultural, industrial, and municipal water supplies. Excessive withdrawal can result in lowered water tables, diminished local aquifer discharges, and reductions in water available to sustain stream base flows, spring discharges, isolated wetlands, karst environments, and seepage communities.

14. Incompatible Forestry Practices

Involves poor forestry practices that impact species of concern. This includes failure to follow BMPs and site management activities that result in altered structure and composition of adjacent natural habitats or degraded stream or wetland habitats.

15. Global Warming/Climate Change

Defined as consistent, directed change in climatic conditions at regional scales. Such changes may include increases or decreases in average temperatures, changes in the rates, distribution, frequency, or timing of precipitation, and frequency and intensity of storm events. Local effects are often difficult to quantify.

18. Incompatible Mining/Mineral Extraction

Includes extraction of minerals, oil, or gas or similar activities that result in the disturbance or destruction of natural habitats as well as secondary impacts such as sedimentation or releases of toxins. Impacts may include increased sediment loads, downstream scouring, habitat destruction and disturbance, fragmentation, and creation of migration routes for invasive exotic species.

19. Incompatible Road/Utility Management

Includes management of roads or utility corridors that results in excessive releases of sediment or provides access for non-native species, as well as vegetation management practices that are environmentally "unfriendly" (e.g. indiscriminant use of herbicides).

20. Industrial/Municipal Pollution

Includes toxins and air-borne pollutants, thermally altered effluent, and other point source pollutants derived from industrial/commercial land uses in an urban or suburban setting. Involves direct impacts in the form of chemical or thermal stresses to species or natural communities.

21. Invasive/Alien Species

Includes exotic species as well as native species that have become invasive due to past habitat alterations (e.g. hardwood encroachment of long leaf pine habitats following fire suppression). Impacts include competition, hybridization, and predation as well as long-term alterations of ecological systems and processes (e.g. hydrologic changes, changes in soil attributes, altered fire regimes).

23. Residential Development

Includes primary and secondary home construction as well as development of associated infrastructure (e.g. subdivision roads and driveways, sewer and stormwater utilities). Impacts may include habitat destruction, disturbance, fragmentation, and introduction of invasive species."

9

Rapid Assessment Method of Flood Damage Using Spatial-Statistical Models

Abdul Hamid Mar Iman[1] & Edlic Sathiamurthy[2]

[1] Sustainable Environment and Conservation Cluster, Faculty of Agro-Based Industry, Universiti Malaysia Kelantan, Malaysia

[2] School of Marine Science and Environment, Universiti Malaysia Terengganu, Kuala Terengganu,Terengganu, Malaysia

Correspondence: Abdul Hamid Mar Iman, Sustainable Environment and Conservation Cluster, Faculty of Agro-Based Industry, Universiti Malaysia Kelantan, 17600 Jeli, Kelantan, Malaysia. E-mail: hamid.m@umk.edu.my; edlic@umt.edu.my

Abstract

Attention to damage assessment is always a priority especially in cases of natural disaster. The state of Kelantan is known to be one of a few Malaysian states with noticeable natural disaster, in particular, flood. In December 2014, an extraordinary magnitude of flood – nicknamed as yellow flood – struck the state causing hundreds of million ringgit of damage to properties. The purpose of this study is to demonstrate a spatial approach to estimating property damage incurred by flood. By selecting a badly affected area, GIS was used to map geo-referenced flood-hit location in Kuala Krai, Kelantan. Flood hazard was modelled and superimposed on estimated property damage. GIS spatial technique was then employed to estimate the flood damage incurred. This study, however, did not make a complete damage assessment of the properties but rather focusing on the methodology of damage assessment to show how it can be implemented. In conclusion, GIS spatial technique can generally be used to provide flood damage rapid assessment method.

Keywords: damage assessment, natural disaster, flood, property damage

1. Introduction

The December 2014's flood has caused huge damage of close to RM 1 billion to the country, exclusive of RM 78 million for cleaning operations in Kelantan. A report quoted that about RM 200 million was estimated for the damage of infrastructure in Kelantan (The Star, 2/2/2015). According to Urban Wellbeing, Housing and Local Government Minister, Datuk Rahman Dahlan, between 2,000 and 3,000 houses in Kelantan were destroyed in the worst flood ever in decades (Azura, 2015). More than 200,000 victims were affected by the massive flood which claimed 21 lives (Anon, 2015).

One of the main concerns of flood is to estimate the extent of damage to properties and other assets. It is an intricate task to perform since damage assessment needs itemized identification and estimate of affected objects. Some studies resort to only assessing flood impact without being able to provide the monetary estimate of the damage (see for e.g. Ab-Jalil and Aminuddin, 2006; Pradan, 2009). Therefore, it is vitally important to devise a rapid assessment method that can provide a reliable method for estimating the monetary loss as soon as flood strikes in a particular location. Flood damage assessment itself is not a new thing; there has been a substantial body of literature dealing with it. However, the techniques are difficult to generalize since they vary and case-to-case.

By applying empirical damage or loss functions meant for compensation, relief, and/or insurance purposes, flood damage rapid assessment method (FD-RAM) seeks to estimate the expected monetary damage as soon as a disaster strikes (Poser and Dransch, 2010). In case of flood, these models calculate the expected damage as a function of inundation depth, building characteristics, and possibly further parameters such as water contamination (Poser and Dransch, 2010).

2. Theoretical Background

2.1 Flood Damage Model

For any property, expected physical damage (EPD) is generally modelled as:

$$EPD = f(SD, CD) \tag{1}$$

where SD is structural damage and CD is content damage. SD comprises damage to land/soil and building while CD can refer to any type and/or amount of 'content' asset. Therefore, 'content' can comprise any moveable asset inside or outside a building such furniture, radio, television, appliance, vehicle, clothes, money, etc. Damage to land/soil is difficult to ascertain. For example, the eroded soil of a land parcel may need to be replaced. Consequently, it incurs re-fill cost. However, the amount of nutrients that is being washed away from a farm as well as re-fill cost are difficult to measure. In the same manner, the number of trees/crop damaged by flood is not easy to quantify.

To overcome the above difficulty, a survey based approach is proposed adopting the model as shown in equation (1). A sample survey needs to be conducted to collect data on the quantum of damage of each property or item at a particular site. For landed properties such as residential, office, and commercial, structural as well as content damages are taken as some percentages of property value. In general, equation (1) can be re-expressed as:

$$EPD = SD + CD = (.p1*ALV + .p2*ABV) + .p3*(ALV + ABV) \tag{2}$$

where ALV = assessed land value; ABV = assessed building value; .p1, .p2, and .p3 = certain defined "proportion" or "percentage" property component's damage in decimal form. ALV, ABV, and any other 'content' asset can be estimated by replacement cost approach. Alternatively, market value (MV) of property can be used in place of ALV and ABV if sales data are available.

For agricultural properties, damage can occur to land/soil (structure) and tree/crop (content). Again, it is difficult to ascertain damage to these elements. For compensation purposes, land/soil damage can be estimated as a percentage of market value of a particular type of agricultural property but tree/crop damage is much more difficult to estimate. The general formula for damage estimation of agricultural properties with immature trees/crop is modified from equation (2) as follows:

$$EPD = SD + CD = land/soil + tree/crop = .q1*MV + n[(c-d)(1+i)^t] \tag{3}$$

where MV = market value of a particular type of agricultural property (alternatively, actual replacement cost can be used); .q1 = a defined proportion in decimal form; c = cost of replacement new of the tree/crop; i = discounting rate; t = age of immature crop; n = number of damage trees/crop.

However, this formula cannot be used directly without modification based on the type of agricultural property under view. For example, damage to annual and perennial crop such as banana, maize, rubber, oil palm, cocoa, and orchard trees need to be estimated by "individual" tree counting – a daunting, if not impossible, task in FD-RAM. As another example, the immaturity period is different for different crops. For instance, the immature period for oil palm is four years, rubber five years, while for some orchard trees, this period may be up to seven years.

A sample survey in the disaster area is needed in order to compute the reasonable figures of all the above damage components. Specifically, a priori information is needed to compute .p1, .p2, and .p3.

2.2 Rapid Damage Assessment Procedure

The whole procedure of rapid assessment of flood damage is part of the general concept of decision support system promoted by Malczewski (1997). Ideally, it should become part of national disaster management programs of any country troubled by the disaster. The actual implementation of flood damage rapid assessment method is rather complex. It has two main components, namely mapping component and spatial modelling component.

The mapping component has the following mapping activities: boundary of study area; and distribution of poor population; sampling points to compute asset value, particularly building and land value. Geographic Information System (GIS) is a standard method for flood mapping through various kinds of software such as ArcGIS, MapInfo, Idrisi, etc. One of the most widely used GIS software is Environmental System Research Institute's (ESRI) ArcGIS 10.x.

The spatial modelling process has the following modelling activities: flood inundation coverage/flood modelling based on rainfall-runoff method; spatial flood damage-estimating model; and general damage estimate. Fundamentally, we can specify flood damage-estimating model in a number of ways (Messner et al., 2007; Merz *et al.*, 2010; Green *et al.*, 2011). Factors such as flood depth, velocity, duration, water contamination, precaution, and warning time can be included. However, inclusion of flood factors cannot be generalized and is very much determined by data availability.

One potential spatial flood damage-estimating model is Geographically Weighted Regression (GWR) originally developed by Fotheringham *et al.* (2000; 2002; 2005). Suppose we had some location in the study area, perhaps one of the data points, where (x,y) are the coordinates of its position. We can rewrite the model, in vector form as:

$$V(x,y)W = W(x,y)a + W(x,y)Z + W(x,y)e \qquad (4)$$

where V is value of damage, a is regression's intercept, Z represents hydrological, physical, environmental, and socio-economic variables/factors, W is spatial weight matrix, e is error term, and is some measure of spatial component of data points. This relationship is fitted by least squares to give an estimate of the parameters at the location (x,y) and a predicted value. This is achieved through the implementation of the geographical weighting scheme. The weighting scheme is organized such that data nearer (x,y) is given a heavier weight in the model than data further away.

Using OLS, the parameters for a linear regression model is obtained by solving:

$$\beta = (Z^{T}Z)^{-1}Z^{T}V \qquad (5)$$

The parameter estimates for GWR are solved using a weighting scheme:

$$\beta(g) = (Z^{T}W(g)X)^{-1}Z^{T}W(g)V \qquad (6)$$

The weights are chosen such that those observations near the point in space where the parameter estimates are desired have more influence on the result than observations further away. Two functions we have used for the weight calculation have been (a) bi-square and (b) Gaussian. In the case of the Gaussian scheme, the weight for the i_{th} observation is:

$$w_i(g) = \exp(-d/h)^2 \qquad (7)$$

where d is the Euclidean distance between the location of observation i and location g, and h is a quantity known as the bandwidth. (There are similarities between GWR and kernel regression). One characteristic that is not immediately obvious, is that the locations at which parameters are estimated need not be the ones at which the data have been collected.

The resulting parameter estimates are mapped in order to examine local variations in the parameter estimates. One might also map the standard errors of the parameters estimates as well. Hypothesis tests are possible - for example one might wish to test whether or not the variations in the values of a parameter in the study area are due to chance. The bandwidth may be either supplied by the user, or estimated using a technique such as cross validation technique. The (x,y)s are typically the locations at which data are collected. This allows a separate estimate of the parameters to be made at each data point. The resulting parameter estimates can them be mapped.

Flood Loss Estimation Model for the private sector (FLEMOps) on the meso scale (Thieken et al., 2008) is applied with some adaptation to the location situations. This model calculates the damage ratio for residential buildings as a function of inundation depth classified into five classes and building characteristics, i.e. three buildings types and two building qualities. To be applicable on the meso scale, mean building composition and the mean building quality per municipality were derived and the resulting damage ratios are multiplied by total asset values disaggregated to land use units (Thieken et al., 2005).

Spatially assessed flood damage by kriging technique is used in performing data analysis. A modified Ordinary Least Squares technique, kriging adopts weights to the surrounding measured values to derive a prediction for an unmeasured location. The general formula for both interpolators is formed as a weighted sum of the data:

$$\hat{Z}_{(S_0)} = \sum_{i=1}^{N} \lambda_i Z(S_i) \qquad (8)$$

where $\hat{Z}_{(S_0)}$ = weighted sum of values; $Z(S_i)$ = the measured value at the i*th* location; λ_i = an unknown weight for the measured value at the ith location; s_0 = the prediction location; N = the number of measured values.

In the kriging technique, the weights (represented by λ_i) are based on both the distance between the measured points and the prediction location and also the overall spatial arrangement of the measured points. To use the spatial arrangement in the weights, the spatial autocorrelation must be quantified.

In the ordinary kriging, the weight, λ_i depends on a fitted model to the measured points, the distance to the prediction location, and the spatial relationships among the measured values around the prediction location. The following section briefly discusses how the ordinary kriging formula is used to create a map of the prediction surface and a map of the accuracy of the predictions.

There are a number of kriging techniques discussed in the literature. However, to avoid cumbersome discussion, we would only adopt ordinary kriging in this study. Ordinary kriging estimates the unknown value using weighted linear combinations of the available sample (Isaaks and Srivastava, 1989):

$$\hat{v} = \sum_{j=1}^{n} w_j * v \quad \sum_{i=1}^{n} w_i = 1 \tag{9}$$

The error of ith estimate, r_i, is the difference of estimated value and true value at that same location:

$$r_i = \hat{v} - v_i \tag{10}$$

The average error of a set of k estimates is:

$$m_\tau = \frac{1}{k}\sum_{i=1}^{k} r_i = \frac{1}{k}\sum_{i=1}^{k} \hat{v}_i - v_i \tag{11}$$

The error variance is:

$$\delta_R^2 = \frac{1}{k}\sum_{i=1}^{k}(r_i - m_R)^2 = \frac{1}{k}\sum_{i=1}^{k}\left[\hat{v}_i - v_i - \frac{1}{k}\sum_{i=1}^{k}(\hat{v}_i - v_i)\right]^2 \tag{12}$$

However, we cannot use the equation because we do not know the true value $V_1,...,V_k$. In order to solve this problem, we apply a stationary random function that consists of several random variables, $V(x_i)$. X_i is the location of observed data for $i > 0$ and $i \leqq n$. (n is the total number of observed data). The unknown value at the location X_0 we are trying to estimate is $V(x_0)$. The estimated value represented by random function is:

$$\widetilde{V}(x_0) = \sum_{i=1}^{n} w_i * V(x_i)$$

$$R(x_0) = \widetilde{V}(x_0) - V(x_0) \tag{13}$$

The error variance is:

$$\widetilde{\delta}_R^2 = \widetilde{\delta}^2 + \sum_{i=1}^{n}\sum_{j=1}^{n} w_i w_j \widetilde{C}_{ij} - 2\sum_{i=1}^{n} w_i \widetilde{C}_{i0} + 2\mu(\sum_{i=1}^{n} w_i - 1) \tag{14}$$

$\widetilde{\delta}^2$ is the covariance of the random variable $V_{(X0)}$ with itself and we assume that all of our random variables have the same variance while μ is the Lagrange parameter.

In order to get the minimum variance of error, we calculate the partial first derivatives of the equation (11) for each w and setting the result to 0. The example of differentiation with respect to w is:

$$\frac{\delta(\widetilde{\sigma}_R^2)}{\delta w_1} = 2\sum_{j=1}^{n} w_j \widetilde{C}_{1j} - 2\widetilde{C}_{10} + 2\mu = 0 \quad \sum_{j=1}^{n} w_j \widetilde{C}_{1j} + \mu = \widetilde{C}_{10} \tag{15}$$

All of weight w_i can be represented as:

$$\sum_{j=1}^{n} w_i \widetilde{C}_{ij} + \mu = \widetilde{C}_{i0} \tag{16}$$

For each i, $1 \leq i \leq n$

We can get each weight W_i through equation (13). After getting the value, we can estimate the value located in X_0. We can use variogram instead of covariance to calculate each weight of equation (12). The variogram and minimized estimation variance are:

$$\gamma_{ij} = \tilde{\delta}^2 - \tilde{C}_{ij} \tag{17}$$

$$\tilde{\delta}_R^2 = \sum_{i=1}^{n} w_i \gamma_{i0} + \mu$$

The kriging module includes two variogram models:

Spherical

$$\tilde{\gamma}(h) = \begin{cases} C + C\left(1.5\dfrac{h}{a} - 0.5\left(\dfrac{h}{a}\right)\right)^3 & \text{if } |h| \leq a \\ C_0 + C_1 & \text{if } |h| > a \end{cases} \tag{18}$$

Exponential

$$\tilde{\gamma}(h) = \begin{cases} 0 & \text{if } |h| = a \\ C_0 + C_1\left(1 - \exp\left(\dfrac{-3|h|}{a}\right)\right) & \text{if } |h| > a \end{cases} \tag{19}$$

Nugget effect **(c_0)**

Though the value of the variogram for $h = 0$ is strictly 0, several factors, such as sampling error and short scale variability, may cause sample values separated by extremely small distances to be quite dissimilar. This causes a discontinuity at the origin of the variogram. The vertical jump from the value of 0 at the origin to the value of the variogram at extremely small separation distances is called the nugget effect (Isaaks and Srivastava, 1989).

Range **(a)**

As the distance of two pairs increases, the variogram of those two pairs also increases. Eventually, the increase of the distance cannot cause the variogram to increase. The distance which causes the variogram to reach plateau is called range (see Figure 1).

Sill **($C_0 + C_1$)**

It is the maximum variogram value which is the height of plateau (see Figure 1).

Distance **h**

It is the distance between estimated location and observed location.

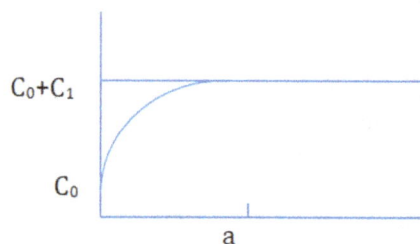

Figure 1. An Example of Exponential Variogram Model

Equation (16) can be written in matrix notation as **V * W = D** where V is (n+1) x (n+1) matrix which contains the variogram of each known data. The components of last column and row are 1 and the last component of the matrix is 0; W is (n+1) matrix which contains the weight corresponding to each location. the last of component of matrix is Lagrange parameter; and D is (n+1) matrix which contains the variogram of known data and estimated data. The last component of the matrix is 1.

Since V and D is known, we can get the unknown matrix W by $\mathbf{W = invert(V) * D}$. Applying equation (13), we can get the estimated value on a specific location. We also can get the error variance from the square root of equation (17).

3. Methodology

A sample-based flood damage survey was conducted in early 2015 in Kuaka Krai and Dabong. This study area was chosen because it was the most severely-hit sub-region of the state of Kelantan. Furthermore, state-wide FD-RAM was not possible due to data and financial limitations. Sample-based field inspections were conducted to estimate flood damage to buildings, trees, and other items. Since it was very difficult to account for each item damaged by flood, this study was confined only to estimating damages of residential and agricultural properties. Some moveable assets categorised as "contents" (e.g. furniture, house appliance, equipment), were, however, accounted for. As many as 336 geo-referenced sites (longitude and latitude in metres) within the flood inundated river corridors were sampled and mapped as "survey points" shape file (see **Figure 2**).

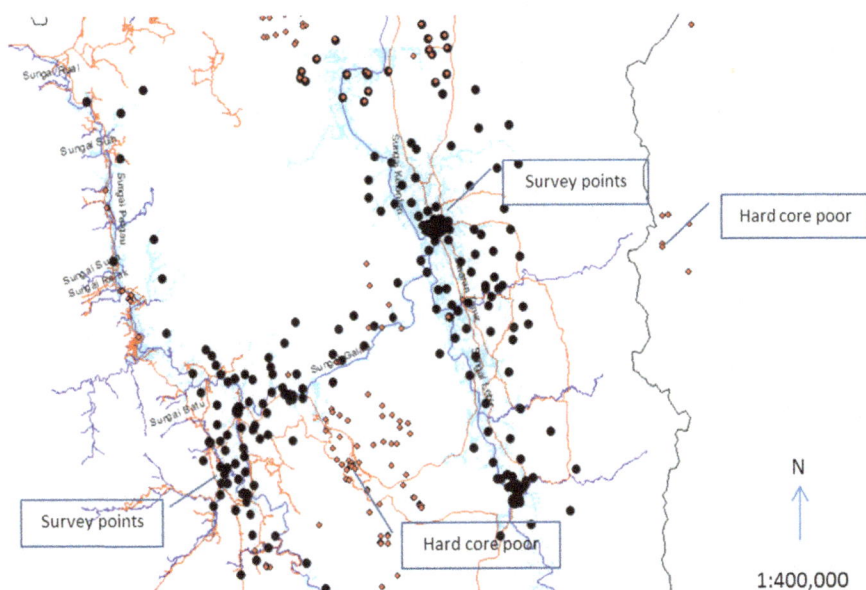

Figure 2. Survey points shape file in the selected study area. These survey points include locations of some hard core poor's homes (smaller dots)

Data on flood-related factors were collected at each sampled location, namely land value (asking price) (RM/acre); building value (replacement cost new) (RM/unit); proportion of structural damage (%); proportion of content damage (%); current use (forest, agriculture, natural vegetation, urban, transport, built-up); use activity (rubber, oil palm, orchard, water body, road, vacant, residential); structural type (soil, building); content type (tree, building, miscellaneous items); and flood depth (feet). All of the information was formatted as attribute table of the "survey points" shape file in ArcGIS software. The purpose of this shape file was to enable spatial modelling of flood damage using Geographically Weighted Regression (GWR) technique based on the following specification:

$$TotDmg = f(Curuse, Activ, Structy, Contyp, Flo\_dep) \qquad (20)$$

where TotDmg = Total flood damage (RM); Curuse = Current use; Acti = Use activity; Structy = Structural type; Contyp = Content type; and Floo_dep = Flood depth (feet).

Damage estimation according to of property type is given as in equations (1) and (2) above. Spatially assessed flood damage by kriging technique was used in performing data analysis.. Flood damage was calculated as follows:

TotDmg = ContDmg1 + ContDmg2 + StrDmg1 + StrDmg2

where ContDmg1 = CD_P x Buildv x ef; ContDmg2 = CD_P x Landv x ef; StrDmg1= SD_P x Buildv x ef; StrDmg2 = SD_P x Landv x ef. [ef = 1 IF sampled point = Residential/building; ef = 0 IF sampled point =

Agriculture/forest]

where CD_P = % of content damage; SD_P = % of structural damage; Landv = land value (RM/unit); and Buildv = building value (RM/unit).

In the damage assessment process, the following guide was used:

ContDmg1 = content damage for residential/building

ContDmg2 = content damage for agricultural crop/forest

StrDmg1 = structural damage for Residential/building

StrDmg2 = structural damage for agricultural crop/forest

Building value was estimated based on replacement cost new (RCN) of the original building. This was a challenging process since RCN cannot easily and accurately be estimated. Although the ideal method was to base value estimates on official government valuation, this was not possible due to resource constraints. The regression procedure for the above specification followed the steps as outlined in equations (4) through (17). GWR was run to relate flood damage (content & structural) with their influencing factors, namely current land use (Curuse), land use activity (Activ), property structural type (Structy), and flood depth (Flo_dep).

Once outputs were generated, superimposition process was performed whereby land use map was overlaid on modelled flood, and GWR-kriged flood damage estimate. A manual process of identifying, listing, and estimating damages of various types of properties was carried out using this superimposed map. (See example in Figure 3.)

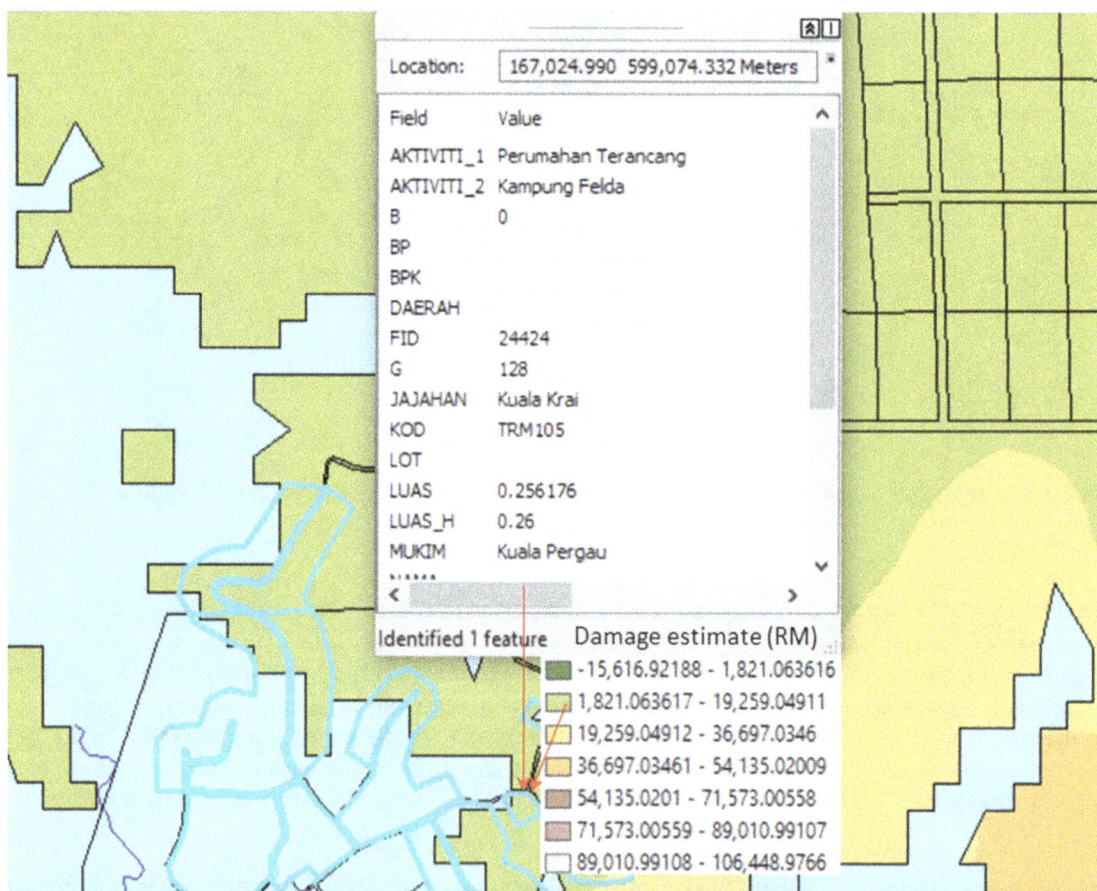

Figure 3. Screen shot of an overly of land use, modelled flood, and GWR-kriged flood damage estimate

4. Results and Discussion

Figure 4 shows flood hazard superimposed on GWR-kriged flood damage map over the study area.

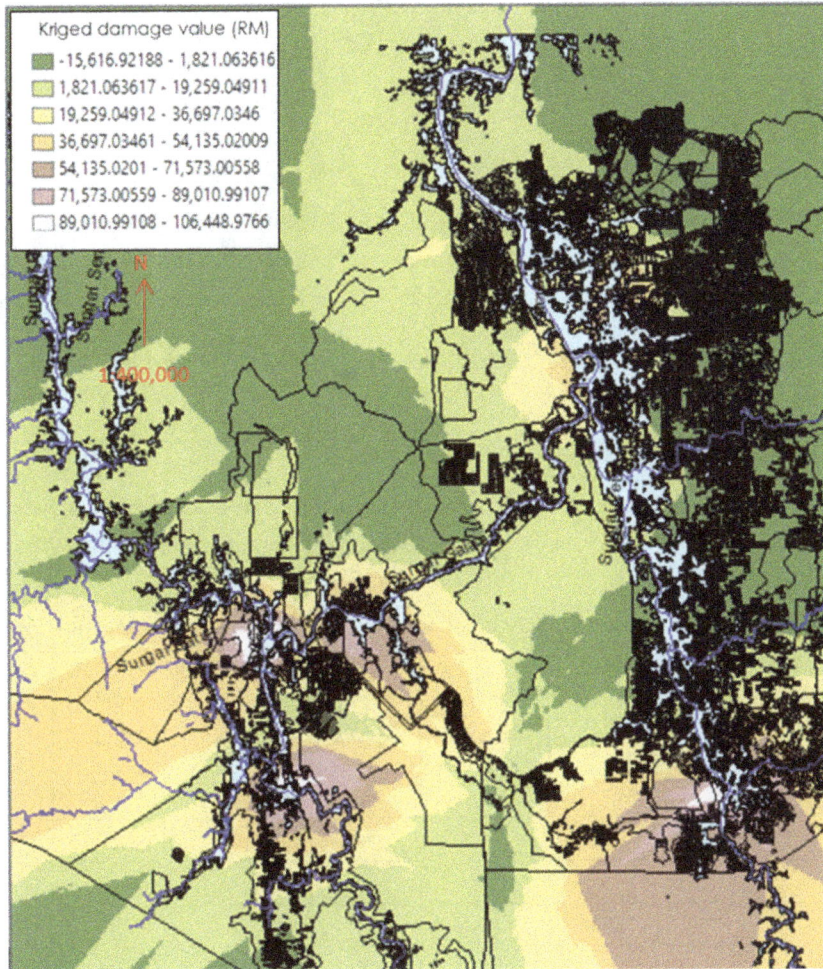

Figure 4. Flood hazard superimposed on GWR-kriged flood damage map

Flood damage and, thus, flood risk is higher in densely populated locations such as urban or residential areas. Besides, sites closer to river banks (say, less than 1 km) were mostly exhibited greater flood depth. Other factors also contribute to the magnitude of damage.

The regression results are shown in Table 1. The performance of the GWR was very modest with a local R^2 of only 0.58. This reflects the shortcoming in modelling spatial relationship of flood damage since, apart from land use factors, many other hydrological and geomorphological factors were not included in the model specification due to data limitation.

From **Table 1**, flood depth was found to be significantly influencing flood damage. Content type was also significant to property damage while other land use factors did not show statistical significance. With respect to content type, miscellaneous contents of moveable property such as furniture and appliances could have incurred damage of RM 31,221 more than other types of contents such as trees and vegetation whenever there was flood inundation in the study area.

Table 1. General statistical results of Geographically Weighted Regression (GWR)

VARNAME	VARIABLE	DEFINITION
Bandwidth	26,115.340592	
Residual Squares	506,075,220,856.614	
Effective Number	15.870035	
Sigma	3,9697.86226	
AICc	8,105.709384	
Dependent: TotDmg		Total flood damage (RM)

Local R^2	0.58					
R^2 Adjusted	0.56					
Residual	1,480.77					
Standard Error	38,692.17					
Std. Residual	0.03					
Sample size	336					
	Coefficient	Std. error	t-value	Min	max	95% confidence
Intercept	-26,585.93	5,681.39	-4.68	-31,286.39	-16,581.87	332.69
Current use (Curuse)	24,545.44	21,285.12	1.15	18,093.14	42,160.54	574.24
Activity (Acti)	13,029.86	17,976.61	0.72	-4,936.35	21,905.10	523.05
Structural type (Structy)	8,150.71	19,128.44	0.43	-5,129.26	32,765.70	1,103.62
Content type (Contyp)	31,221.04	18,104.81	1.72	15,234.98	36,765.76	396.62
Flood depth (Floo_dep)	5,547.52	873.03	6.35	3,659.86	6,808.20	97.33

By manually using the GIS map, various types of properties were identified and listed together with their corresponding damage (see Table 2). Many places were severely inundated, more than 70% in some cases.

Table 2. Flood inundation over some selected land uses in the study area – GIS analysis

Land use	Total area (ha.)	Total inundated area (ha.)	Approx (%)	Structural Damage (%)	Content Damage (%)	Area Affected (structural) (ha.)	Area Affected (content) (ha.)
Kediaman:							
Kampung Felda	310.16	92.26	30	0	45	0.00	1.45
Kampung Setinggan	0.67	0.18	27				
Kampung Tersusun	112.36	90.33	80	55	61	0.76	0.85
Kampung Tradisi	147.21	128.23	87	70	72	6.00	6.18
Perumahan Strata	0.03	0.03	100				
Perumahan Bukan Strata	56.89	43.01	76	80	94	1.00	1.17
Perumahan Kakitangan	10.54	10.27	97				
Perumahan Ladang/Estet	42.11	11.52	27				
Perniagaan dan Perkhidmatan:							
Perniagaan Terancang	31.71	20.91	66	50	60	0.02	0.02
Perniagaan Tidak Terancang	30.26	22.17	73				
Pertanian:							
Getah	74233.19	46415.61	63	30	7	5326.82	1242.92
Kelapa Sawit	8825.22	5947.63	67	0	11	0.00	354.53
Padi	373.67	172.03	46				
Dusun	5671.7	2795.32	49	63	90	3.71	5.30
Tanah Terbiar (Pertanian tidak diusahakan)	746.82	643.37	86	0	5	0.00	0.39
Industri:							
Industri Terancang	94.44	66.68	71				

Industri Tidak Terancang	71.77	51.42	72				
Infrastruktur dan Utiliti:							
Bekalan Air	8.38	7.1	85				
Bekalan Elektrik	352.68	270.79	77				
Pengairan dan Perparitan	12.4	6.87	55				
Telekomunikasi	2.06	1.11	54				
Institusi dan Kemudahan Masyarakat:							
a) Keagamaan							
Masjid	21.45	10.46	49				
Surau	2.6	1.31	50				
Tokong	0.36	0.36	100				
Kuil	0.27	0.16	59				
b) Kegunaan Kerajaan/Badan Berkanun:							
Pejabat Kerajaan/Agensi Kerajaan	46.15	35	76	50	60	0.03	0.03
Badan Berkanun	37.5	37.4	100	50	60	0.03	0.03
c) Keselamatan							
Balai Polis	5.02	5.02	100				
Balai Bomba	0.4	0.4	100				
Pondok Polis	4.71	3.89	83				
Kem Tentera	7.3	7.3	100				
d) Kesihatan							
Klinik Kesihatan	9.07	9.07	100				
Klinik Desa	6.3	2.23	35	0	0	0.00	0.00
Hospital	4.23	4.23	100				
e) Pendidikan							
Tadika	4.82	2.02	42				
Sekolah Rendah	130.07	91.36	70	30	50		
Sekolah Menengah	55.8	49.24	88				
Sekolah Agama	13.76	9.31	68				
Institut Latihan	0.33	-					
f) Perkuburan							
Islam	46.3	34.75	75	60	65		
Cina	11.06	11.06	100				
Hindu/Sikh	0.58	0.58	100				
g) Lain-lain Kemudahan Masyarakat							
Balai Raya	1.75	1.25	71				
Dewan Serbaguna Awam	0.79	0.26	33				

Dewan Orang Ramai	0.95	0.67	71				
Perpustakaan Awam	1.66	0.39	23				
Lain-lain:	0.85	0.21	25				
Pusat Rukun Tetangga Kuala Krai							
Pusat Aktiviti Rukun Tetangga							
Pusat Sumber KEMAS / Pusat Literasi Komputer							
Pusat Kominiti Desa							
Dewan Rukun Tetangga Taman Gucil Jaya							
Pengangkutan:							
Jalan	1338.65	1082.19	81	60	56		
Stesen Bas	0.29	0.29	100				
Stesen Keretapi	4.36	4.36	100				
Penternakan dan Akuakultur	44.67	2.37	5				
Tanah Kosong	836.17	633	76	63	46		
Hutan	28916.39	70860.78	55	0	0		
Tanah Lapang dan Rekreasi	900.69	835.07	93	55	54		

* Expressed as number of units rather than area of land (ha.)

　　　　　\No data were available on the map

To further illustrate the use of FD-RAM, Figure 5 took a group of hard core poor people as a case. The map indicates that the hard core poor group experienced low to severe flood damage. Most of them experienced a total flood damage of about RM 10,000/household. This a was quite small figure and was not surprising as many of them did not own high-value property. Nonetheless, this damage was about 26 times their monthly income and can be considered a huge suffering for a hard core poor family. The model, however, suffered from prediction inaccuracy and, thus, overstressing on damage figure may not be desirable due to possible over- or underestimation in the assessment process.

Not all of hard core poor in the study area were affected by flood and, thus, those hit must be identified. This was done by picking the affected hard core poor's homes from the map via clipping menu available in ArcGIS. In this case, modelled "flood polygon" layer was clipped onto "survey points" layer. The resulting clipped layer was then superimposed on another layer, namely kriged estimated total flood damage (ETFD). Figure 5 shows the locational distribution of hard core poor which was superimposed over kriged values of estimated total flood damage (ETFD) modelled using Geographically Weighted Regression based on equation (17). By this way, the hydrological and physical aspects of flood were factored into flood damage-estimating model.

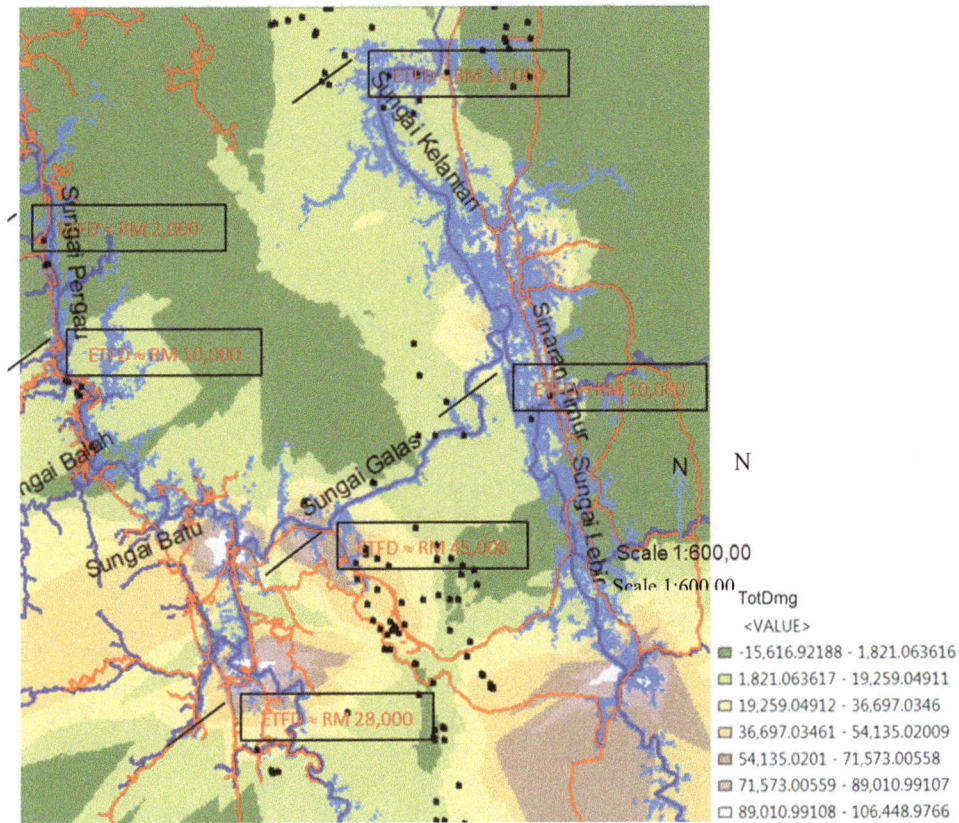

Figure 5. Kriged value of estimated total flood damage (ETFD) based on Geographically Weighted Regression among hard core poor (black dots) in the study area. Figures shown are middle-values of ETFD.

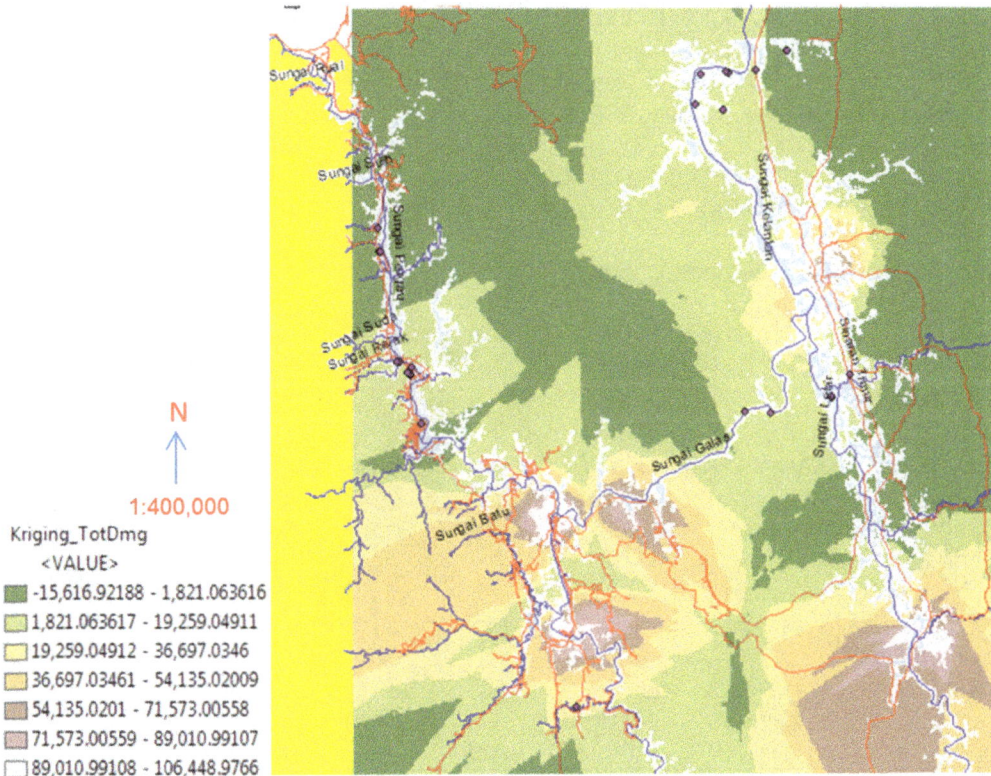

Figure 6. Identification of flood-hit hard core poor by ArcGIS procedure

Table 3. Estimated total flood damage (ETFD) incurred by the hard core poor in the study area

No	Name	Address	Lat	Long	District	Occupation	ETFD (RM)
1.	JAHARAH BT SALLEH	KG LALANG JENAL,KUALA GRIS DAB 18200	5.2477	102.0250	DABONG	0	27,978
2.	MOHD RONI B ZAKARIA	KG KUALA BALAH KUALA BALAH 17610	5.4445	101.9145	KUALA BALAH	0	10,540
3.	MARILA BINTI ISMAIL	KG. AIR BELAGA, MACHNAG 18500	5.4524	102.1638	ULU SAT	0	10,540
4.	SALLAH BIN DAUD	KAMPUNG BUKIT TIU 18500 MACHANG 18500	5.4526	102.1447	ULU SAT	0	10,540
5.	RAHIMAH BINTI SULAIMAN	KG. TELOSAN 16800	5.4632	102.2081	JERAM	0	10,540
6.	ZAKIAH BINTI DOLLAH	KG. TELOSAN 16800	5.4636	102.2083	JERAM	5	10,540
7.	MAJID BIN DAUD	KG BUKIT JERING KUALA BALAH 17610	5.4780	101.9062	KUALA BALAH	4	10,540
8.	ZABIDAH BINTI IBRAHIM	KG. BUKIT JERING KUALA BALAH 17610	5.4787	101.9064	KUALA BALAH	4	10,540
9.	ABDUL MALIK BIN ISMAIL	KG. JERAM 16800	5.4791	102.2211	JERAM	0	10,540
10.	MUHAMMAD BIN ALI	KG. BKT JERING KUALA BALAH 17610	5.4800	101.9048	KUALA BALAH	1	10,540
11.	HALIMAH BINTI MOHAMAD	KG. JERIMBONG KUALA BALAH 17610	5.4835	101.9075	KUALA BALAH	1	10,540
12.	MA KALSOM BINTI OMAR	KG. BUKIT SELAR KUALA BALAH 17610	5.4871	101.8987	KUALA BALAH	1	10,540
13.	HASLI BIN IBRAHIM	KG. RELAK KUALA BALAH 17610	5.4876	101.8970	KUALA BALAH	4	10,540
14.	MOHD ABU BAKAR BIN MAT JIDIN	NO.117, KG. LUBOK BONGOR KUALA BALAH 17610	5.5635	101.8837	KUALA BALAH	0	2,000
15.	YAAKUB BIN MAT MIN	NO. 134, KG. LUBOK BONGOR KUALA BALAH 17610	5.5640	101.8848	KUALA BALAH	4	2,000
16.	MAT YAAKOB BIN SALLEH	KG SG RENYUK KUALA BALAH 17610	5.5799	101.8827	BALAH	0	2,000
17.	MOHAMAD BIN SAHAK	LEPAN PERINGAT 18400	5.6628	102.1289	TEMANGAN	3	10,540
18.	MEK NABLOH BINTI ABDULLAH	KAMPUNG KERILLA 18500	5.6664	102.1090	TEMANGAN	0	10,540
19.	NAZMIAH BT HARON	KG. PASIR SENOR TEMANGAN 18400	5.6874	102.1127	TEMANGAN	0	10,540
20.	ABDULLAH BIN AWANG HAMAT	KAMPUNG TEMANGAN LAMA 18400	5.6878	102.1322	TEMANGAN	0	10,540
21.	ZAINI BIN CHE THE	KG KERILLA 18500	5.6891	102.1305	TEMANGAN	0	10,540
22.	FATIMAH BINTI HASSAN	KAMPUNG PAUH TEMANGAN 18400	5.6902	102.1522	TEMANGAN	0	10,540

Note: Some of data columns were removed to save space

5. Conclusion

Although accurate estimate was not the focus of this study, being able to derive some initial figure of flood damage is an important aspect of emergency relief and recovery program by the authority. The ability of knowing the 'possible' amount of damage at a specific site is an additional useful piece of information to the government.

The usefulness of rapid damage assessment of flood disaster largely depends on the completeness of data and accuracy of damage-estimating model. The correct GWR model specification that will result in satisfactory results was rather difficult and the available body of literature was not that useful to identify all the correct variables to include. Trial and error specification and test of the candidate variables such as those of geomorphological, hydrological, physical demanded a lot of data collection that was not possible due to resource constraint.

Accurate identification of 'itemised objects' affected by flood is always a problem of flood damage estimation. In this study, only content and structural damage of certain types of property/asset were quite conveniently accounted for their respective owners their respective owners their respective owners. Moveable assets such as vehicle, machinery, agricultural tools, etc. were not easily taken into account for various technical reasons. Assignment of damages of crops and animals to their respective owners was also difficult especially for those whose properties/assets were located on different sites away from their living premise.

Estimating flood damage was very challenging particularly in choosing the most appropriate approach of valuation. Cost, market and investment approaches are legitimate bases of asset valuation but none can be suitable for all situations and for all property types. Detailed examination of the property is thus necessary before deciding on the appropriate approach to valuation. This was simply not possible in rapid damage assessment procedure.

References

Abd, J. H., & Aminuddin, A. G. (2006). Development of flood risk map using GIS for Sg. Selangor Basin. Retrieved from http://www.redac.eng.usm.my.html

Azura, A. (2015). RM78mil to clean post-flood Kelantan. *New Straits Times*, 7(January).

Fotheringham, A. S., Brunsdon, C., & Charlton, M. (2002). *Geographically Weighted Regression the Analysis of Spatially Varying Relationship*. John Wiley & Sons, LTD.

Fotheringham, A. S., Brunsdon, C., & Charlton, M. (2005). *Geographically Weighted Regression*. ESRC National Centre for Research Methods, June, University of Leeds.

Fotheringham, A. S., Brunsdon, C., & Charlton, M. E. (2000). *Quantitative Geography*, London: Sage.

Green, C. H., Viavattene, C., & Thompson, P. (2011). *Guidance for Assessing Food Losses*. CONHAZ Report, Flood Hazard Research Centre – Middlesex University, Middlesex, WP6 Report.

Isaaks, E. H., & Srivastava, R. H. (1989). *An Introduction to Applied Geostatistics*.

Malczewski, J. (1997) Spatial Decision Support Systems, NCGIA Core Curriculum in GIScience. Retrieved from http://www.ncgia.ucsb.edu/giscc/units/u127/u127.html

Merz, B., Kreibich, H., Schwarze, R., & Thieken, A. (2010). Assessment of Economic Flood Damage. *Natural Hazards and Earth System Science, 10*, 1697–1724.

Messner, F., PennningRowsell, E. C., Green, C., Meyer, V., Tunstall, S. M., & van der Veen, A. (2007). *Evaluating flood damages: Guidance and recommendations on principles and methods*, FLOODsite, Wallingford, UK, T09-06-01.

Poser, K., & Dransch, D. (2010). Volunteered Geographic Information for Disaster Management with Application to Rapid Flood Damage Estimation. *Geomatica, 64*(1), 89-98.

Pradhan, B. (2009). Flood Susceptible mapping and risk area delineation using logistic regression, GIS and remote sensing. *Journal of Spatial Hydrology, 9*(2), 1–18.

Thieken, A. H., Muller, M., Kreibich, H., & Merz, B. (2005). Flood damage and influencing factors: New insights from the August 2002 flood in Germany. *Water Resources Research, 41*(12), W12430+.

Thieken, A. H., Olschewski, A., Kreibich, H., Kobsch, S., & Merz, B. (2008). Development and evaluation of FLEMOps a new Flood Loss Estimation Model for the private sector. In D. Proverbs, C. A. Brebbia, and E. Penning-Rowsell, editors, *Flood Recovery, Innovation and Response I*.

10

Assessing the Impact of Projected Climate Change on Zoo Visitation in Toronto (Canada)

Micah J. Hewer[1] & William A. Gough[2]

[1] Department of Geography, University of Toronto, Toronto, Ontario, Canada

[2] Department of Physical and Environmental Sciences, University of Toronto Scarborough, Scarborough, Ontario, Canada

Correspondance: Micah Hewer, Department of Geography, University of Toronto, Toronto, ON, M5S 3G3, Canada. E-mail: micah.hewer@mail.utoronto.ca

This research did not receive any external funding

Abstract

Weather and climate have been widely recognised as having an important influence on tourism and recreational activities. However, the nature of these relationships varies depending on the type, timing and location of these activities. Climate change is expected to have considerable and diverse impacts on recreation and tourism. Nonetheless, the potential impact of climate change on zoo visitation has yet to be assessed in a scientific manner. This case study begins by establishing the baseline conditions and statistical relationship between weather and zoo visitation in Toronto, Canada. Regression analysis, relying on historical weather and visitation data, measured at the daily time scale, formed the basis for this analysis. Climate change projections relied on output produced by Global Climate Models (GCMs) for the Intergovernmental Panel on Climate Change's 2013 Fifth Assessment Report, ranked and selected using the herein defined Selective Ensemble Approach. This seasonal GCM output was then used to inform daily, local, climate change scenarios, generated using Statistical Down-Scaling Model Version 5.2. A series of seasonal models were then used to assess the impact of projected climate change on zoo visitation. While accounting for the negative effects of precipitation and extreme heat, the models suggested that annual visitation to the zoo will likely increase over the course of the 21st century due to projected climate change: from +8% in the 2020s to +18% by the 2080s, for the least change scenario; and from +8% in the 2020s to +34% in the 2080s, for the greatest change scenario. The majority of the positive impact of projected climate change on zoo visitation in Toronto will likely occur in the shoulder season (spring and fall); with only moderate increases in the off season (winter) and potentially negative impacts associated with the peak season (summer), especially if warming exceeds 3.5 °C.

Keywords: weather sensitivity, climate change impacts, tourism and recreation, zoo visitation, tourism seasonality, tourist behaviour

1. Study Context

International tourism accounts for 30% of the world's exports of services and 6% of world's total exports, according to the United Nations World Tourism Organization (UNWTO, 2015). As an export category, it ranks fourth worldwide, after fuels, chemicals and food, but notably ahead of automotive products (UNWTO, 2015). Furthermore, compared to fuels, earnings from international tourism benefit a larger number of exporters and the sector also tends to generate more employment (UNWTO, 2015). In 2014, 1.3 billion people travelled internationally, up 4.4% from 2013, marking a fifth year of consecutive growth since the global economic crisis in 2009 (UNWTO, 2015).

According to the Canadian Tourism Commission (CTC, 2015), Canada's tourism industry is one of the country's fundamental economic drivers and is the country's number one service export. Total overnight international visitors to Canada increased to 17.1 million in 2014, up 3.2% from 2013 (CTC, 2015). In 2014, the Canadian tourism industry supported nearly 627,000 jobs, being comprised of over 170,000 tourism business establishments (CTC, 2015).

Based on the most recent figures released by the Ontario Ministry of Tourism, Culture and Sport (OMTCS,

2015), Ontario received $28.1 billion in total tourism receipts in 2012; representing 4.2% of provincial GDP that year. In the same year, tourism in Ontario created 359,000 jobs, representing 5.2% of provincial employment in 2012 (OMTCS, 2015). Tourism in Ontario also generated $4.9 billion in tax revenue for 2012, representing 4.2% of provincial revenues that year (OMTCS, 2015).

In 2013, 40 million people visited the greater Toronto region (13.7 million overnight visitors), generating visitor expenditures of approximately $6.5 billion (Tourism Toronto, 2014). The Toronto Zoo is marketed as a world-class zoo and is a highly recognised and promoted tourism attraction in the region (Tourism Toronto, 2014). The Toronto Zoo is therefore an important element of the tourism industry in this region, recording 1.43 million visitors in 2013 (Toronto Zoo, 2014). The zoo is a municipally owned and operated organization with a mission to protect wildlife populations and their habitats by facilitating conservation programs and offering wildlife education opportunities (Toronto Zoo, 2014). In 2013, the Toronto Zoo (2014) declared operating expenses of nearly $53 million. Approximately 73% of the zoo's operating expenses were covered by admission fees, membership sales and other visitor services; the remaining expenses were funded by the City of Toronto (Toronto Zoo, 2014). Since visitor revenues are essential for funding conservation and education programmes at the zoo, it is vital for operations management and city planners to understand external factors that may influence visitation trends.

Weather and climate are two of the critical elements for the natural resource-base of recreation, and are exploited by the tourism industry (de Freitas, 2003). According to Butler (1998), climate and weather also determine the length and quality of certain recreational seasons, by controlling when certain activities are available for participation. Since tourism and recreation are voluntary activities undertaken for personal satisfaction and pleasure (Yukic, 1970); it was argued by de Freitas (2003) that participation will only occur if the participant perceives conditions to be suitable. Weather and climate directly affect tourist satisfaction (Williams et al. 1997); satisfaction in turn affects participation (Butler, 1998), which can be considered a measure of demand for a climatic resource (de Freitas, 2003). The relationship between satisfaction and participation is therefore a reflection of climate and weather related decision-making among tourists (de Freitas, 2014).

Climate was once considered one of the more stable properties of tourism destinations (Abegg et al. 1997). However, this position has been recently abandoned given the increasing evidence of global climate change (Moreno & Amelung, 2009). According to the Intergovernmental Panel on Climate Change (IPCC, 2013), warming of the climate system is unequivocal, and since the 1950s, many of the observed changes are unprecedented over decades to millennia. The earth's atmosphere and oceans have warmed, the amounts of snow and ice have diminished, sea level has risen, and the concentrations of greenhouse gases (GHGs) have increased (IPCC, 2013). Each of the last three decades has been successively warmer at the Earth's surface than any preceding decade since 1850 (IPCC, 2013). In the Northern Hemisphere, 1983–2012 was likely the warmest 30-year period of the last 1400 years (IPCC, 2013). By the end of the 21st century, a warming of between 1.7 to 4.8 °C is expected for global annual mean surface air temperatures (IPCC, 2013). Continued emissions of GHGs will cause further warming and changes in all components of the climate system (IPCC, 2013). Most aspects of climate change will persist for many centuries even if emissions of CO_2 are stopped (IPCC, 2013). Numerous assessments have suggested that climate change is likely to affect tourist destination choice, activity selection, seasonality and tourism demand; from which multiple climate change and tourism literature reviews have recently emerged (Scott et al. 2012; Becken, 2013; Kajan & Saarinen, 2013; Pang et al. 2013; Rossello-Nadal, 2014; Njoroge, 2015).

However, the precise relationship between weather and recreation is not universal for all forms of tourism, as certain activities have different climatic requirements (Morgan et al. 2000; Scott et al. 2008; Rutty & Scott, 2010; Hewer et al. 2014). For this reason, climate change impact assessments that focus on international tourism arrivals without distinguishing between the various tourism activities involved in certain national tourism economies are considered conceptually unsound (de Freitas et al. 2008; Scott et al. 2008; Scott et al. 2012). It is therefore essential to examine each tourism segment individually, in order to effectively determine its relative weather sensitivity (de Freitas et al. 2008); from which more informed climate change impact assessments can be conducted.

In a Canadian context, a number of climate change impact assessments have been conducted on different tourism and recreation activities that were perceived to be vital to the national tourism economy and vulnerable to climatic change and variability. Climate change impact assessments in a Canadian context have focused on the ski industry (Scott et al. 2002, 2003, 2006, 2007a); parks and protected areas (Jones & Scott, 2006a, 2006b; Scott et al. 2007b; the golf industry (Scott & Jones, 2006, 2007); as well as general tourism activities such as sightseeing and shopping (Scott et al. 2004). Overall, climate change is expected to have a significant impact on tourism seasonality and tourist visitation in Canada. In regard to warm-weather recreational activities, climate change is expected to cause conditions in Canada to become more suitable for tourism. Inversely, climate change is expected to have negative impacts on winter-season recreational activities, especially those dependent on snow or ice as the main climatic resource. Although our understanding of climate change impacts and tourism in Canada has improved substantially, especially since the turn of the century, there is still a great need for further research in this field. A number of important tourism activities and attractions have yet to be assessed in relation to their respective weather sensitivity or the subsequent impacts of projected climate change. Climate change is very likely to present a number of different opportunities and threats for the tourism industry in Canada, much of which has yet to be explored in a formal scientific manner. As a result, the on-going assessment of the weather sensitivity and potential climate change impacts associated with tourism activities in Canada is an important area of future research, in order to foster more informed climate change adaptation strategies across the industry.

There has been only limited research conducted in regard to the relationship between zoos and aquariums with climate and climatic change. Junhold and Oberwemmer (2011) summarised the way climate change might directly impact zoos and aquariums in regard to their conservation planning. Furthermore, Pearce-Kelly et al. (2013) discussed the increasing demand on zoos and aquariums due to the threat of projected climate change on global biodiversity; illustrating the likelihood that more species will require ex-situ conservation efforts due to harmful climate and climate-induced environmental change. The significant climate change-related research potential of aquariums and zoos has been highlighted by Barbosa (2009). Aylen et al (2014) used multivariate regression analysis to determine the statistical relationship between climate and weather with average monthly and total daily visitation to a particular zoo in Manchester, England. The study reported that zoo attendance is sensitive to weather and climate variability, but concluded that projected climate change is unlikely to have a major impact on zoo visitation in that region (Aylen et al. 2014). However, no research to date has formally assessed the impact of projected climate change on zoo visitation numbers, patterns of seasonality, or the implications for revenue generation and management responses.

2. Establishing the Baseline

The first step in a formal climate change impact assessment is to understand the baseline conditions relating to the climatic variables of interest (temperature and precipitation, in this case) as well as for the exposure unit (zoo visitation). A climatic baseline, or climate normal, is generally understood as a period of at least 30 years (IPCC, 2007). Taking the most recent 30 year period, the baseline period for this study is therefore 1981 to 2010. Climate data for this period was retrieved from the closest Environment Canada meteorological station in proximity to the Toronto Zoo with the required data for the variables of interest (Richmond Hill weather station, in this case). Data for the exposure unit was obtained directly from the Toronto Zoo. However, visitation data was not available for the entire baseline period (1981-2010); but rather, only for the period from 1999-2013. The climatic and visitation data retrieved for this study were all measured at the daily time scale, which enables a more accurate and precise investigation of the relationship between weather and visitation, as well as the prediction of future visitation under a changed climate, since visitors respond to the current and expected atmospheric conditions on a given day (de Freitas, 2003). Monthly and annual averages are not relative to actual visitor decision-making (de Freitas, 2003, de Freitas et al. 2008); therefore, using these course aggregate measures to model the relationship between climate and visitation and then predict future patterns of tourist behaviour (Jones and Scott 2006a, 2006b; Loomis and Richardson, 2006; Scott et al. 2007b; Fisichelli et al. 2015), could be problematic.

Nonetheless, in order to understand the current trends in these variables of interest over time, it was necessary to reduce the daily weather and visitation data to its annual aggregate form (Figure 1). Fig 1 shows the change in average annual temperature (both minimum and maximum, °C), as well as total annual precipitation (mm), over the baseline period from 1981 to 2010. Figure 1 also records the equation for the slope of the linear trend line and the results from a linear regression analysis, to illustrate the statistical strength of the relationship.

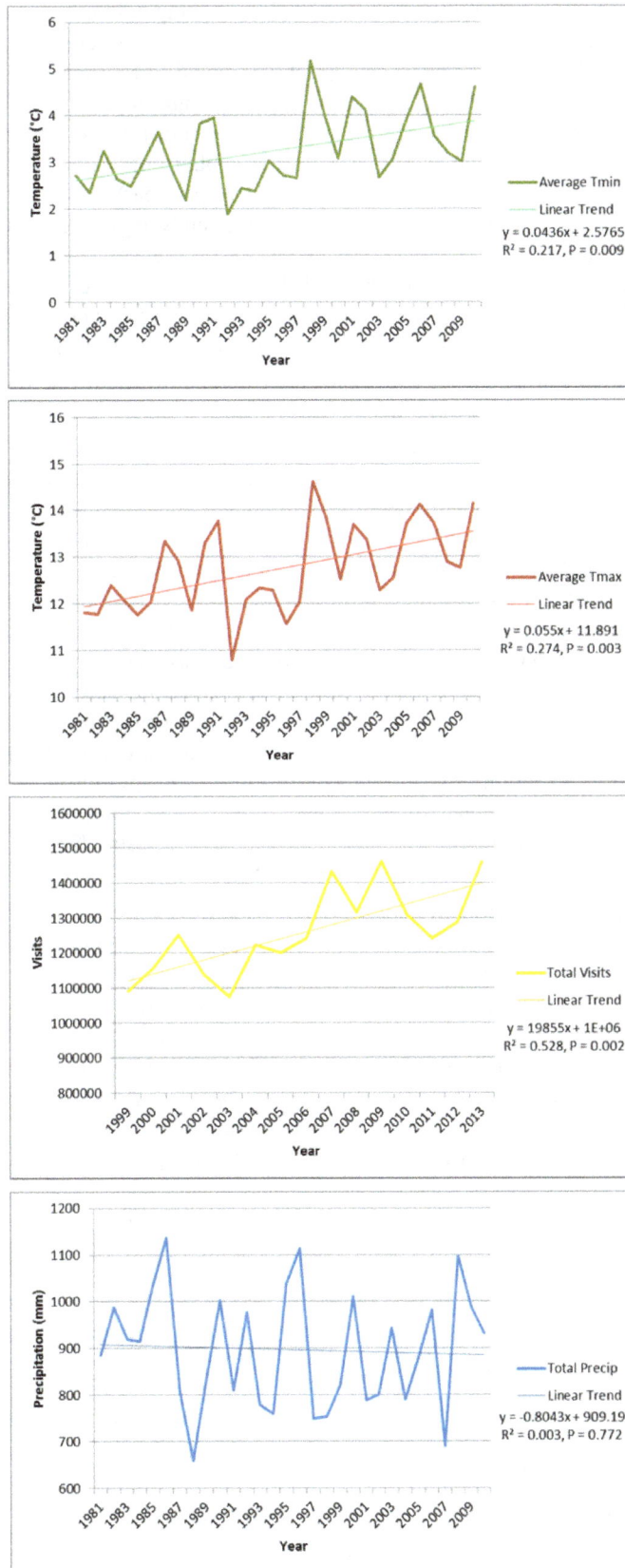

Figure 1. Linear Trends in Baseline Climate Conditions for Richmond Hill from 1981-2010 and Visitation for the Toronto Zoo from 1999-2013 (Top left: annual average for daily maximum temperatures; top right: annual average for daily minimum temperatures; bottom left: annual sum for total daily precipitation; and bottom right: annual sum for total daily visits)

From this preliminary analysis, it is evident that both measures of temperature have been increasing over time within the study region. Average annual maximum temperatures (Tmax) have been increasing at a rate of 0.06 °C per year, over the course of the baseline period, this trend was found to be statistically significant at the 95% confidence interval ($R^2 = 0.274$, P = 0.003). Average annual minimum temperatures (Tmin) also displayed a positive linear trend, but to a lesser degree, increasing by 0.04 °C per year, yet still being statistically significant ($R^2 = 0217$, P = 0.009). Total annual precipitation, was most variable from year to year, and although there was some suggestion of a negative linear trend (decreasing by 0.8 mm per year), the relationship between time and precipitation was not found to be statistically significant for the study region ($R^2 = 0.003$, P = 0.772). Even though the data set for total annual zoo visitation had the smallest sample size since there were fewer years of available data (15 years rather than 30), it still recorded the strongest statistical relationship with time ($R^2 = 0.528$, P = 0.002). According to the slope of the linear trend line, total annual zoo visitation has been increasing at the rate of nearly 20,000 visitors each year, over the course of the baseline period from 1999 to 2013.

What we learn from this preliminary analysis is that temperatures have already been increasing within the study region. Some of this may be attributable to global climate change; however, a greater percentage in likely the effect of the Urban Heat Island (UHI) that characterises this region (Gough & Rosanov, 2001; Mohsin & Gough, 2012). Nonetheless, it is reasonable to conclude that temperatures will continue to increase in this region due to the combined effects of global climate change and the UHI. Therefore, even without the use of sophisticated Global Climate Models (GCMs), zoo managers can reasonably anticipate a continued warming trend, which if the current trend continues, could result in annual average temperatures increasing by 4.5 °C come the end of the 21st century. There was no observable trend in precipitation over the baseline period for the study region; therefore, uncertainty remains concerning the role of precipitation under a changed climate. Although there were only 15 years of data to analyse, zoo visitation showed the strongest positive trend, which suggests that visitation is likely to continue to increase, regardless of future climate change impacts. This gradual increase in zoo visitation is most likely due to population growth in the region, accompanied by subsequent zoo expansion. At the current rate of increase, average annual zoo visitation is on track to nearly triple by the end of the 21st century (increasing from just over one million annual visitors to nearly three million by the year 2100). However, maintaining this rate of increase is conceptually unreasonable, as at some point in the future population growth in the region will likely level off and more importantly, the physical infrastructure of the zoo and its ability to further expand will reach their limits as well, at which point annual visitation will likely plateau.

Although monthly averages for weather and visitation are not appropriate measures for use in predictive models designed to project future behaviour, they still do have application when trying to understand the natural and institutional seasonality associated with both climate and participation in a given tourism and/or recreation context. In order to visually present and compare the patterns of seasonality associated with the two climate variables (temperature and precipitation) as well as the exposure unit (zoo visitation), the raw data was aggregated into monthly averages for the period from 1999-2013 and then the standardised scores were equated for each variable. Standardised scores, or Z scores, for each variable were determined by subtracting the average of all twelve months from the average for each month, then dividing by the standard deviation of the 12 monthly averages. The resulting Z scores from each variable, for each month, are presented in Figure 2.

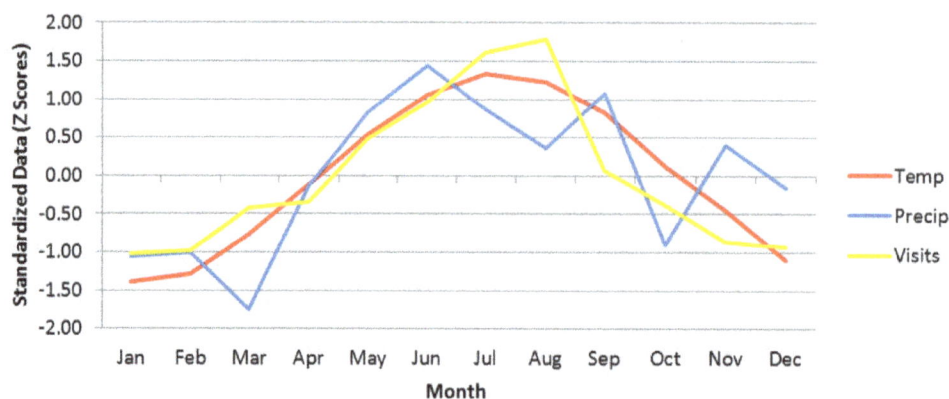

Figure 2. Seasonality of Temperature, Precipitation and Zoo Visitation (Monthly averages for temperature, precipitation and zoo visitation from 1999 to 2013 computed into standardized data (Z scores) the plotted, in order to compare variables recorded in different units of measurement)

What becomes apparent from this illustration is that although temperature increases gradually from the winter low, continuing to rise through the spring and then peaks in the summer before beginning its decline in the fall; visitation does not mimic this pattern precisely. It is apparent that visitation seems to follow the pattern of temperature, in general, but there are some irregularities. Visitation is somewhat level and non-responsive to changes in temperature from November to February, then visitation begin to increase rapidly in March and follows the increases in temperature until the summer months of July and August when visitation in even higher than what can be expected due to the summer increase in temperature alone. Finally, temperatures gradually decline from the summer, through the fall and into the winter but visitation's decline is much more abrupt and drops sharply once August ends.

The pattern for precipitation is much harder to follow and seemingly acts counter to that of temperature and visitation, with the wettest months also being quite warm with high levels of visitation. Since the relationship between temperature and visitation does not appear to be constant across the different seasons, it makes intuitive sense that it may be necessary to model each season separately. For example, since temperatures are quite cold, on average, from November to February, day to day temperature variability is unlikely to have much of an effect on visitation patterns. However, once temperatures rise into a more comfortable range, such as from March to June, as well as September and October, it is more likely that visitation will be more sensitive to temperature variability. Finally, it is apparent that something other than temperature is driving the exceptionally high levels of visitation in July and August. The most likely explanation is the fact that many schools are closed during these months and parents are more likely to take time off work and participate in tourism and recreation activities with their children during this time; therefore these months should be modelled separately as well.

Given that fact that the data for this study was measured at the daily time scale and that the predictive models are designed to project future visitation based on changes in temperature and precipitation at the same scale, it is important to understand the day-to-day variability of the exposure unit. Certain social features that will likely remain constant in the future have considerable influence on the day-to-day variability of zoo visitation and therefore will be useful inclusions when creating predictive weather-visitation models. For example, weekdays (Monday to Friday) are associated with considerably lower levels of visitation than weekends (Saturday and Sunday). Furthermore, certain months contain institutional holidays that result in visitation levels which are even higher than those experienced on weekends during that month (Figure 3). All months except June and November have some form of holiday that is characterised by higher than average visitation. Most of these holidays result in what is referred to as a "long weekend" and the increased visitation usually occurs on either the Friday or Monday, depending on the timing of the holiday. Two exceptions are the months of March and December, in which there is a full week when holiday like visitation occurs associated with times that many schools in the region go on breaks.

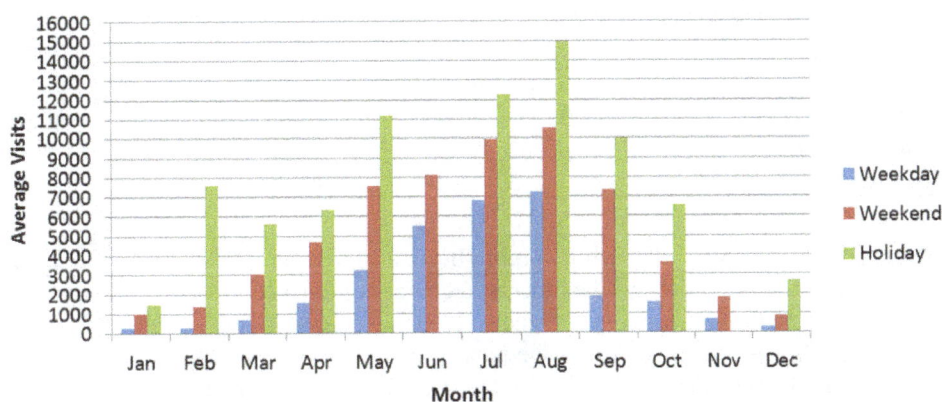

Figure 3. Day-to-Day Variability associated with Zoo Visitation due to Institutional Factors (Average visitation recorded for weekdays, weekends and holidays within each month)

3. Creating Predictive Weather-Visitation Models

Based on the preliminary baseline analyses, informed construction of predictive weather-visitation models was possible. Since the relationship between weather and visitation seemed to vary across seasons, three different models were created. An off-season model was created for the months from November to February, a shoulder season model for March to June, along with September and October; then finally, a peak season model was created for July and August. In addition to the selected climatic variables, a number of social variables were also included. A yearly rank variable was created and included to control for the observed positive linear trend in total annual zoo visitation, which serves as a proxy for population growth and zoo expansion. Binary variables were created and included to control for the effect of weekends and holidays on the day-to-day variability of zoo visitation. In an effort to normalise the data and meet the assumptions of regression analysis, the natural logarithm transformation was employed of zoo visitation data for both the off and shoulder season models. Daily precipitation data was transformed into a daily precipitation scale: 0 mm (no rain); 0 to 2 mm (trace rain); 2 to 5 mm (light rain) 5 to 10 mm (rain); 10 to 20 mm (heavy rain); and more than 20 mm (very heavy rain). Finally, a binary temperature threshold variable was included in the shoulder season model since it was hypothesised that at some critical point the positive linear trend between temperature and visitation would reach a break point and conditions would become too hot, causing visitation to decline. After testing a series of temperature thresholds within the model, it was determined that the binary variable which triggered when temperatures exceeded 26 °C had the strongest influence. Table 1 records the summary statistics for each of the three seasonal weather-visitation models.

Table 1. Seasonal weather-visitation predictive regression models

Predictor Variable	Beta	Part. Corr.	B	t	P
OFF SEASON MODEL (R^2=0.597, F=525.15, P<0.001, SEE=0.654, N=1777, d=1.46)					
Intercept (LN of Visits)			5.553	151.404	<0.001
Year Rank (1-15)	0.099	0.099	0.024	6.545	<0.001
Weekend Binary (0,1)	0.437	0.437	0.995	28.955	<0.001
Holiday Binary (0,1)	0.395	0.393	1.691	26.079	<0.001
Maximum Temperature (°C)	0.502	0.499	0.081	33.102	<0.001
Precipitation Scale (0-5)	-0.206	-0.206	-0.168	-13.633	<0.001
SHOULDER SEASON MODEL (R^2=0.631, F=768.06, P<0.001, SEE=0.623, N=2698, d=1.20)					
Intercept (LN of Visits)			6.141	162.929	<0.001
Year Rank (1-15)	0.042	0.042	0.010	3.559	<0.001
Weekend Binary (0,1)	0.443	0.438	1.005	37.458	<0.001
Holiday Binary (0,1)	0.343	0.335	1.665	28.617	<0.001
Maximum Temperature (°C)	0.680	0.559	0.082	47.751	<0.001
Threshold (Tmax > 26 °C)	-0.121	-0.101	-0.379	-8.596	<0.001
Precipitation Scale (0-5)	-0.174	-0.174	-0.130	-14.825	<0.001
PEAK SEASON MODEL (R^2=0.567, F=241.33, P<0.001, SEE=2077.3, N=926, d=1.81)					
Intercept (Visits)			12259.01	34.222	<0.001
Year Rank (1-15)	0.186	0.185	135.08	8.525	<0.001
Weekend Binary (0,1)	0.445	0.441	3103.92	20.352	<0.001
Holiday Binary (0,1)	0.351	0.348	6345.14	16.051	<0.001
Minimum Temperature (°C)	-0.382	-0.374	-362.52	-17.233	<0.001
Precipitation Scale (0-5)	-0.253	-0.248	-570.15	-11.427	<0.001

Each of the three models had similar explanatory power, being able to explain between 56 and 63 percent of the observed variability in daily zoo vitiation. The order of the rank of influence for each of the predictor variables remained the same between the off season and shoulder season models, with some variation in the magnitude of influence for each variable and the inclusion of the temperature threshold variable in the shoulder season. For both the off season and the shoulder season, temperature was the strongest predictor variable, followed by the social effects of weekends and then holidays, next was precipitation and then the yearly rank variable.

The peak season was certainly unique among the models with weekends having the strongest effect on visitation, followed by the negative effect of temperature, then the effect of holidays, followed by the negative effect of precipitation and then the yearly rank variable. It is also worth noting that both precipitation and the yearly rank variables had the strongest effect on visitation during the peak season, compared to the off and shoulder seasons. Furthermore, although the coefficient of determination (R^2) was lowest during the peak season; this model also had the fewest data points available for inclusion. Therefore, the slightly lower R^2 value for the peak season model, compared to the other two models, is not necessarily indicative of decreased weather sensitivity, but is more likely due to the diminished statistical power of the sample size. Each of the predictor variables included in all three models were found to be statistically significant, with very low P values and excellent partial correlations, not much lower than the recoded Beta values themselves. Finally, there was no evidence of serial correlation within the data for any of the models, as evidenced by the related Durban-Watson tests, since none of the d values were greater than two (negative serial correlation) nor were they less than one (positive serial correlation).

4. Ranking and Selecting Global Climate Models

Climate models have improved since the IPCC's 2007 Forth Assessment Report (AR4). The IPCC's 2013 Fifth Assessment Report (AR5) models reproduce observed continental-scale surface temperature patterns and trends over many decades, including the more rapid warming since the mid-20th century and the cooling immediately following large volcanic eruptions. According to the IPCC (2013), the scientific community has defined a set of four new scenarios for the AR5, denoted Representative Concentration Pathways (RCPs). They are identified by their approximate total radiative forcing in year 2100 relative to 1750: 2.6 W m^{-2} for RCP 2.6, 4.5 W m^{-2} for RCP 4.5, 6.0 W m^{-2} for RCP 6.0, and 8.5 W m^{-2} for RCP 8.5. As described by the IPCC (2013), these four RCPs include one mitigation scenario leading to a very low forcing level (RCP 2.6), two stabilization scenarios (RCP 4.5 and RCP 6), and one scenario with very high GHG emissions (RCP 8.5). It is explained by the IPCC (2013) that the RCPs represent a range of 21st century climate policies, compared with the no-climate policy approach of the Special Report on Emissions Scenarios used in the IPCC's 2001 Third Assessment Report (TAR) and 2007 AR4.

Global Climate Model output from the IPCC's 2013 AR5 can be obtained from the Coupled Model Intercomparison Project of the World Climate Research Programme. However, there is a wide selection of global climate models available to provide projections of future climate change, forty in total from the most recent assessment. Furthermore, each of the forty modelling centres provide future projections for a number of the four different RCPs, which describe how GHG concentrations could evolve over the next 100 years and thereby influence global climate. There are many approaches that have been developed in order to provide some direction for determining which of the future projections of climate available for impact assessments should be used in planning (Fenech et al., 2007). Compared against historical observed gridded data, climate projections using the ensemble approach have been shown to come closest to replicating the historical climate (IPCC, 2010). This approach suggests that it is best to plan for the average climate change from all the climate model projections by using a mean of all the models to reduce the uncertainty associated with any individual model. In effect, the individual model biases seem to offset one another when considered together. It is generally accepted that climate models can be evaluated based on their ability to reproduce baseline conditions (IPCC, 2013). Some climate models perform better in certain regions than they do in others. For this reason, it seems unreasonable to create a "full" ensemble including all of the available GCMs when it is evident that some models are unable to reproduce past climate for the study region. Based on this logic, it would seem more appropriate to evaluate each model individually, based on its ability to reproduce past climate, and then rank and select the best three models to create a "selective" ensemble from these top performing models. Tables 2 and 3 present the resulting models identified from this Selective Ensemble Approach, as well as the statistics required to formulate the ranking, referred to as the Gough-Fenech confidence index (GFCI) (Fenech, 2009).

Table 2. Selected seasonal global climate models for precipitation (GCM output from the IPCC's 2013 AR5 for precipitation at Richmond Hill based on the 1981 to 2010 baseline)

Season	Model - Experiment	OBS Mean	OBS StDev	MOD Mean	GFCI Score
Winter (DJF)	CESM1-CAM5 (Mean) RCP2.6	2.132	4.387	2.120	0.003
	CESM1-CAM5 (Mean) RCP8.5	2.132	4.387	2.072	0.014
	FGOALS-g2 (Run 1) RCP2.6	2.132	4.387	2.173	0.009
	FGOALS-g2 (Run 1) RCP8.5	2.132	4.387	2.209	0.018
	GISS-E2-R (Run 1) RCP2.6	2.132	4.387	2.082	0.011
	GISS-E2-R (Run 1) RCP8.5	2.132	4.387	2.083	0.011
Spring (MAM)	IPSL-CM5A-MR(Run 1) RCP2.6	2.286	5.076	2.138	0.029
	IPSL-CM5A-MR(Run 1) RCP8.5	2.286	5.076	2.190	0.019
	MIROC5(Mean) RCP2.6	2.286	5.076	2.320	0.007
	MIROC5(Mean) RCP8.5	2.286	5.076	2.367	0.016
	NorESM1-ME(Run 1) RCP2.6	2.286	5.076	2.317	0.006
	NorESM1-ME(Run 1) RCP8.5	2.286	5.076	2.327	0.008

Season	Model - Experiment	OBS Mean	OBS StDev	MOD Mean	GFCI Score
Summer (JJA)	bcc-csm1-1(Run 1) RCP2.6	2.788	7.263	2.683	0.014
	bcc-csm1-1(Run 1) RCP8.5	2.788	7.263	2.807	0.003
	BNU-ESM(Run 1) RCP2.6	2.788	7.263	2.889	0.014
	BNU-ESM(Run 1) RCP8.5	2.788	7.263	2.933	0.020
	FGOALS-g2(Run 1) RCP2.6	2.788	7.263	2.716	0.010
	FGOALS-g2(Run 1) RCP8.5	2.788	7.263	2.687	0.014
Autumn (SON)	BNU-ESM(Run 1) RCP2.6	2.652	6.162	2.644	0.001
	BNU-ESM(Run 1) RCP8.5	2.652	6.162	2.632	0.003
	HadGEM2-AO(Run 1) RCP2.6	2.652	6.162	2.681	0.005
	HadGEM2-AO(Run 1) RCP8.5	2.652	6.162	2.718	0.011
	MPI-ESM-MR(Run 1) RCP2.6	2.652	6.162	2.707	0.009
	MPI-ESM-MR(Run 1) RCP8.5	2.652	6.162	2.623	0.005

Table 3. Selected seasonal global climate models for temperature (GCM output from the IPCC's 2013 AR5 for temperature at Richmond Hill based on the 1981 to 2010 baseline)

Season	Model - Experiment	OBS Mean	OBS StDev	MOD Mean	GFCI Score
Winter (DJF) Max Temp (°C)	FIO-ESM (Mean) RCP2.6	-0.744	5.601	-0.679	0.012
	FIO-ESM (Mean) RCP8.5	-0.744	5.601	-0.568	0.031
	GFDL-CM3 (Run 1) RCP2.6	-0.744	5.601	-0.667	0.014
	GFDL-CM3 (Run 1) RCP8.5	-0.744	5.601	-0.514	0.041
	MIROC-ESM-CHEM (Run 1) RCP2.6	-0.744	5.601	-0.584	0.029
	MIROC-ESM-CHEM (Run 1) RCP8.5	-0.744	5.601	-0.643	0.018
Spring (MAM) Max Temp (°C)	bcc-csm1-1-m (Run 1) RCP2.6	11.953	8.562	11.708	0.029
	bcc-csm1-1-m (Run 1) RCP8.5	11.953	8.562	11.924	0.003
	GISS-E2-H (Run 1) RCP2.6	11.953	8.562	11.857	0.011
	GISS-E2-H (Run 1) RCP8.5	11.953	8.562	11.795	0.018
	HadGEM2-ES (Mean) RCP2.6	11.953	8.562	12.065	0.013
	HadGEM2-ES (Mean) RCP8.5	11.953	8.562	11.977	0.003
Summer (JJA) Max Temp (°C)	CSIRO-Mk3-6-0 (Mean) RCP2.6	25.599	4.111	25.770	0.042
	CSIRO-Mk3-6-0 (Mean) RCP8.5	25.599	4.111	25.795	0.048
	HadGEM2-ES (Mean) RCP2.6	25.599	4.111	25.677	0.019
	HadGEM2-ES (Mean) RCP8.5	25.599	4.111	25.620	0.005
	IPSL-CM5A-MR (Run 1) RCP2.6	25.599	4.111	25.476	0.030
	IPSL-CM5A-MR (Run 1) RCP8.5	25.599	4.111	25.487	0.027
Summer (JJA) Min Temp (°C)	bcc-csm1-1-m (Run 1) RCP2.6	14.743	3.715	14.518	0.061
	bcc-csm1-1-m (Run 1) RCP8.5	14.743	3.715	14.312	0.116
	CCSM4 (Mean) RCP2.6	14.743	3.715	15.079	0.091
	CCSM4 (Mean) RCP8.5	14.743	3.715	15.068	0.087
	GFDL-CM3 (Run 1) RCP2.6	14.743	3.715	15.037	0.079
	GFDL-CM3 (Run 1) RCP8.5	14.743	3.715	15.079	0.090
Autumn (SON) Max Temp (°C)	bcc-csm1-1-m (Run 1) RCP2.6	13.935	7.646	14.164	0.030
	bcc-csm1-1-m (Run 1) RCP8.5	13.935	7.646	14.129	0.025
	HadGEM2-AO (Run 1) RCP2.6	13.935	7.646	13.784	0.020
	HadGEM2-AO (Run 1) RCP8.5	13.935	7.646	13.871	0.008
	MIROC-ESM (Run 1) RCP2.6	13.935	7.646	13.746	0.025
	MIROC-ESM (Run 1) RCP8.5	13.935	7.646	13.797	0.018

Of the forty different GCMs available from the IPCC's 2013 AR5, only those with projections for RCP 2.6 and RCP 8.5 when considerable for inclusion within this selective ensemble approach. Previously, researchers in field of climate change and tourism had advocated for selecting models in an effort to capture the full range of model uncertainty by selecting the models whose projections were positioned on the outer-limits of a scatterplot (Jones & Scott, 2006a, 2006b; Scott & Jones, 2006, 2007; Scott et al. 2003, 2006, 2007a). However, this approach does not consider model accuracy and may actually result in the selection of models that cannot reproduce past climate for the study region. Nonetheless, recognising the merit of capturing a full range of uncertainty (in regard to future global emissions, GHG concentrations and subsequent radiative forcing), projections were based on the least (2.6) and greatest (8.5) change RCPs, while still employing the selective ensemble approach to improve climate model accuracy. The GFCI (Fenech, 2009) was employed to evaluate the ability of the available AR5 GCMs to reproduce past climate conditions for temperature and precipitation in each of the four different seasons. This approach takes the absolute value of the observed average baseline condition (either temperature or precipitation), subtracted by the modelled average baseline condition, and then divided by the standard deviation of the observed baseline conditions. From this process, a GFCI score was calculated for each model and for each RCP followed by that model. The closer to 0 that the GFCI score was, the better the model was able to reproduce past climate, and therefore a greater likelihood of generating more accurate climate change projections. The top three models were selected based on the combined GFCI score for the two different RCPs.

5. Downscaling Global Climate Model Output

Not only is GCM output coarse in relation to its spatial scale, representing a general region and not a particular location; it is also coarse in relation to its temporal scale, typically conveying only monthly change and not able to produce daily climate change scenarios (Pielke & Wilby, 2012). For these reasons, the GCM output generated through the selective ensemble approach (Table 4) was used to guide the production of local point, daily climate change scenarios in the Statistical Down-Scaling Model (SDSM) Version 5.2. This software is a stochastic weather generator but also uses data from the National Centers for Environmental Prediction and the National Center for Atmospheric Research, which represents the state of the Earth's atmosphere during the baseline period, to identify the strongest atmospheric predictor variables and thereby create models more capable of generating synthetic weather data that is able to reproduce past climate for a particular location (Wilby et al. 2001). The SDSM process involves an initial data quality control stage; then the evaluation and selection of atmospheric predictor variables; then a model calibration step; followed by a historical weather generation stage; then a model validation process is completed by comparing observed and modelled baseline conditions; finally, future climate change scenarios are generated, guided by the GCM output from the selective ensemble approach. This decision-making tool has been widely applied within the field of applied climatology and for climate change impact assessments (Wilby & Dawson, 2012).

Table 4. Seasonal climate change projections for Richmond Hill.

Season	2020s (2011-2040)	2050s (2041-2070)	2080s (2071-2100
Temperature Change in Degrees Celsius for Daily Maximum (Daily Minimum during Summer)			
Winter (JFD)	1.5 to 1.7	2.0 to 3.5	2.0 to 5.8
Spring (MAM)	1.3 to 1.4	1.6 to 3.1	1.4 to 4.6
Summer (JJA)	1.4 to 1.5 (1.2 to 1.6)	2.1 to 3.8 (1.7 to 3.5)	2.1 to 6.9 (1.8 to 5.3)
Fall (SON)	1.3 to 1.7	2.3 to 3.7	2.4 to 6.4
Precipitation Change in Percentage for Total mm/Day			
Winter (JFD)	1.3 to 6.3	0.5 to 18.5	8.7 to 26.4
Spring (MAM)	5.4 to 6.6	7.2 to 17.1	12.9 to 15.8
Summer (JJA)	-1.2 to -2.9	6.1 to -4.9	2.7 to -4.0
Fall (SON)	-5.7 to -1.2	1.5 to 3.0	-3.0 to -3.7

Note. Temperature and precipitation change based on a selective ensemble of GCM outputs from the IPCC's 2013 AR5, showing the upper and lower bounds of the available RCPs (2.6 and 8.5).

6. Validating the Predictive Models

Before using the predictive weather-visitation models in combination with the daily climate change scenarios to assess the impact on future zoo visitation, it is important to validate the predictive capacity of these models. There is a convenient overlap between the observational zoo visitation record and the modelled baseline climate conditions defined as the period from 1999-2010 (12 years). This period serves as a useful validation tool, a revealing aspect of any modelling study, not often present in tourism climate change impact assessments. Using the synthetic daily weather data generated in SDSM from 1999 to 2010 (taken from the larger data set that covered 1981 to 2010), it was possible to model visitation for these same 12 years using the regression equations that define the weather-visitation models described earlier. The regression equations for each of the seasonal weather-visitation models are as follows:

Off Season Visitation = *EXP*(5.552+0.024*Year+0.995*Weekend+1.691*Holiday+0.081*Tmax

$$-0.168*Precip)*EXP(0.5*POWER(0.654,2))$$

Shoulder Season Visitation = $EXP(6.141+0.010*Year+1.005*Weekend+1.665$
$*Holiday+0.082*Tmax-0.379*TooHot-0.130*Precip)*EXP(0.5*POWER(0.623,2))$

Peak Season Visitation =
$11259.01+135.08*Year+3103.92*Weekend+6345.14*Holiday-362.52*Tmin-570.15*Precip$

Figure 4 illustrates the ability of the predictive models to reproduce the annual trend in zoo visitation from 1999 to 2010. By comparing the slope of the linear trend for total annual zoo visitation from both the observed and modelled data, it is apparent that the predictive models were able to reproduce the magnitude and direction of change in zoo visitation over the validation period. Given the fact that the three models combined were only able to explain an average of 60% of the observed variation in zoo visitation, the two lines are not identical. Zoo visitation increased by an average of 26,000 visitors per year based on the slope of observed linear trend line; however, the modelled record underestimates the rate of increase in zoo visitation, suggesting an increase of only 20,500 visitors each year. Furthermore, the models were not able to fully capture certain annual extremes; such as the visitation low in 2003 or the two spikes in visitation during 2007 and 2009. The conservative nature of the predictive models, as well as their inability to fully capture extremes, is likely explained by other socio-economic variables not included in the model. These sorts of variables cannot be controlled for within the future scenarios (i.e. fuel prices, exchanges rates, disposable income, special attractions), and therefore must be held constant. This issue has been recognised as a limitation to the modelling approach for tourism-based climate change impact assessments (Rosselló-Nadal, 2014). The validation process not only helps identify the limitations of this modelling approach, it also provides assurance in regard to the accuracy of the results, especially in relation to the magnitude and direction of change associated with total annual zoo visitation under projected climate change.

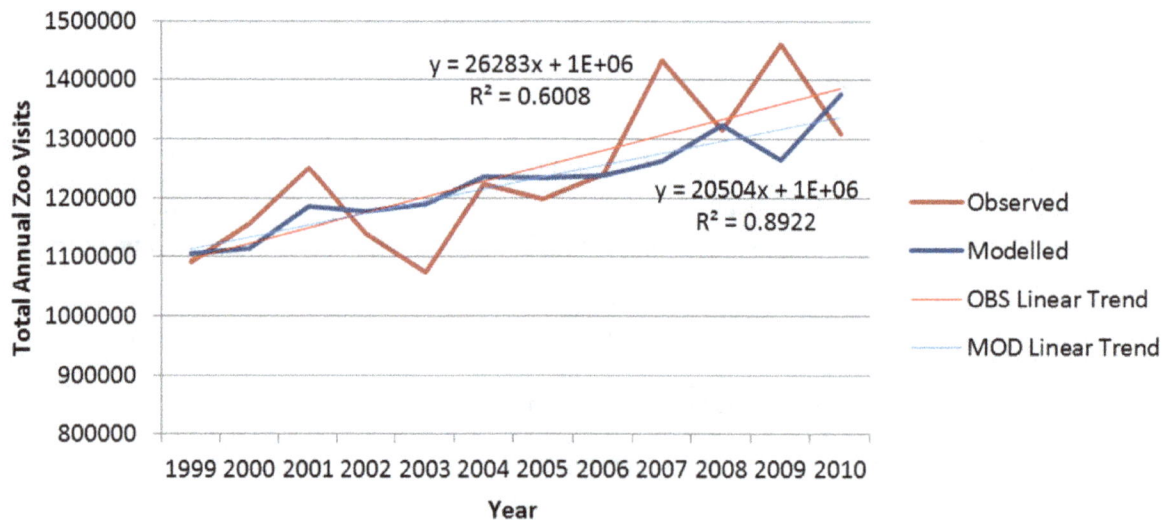

Figure 4. Comparison of observed and modelled total annual zoo visitation from 1999 to 2010 (Linear trend lines have been plotted, including the slope of the line and the co-efficient of determination, describing the strength and direction of the statistical relationship between visitation and time)

Figure 5 illustrates the ability of the predictive models to reproduce patterns of seasonality in zoo visitation over the 12 year validation period from 1999 to 2010. The three seasonal weather-visitation models were able to reproduce patterns of seasonality with greater accuracy than they could reproduce total annual zoo visitation. For 8 out of the 12 months, the plots for average monthly zoo visitation between the observed and modelled data were quite similar. However, the model had difficulty reproducing visitation in the months of May and June as well as September and October. In May and June the model underestimated zoo visitation; whereas, in the September and October the model overestimated zoo visitation. All four of these months were contained within the shoulder season model. Interestingly, there seems to be somewhat of an offsetting effect when considering the overall effect within the model as well as the annual impact since there were two months with lower visitation and two months with higher visitation. Local climate is quite similar when comparing May and June with September and October; however, visitation differs considerably. These monthly discrepancies between modelled and observed visitation is likely due to factors associated with institutional seasonality (Butler, 1998). Although climate conditions are similar either side of the peak season (July & August), zoo visitation seems to be ramping up through the spring (May & June), but then drops sharply in the fall (September & October). Apparently, the models were not able to capture the effects of these socio-cultural factors associated with zoo visitor behaviour (which may be explained by the typical timing of organized school trips to the zoo, which generally tend to be more common in the spring than the fall). However, overall, the predictive models did an excellent job reproducing patterns of seasonality accosted with zoo visitation.

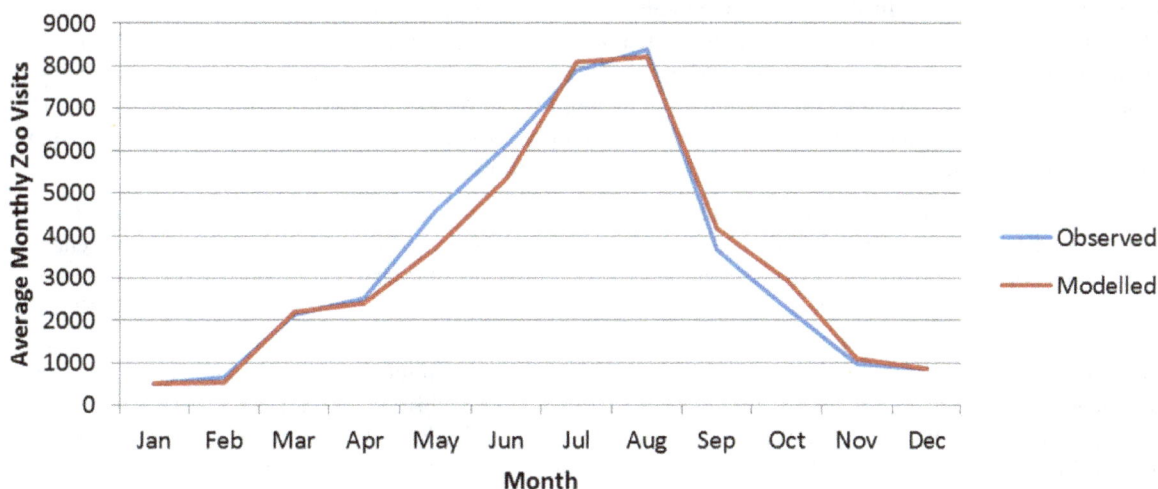

Figure 5. Comparison of observed and modelled patterns of seasonality for zoo visitation (The average number of observed and modelled visitors for 1999 to 2010 plotted for each month)

7. Climate Change Impact Assessment

The seasonal weather-visitation multivariate regression models were used to predict future daily zoo visitation for the remainder of the 21st century. The predictive models were applied to the daily climate change scenarios generated in SDSM, which were guided by GCM outputs taken from the IPCC's 2013 AR5, using the selective ensemble approach. Table 5 displays the magnitude and direction of change associated with the different predictor variables included in the models, as well as the impact on the exposure unit. Although the GCM output was extracted for each climatic season (winter, spring, summer, fall), and zoo visitation was modelled separately for the three different visitation seasons (off, shoulder, peak). The projected changes are summarised based on their annual effects over the three different change periods: the 2020s (2011-2040), the 2050s (2041-2070), and the 2080s (2071-2100).

Table 5. Projected climatic changes and modelled impact on zoo visitation.

Variable	Δ (2020s)	Δ (2050s)	Δ (2080s)
Average Annual Maximum Temperature	1.4 to 1.6 °C	2.0 to 3.5 °C	2.0 to 5.9 °C
Average Annual Number of Days Exceeding 26°C	17 to 19 days	26 to 44 days	26 to 70 days
Average Annual Minimum Temperature	1.2 to 1.6 °C	1.7 to 3.5 °C	1.8 to 5.3 °C
Total Annual Precipitation	-3.0 to -1.3%	1.7 to 10.9%	5.5 to 18.3%
Total Annual Zoo Visits	7.6 to 7.8%	14.0 to 17.3%	18.3 to 33.9%

Note. Climatic change based on SDSM scenarios for Richmond Hill from 2011 to 2100, guided by GCM output (RCP 2.6 to RCP 8.5), selected using the GFCI. Potential impact on visitation to the Toronto Zoo based on predictions from the seasonal weather-visitation multivariate regression models.

The yearly rank variable, whose effect was derived from the increasing trend in annual visitation from 1999-2013, was manipulated in the modelled projections so that each decade represented a year. This was done because it was thought to be unreasonable that visitation would continue to increase at an average rate of 20,000 visitors each year, over the next 90 years; so instead, that rate of increase would only be realised nine times. This was done in an effort to control for limits to population growth and zoo expansion expected to be realised as the 21st century unfolds; while also making the future projections more realistic and further isolating the effect of projected climate change. Based on the daily climate change scenarios, Tmax was projected to increase from +1.4 to +5.9 °C over the course of the 21st century; whereas, Tmin was projected to increase from +1.2 to +5.3 °C. The total number of days when Tmax exceeded 26 °C (not including the peak months of July and August) was projected to increase from 17 to 70 days over the course of the 21st century. Changes in precipitation for the region under projected climate change vary more drastically by season and follow a different pattern than temperature. In the 2020s, precipitation may decrease by -1.3 to -3.0 percent; whereas, from the 2050s and on to the end of the century, precipitation may increase from +1.7 to +18.3 percent. The social variables associated with the effect of weekends and holidays remained constant throughout the modelled projections. Under the low emission, low radiative forcing RCP (2.6), annual visitation was projected to increase from +7.6 to +18.3 percent over the course of the 21st century. Whereas, under the high emission, high radiative forcing RCP (8.5); annual visitation was projected to increase from +7.8 to +33.9 percent.

Figure 6 illustrates the projected rate of increase in total annual zoo visitation for the 21st century under both the selected RCPs. Under the RCP 2.6 climate change scenario, as indicated by the slope of the linear trend line, total annual zoo visitation is projected to increase by approximately 2000 visits each year over the course the 21st century; resulting in approximately 200,000 additional annual visitors by the year 2100. Under RCP 2.6, projected climate change is minimal (unlikely to exceed a 2 °C warming by the end of the century). Therefore, much of the modelled increase in zoo visitation is likely due to continued trends associated with population growth and zoo expansion. Recall that the yearly rank variable derived from the increasing trend in zoo vitiation overtime observed within the observational data, although scaled back for use within the predictive models to account for limits to growth, would likely result in an additional 20,000 annual zoo visitors each decade. Under RCP 8.5, total annual zoo visitation is projected to increase by approximately 4600 visits each year over the course the 21st century; resulting in approximately 400,000 additional annual visitors by the year 2100. Furthermore, the projected impact of climate change on zoo visitation does not vary much between the two RCPs in the 2020s. Whereas, in the 2050s, the impact of projected climate change on annual zoo visitation under RCP 8.5 is noticeably greater than that of RCP 2.6. However, the magnitude of difference in projected impacts between the two RCPs is greatest in the 2080s, where the impact under RCP 8.5 (+33.9%) is nearly twice that of RCP 2.6 (+18.3%).

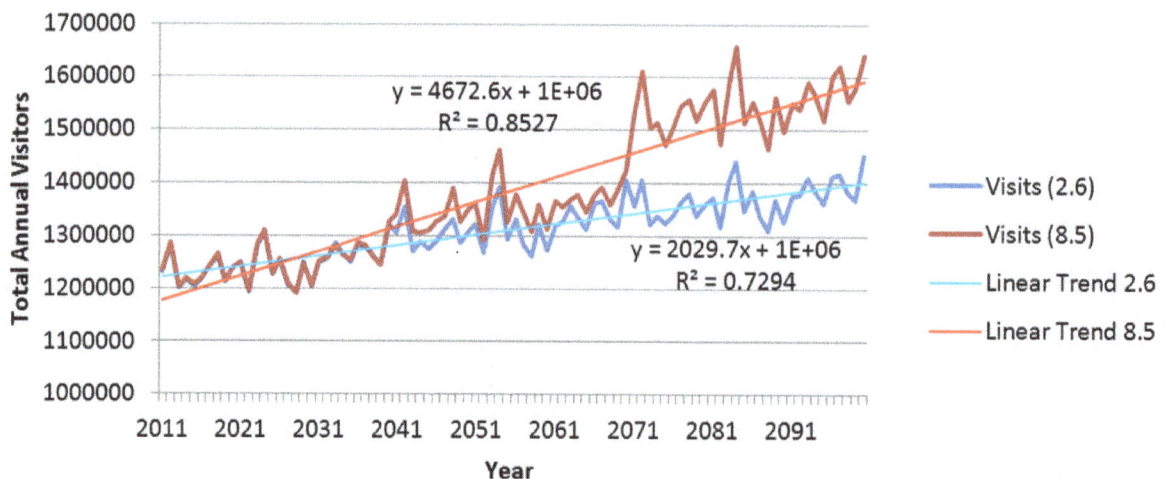

Figure 6. Annual trend in total zoo visitation under projected climate change (Projected total annual zoo plotted across the course of the 21st century for both RCP 2.6 and RCP 8.5 guided climate change scenarios. Linear trend lines have also been plotted, including the slope of the line and the co-efficient of determination, describing the strength and direction of the statistical relationship between predicted visitation and time)

Figure 7 illustrates the impact of projected climate change on the seasonality of visitation to the Toronto Zoo over the course of the 21st century under both RCPs, by graphing monthly averages of total daily visitors for the baseline period (1981-2010) and then for the three future periods: 2020s, 2050s and 2080s. Under RCP 2.6, the model suggests that the patterns of seasonality will remain the same for zoo visitation in Toronto, with visitation expected to increase in each month over the course of the 21st century, but with some months expected to experience greater increases in visitation than others. For example, the summer months are expected to experience greater increases in visitation compared to the winter months, due to the impact of projected climate change under RCP 2.6. However, under RCP 8.5 the patterns of seasonality for zoo visitation in Toronto become reshaped due to the impact of projected climate change. This discrepancy between the impacts of the two RCPs is due mainly to the way visitation responds to projected climate change in the months of July and August. Under RCP 8.5, visitation in July and August declines continually throughout the course of the 21st century, as temperatures continue to warm. The other months are still expected to experience increases in visitation as was seen under RCP 2.6, only to a greater degree. Therefore, as visitation in the shoulder season increases and visitation in the peak season declines, the seasonal patterns of visitation become more rounded and the sharp rise and fall historically associated with the spring and fall seasons are no longer apparent by the end of the 21st century. In the 2080s, November seems to behave more like one of the shoulder season months and may no longer be considered part of the off season; meanwhile as early as the 2050s, June seems to reach peak season levels of visitation.

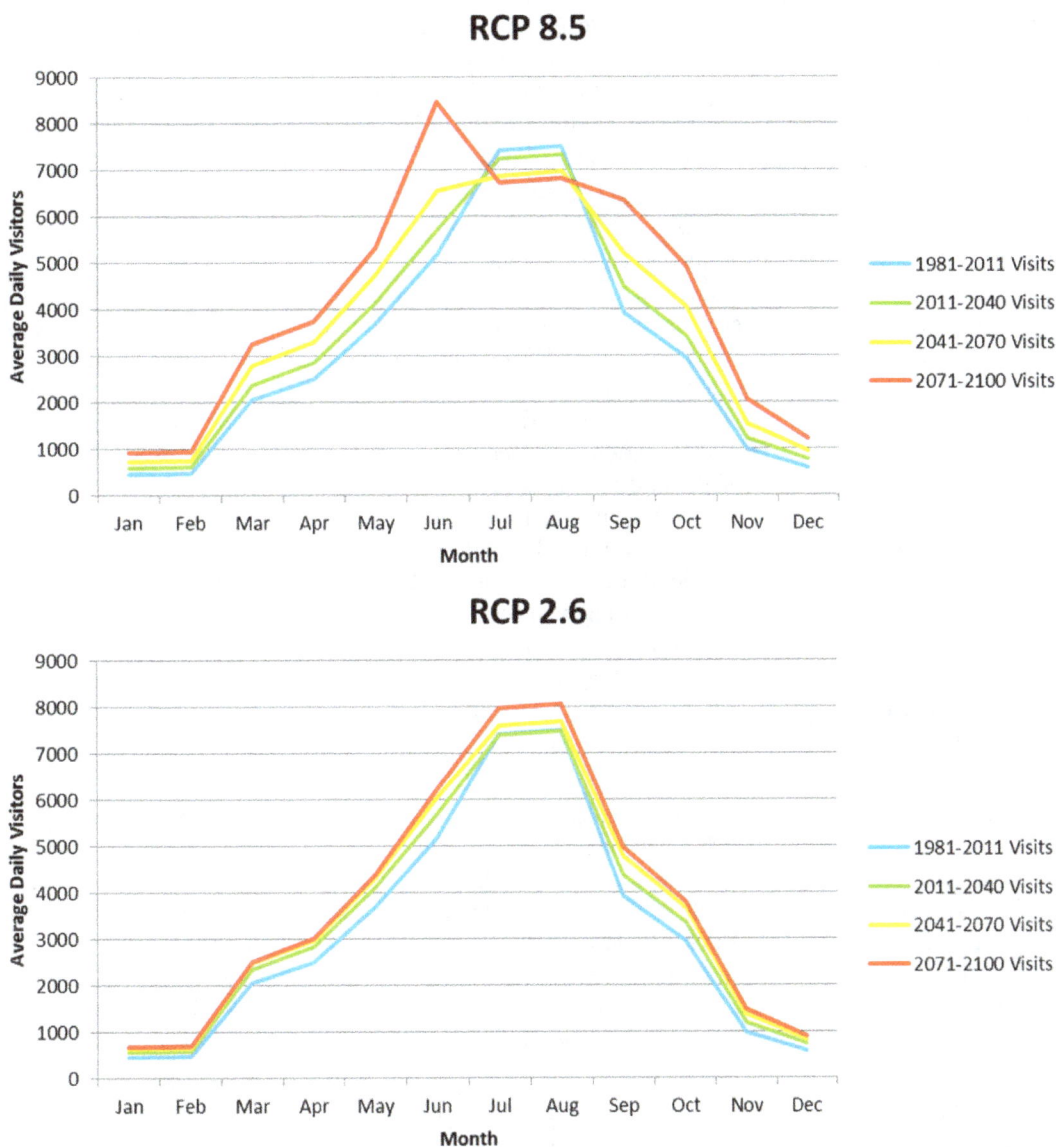

Figure 7. Patterns of seasonality for visitation to the Toronto Zoo under projected climate change (Average visitors for each month have been recorded, plotting the baseline period (1981-2010) and each change period (2020s, 2050s, 2080s), including a graph for both RCP 2.6 and RCP 8.5 climate change scenarios)

There is a clear anomaly associated with the modelled projections under RCP 8.5 in the 2080s for the month of June. The model suggests an unlikely spike in visitation. It is more likely that by the end of the 21st century, visitation in June will be similar to visitation in July and August, and therefore to be modelled correctly would require the use of the peak season model parameters rather than those of the shoulder season model. Recall that the difference between these two models was that the in the shoulder season the strongest predictor of visitation was the positive effect of Tmax; whereas, in the peak season the strongest climatic predictor was the negative effect of Tmin.

8. Discussion and Conclusions

This study conducted a formal climate change impact assessment on visitation to the Toronto Zoo in Ontario, Canada. Multivariate regression was employed to determine the statistical relationship between weather and visitation within the study region, from which predictive models were developed. The GFCI (Fenech, 2009) was used to rank and select a number of top performing GCMs, evaluated based on their ability to reproduce past climate within the region, from which a "selective" ensemble of future climatic change projections was created. This information was then used to generate downscaled, local-point, daily climate change scenarios in SDSM 5.2. Predicted visitation from the models was then validated against observational data, using the synthetic weather generated in SDSM for the period from 1999 to 2010. Satisfied with the validation process, the seasonal weather-visitation models were then employed to assess the potential impact of projected climate change on zoo visitation over the course of the 21st century, based on full range of uncertainty pertaining to future global emissions, GHG concentration and subsequent radiative forcing (RCP 2.6 to RCP 8.5).

The results of this study illustrate that there has already been an observable warming trend within the region. Projected climate change is expected to accelerate that warming process. The regression analysis revealed that daily zoo visitation is highly correlated with weather variability. On average, the seasonal models were able to explain 60% of the observed variability in daily zoo vitiation; with temperature found to be the strongest predictor variable in two out of the three seasonal models. Under a hypothetic global emissions mitigation strategy (RCP 2.6), resulting in lower radiative forcing and subsequent climatic changes, it is projected that the effect of moderate warming over the course of the 21st century (< 2 °C) will result in increased annual zoo visitation (from +7.6% in the 2020s to +18.3 % in the 2080s), without causing any significant changes to the patterns of seasonality. For RCP 8.5, an accelerated emissions and subsequent warming scenario (+3.5 °C by the 2050s and +5.9 °C by the 2080s), annual visitation increased at a much greater rate (from +7.8 % in the 2020s to +33.9 % in the 2080s), with notable shifts in the patterns of seasonality. For RCP 8.5, by the 2050s, there was a notable decline in peak season visitation, accompanied by increases in visitation during the shoulder season months. These changes resulting in a much more rounded pattern of seasonality, compared to the sharp rise and fall associated with visitation on either side of the peak season months of July and August under baseline climate conditions. Furthermore, by the 2080s, this shifting pattern of seasonality was even more pronounced, to such a degree that model apparently failed to predict June visitation for this time period (producing unreasonably high levels of visitation for the month). This modelling limitation was nonetheless revealing, as it suggests that as average annual temperatures exceed 3.5 °C and continue to increase as high as 5.9 °C, visitation in June will start to behave much more like that in July and August and would therefore need to be modelled the way the peak season was in order to produce more reliable predictions.

The results of this study are contrary to the conclusions drawn by Aylen et al (2014), which were derived from their statistical analysis of the relationship between weather, climate and zoo visitation in England. Although no formal climate change impact assessment was conducted, Aylen et al. (2014) suggested that projected climate change in unlikely to have a considerable influence on the total number of visitors to Chester Zoo in England. However, it is reasonable to suggest that the weather sensitivity of zoo visitation varies based on geographic location and climatic conditions. Such disparate results between geographic locations are further emphasised in the findings that suggest zoo visitors in Toronto, Canada found temperatures over 26° C were too hot; whereas, Aylen et al. (2014) reported temperatures exceeding 21° C as being too hot for zoo visitors in Manchester, England.

A number of informed management recommendations can be suggested as an adaptation strategy from the results of this climate change impact assessment of zoo visitation in Toronto. To begin, regardless of the impact of projected climate change, the strongest relationship between visitation and weather in the peak season months of July and August was a negative correlation between Tmin and the total number of daily visitors. Therefore, uncomfortably warm temperatures during the peak season as already a deterrent to zoo visitation. Although this relationship was not expected to cause peak season visitation to decline until additional warming between 3.5 to 5.9 °C was realised (from the mid to late 21st century); it would be in the zoo's immediate best interest to devote further efforts towards mitigating the adverse effects of thermal stress on zoo visitors. Such strategies may include further expansion and promotion of the existing splash pad (water park), as well as increased cooling and hydration stations. Possibly even providing discounted cold beverages and frozen treats when temperatures are perceived by visitors as being too warm (i.e. when Tmax > 26 °C). If the temperatures continue to increase due to the local UHI and global climate change (which is very likely), in additional to threats of declining peak season visitation, there are also opportunities for increased shoulder season visitation. Not only will the potential for increased visitation during the shoulder season months need to be planned for by zoo managers (i.e. additional staff and programmes), but should also be met with increased promotion to take full advantage of this opportunity under projected climate change.

This study was subject to a number of limitations. To begin, the relationship between weather and visitation was determined with participation data that did not reveal the varying characteristics of different zoo visitors. The question that therefore remains is how do different types of zoo visitors perceive and respond to weather (i.e. day users versus season pass holders; families versus couples; the elderly versus young adults). Some of these important questions could be answered by exploring the relationship between weather and zoo visitation using a stated climate preferences approach (Scott et al. 2008; Rutty & Scott, 2010, 2013; Hewer et al. 2014). Furthermore, when validating the predictive models it was acknowledged that the modelled output was not able to capture a number of the annual visitation extremes, which may have been due to prevailing socio-economic conditions or climatic anomalies. In addition, the revealed weather sensitivity and temperature thresholds do not account for the adaptive capacity of tourists (Scott et al. 2012), shifting climate preferences (Gössling et al. 2012) or human acclimatization (Kajan & Saarinen, 2013); concerns that have been raised about using the modelling approach for determining the weather sensitivity and assessing climate change impacts for tourism. This limitation may be overcome by employing a climate analogue approach to better understand how zoo visitation responds during seasons and years when climate conditions are anomalously warm or wet. The use of temporal climate analogues to enable a better understanding of the relationship between weather and tourism and to more accurately assess the impact of climate change on tourism has been strongly encouraged within some of the recent tourism-climate literature reviews (Gössling et al. 2012; Scott et al. 2012), while also being well-defined in its application to the ski industry (Steiger, 2011; Dawson et al. 2009). Nonetheless, the current study has revealed that zoo visitation in Toronto is a highly weather-sensitive tourism and recreation activity which will most likely be impacted by projected climate change, making future research in this regard an important task to enable more informed planning and management.

Acknowledgements

The authors would like to thank the Toronto Zoo for their willingness to support this research by providing the necessary daily zoo visitation data. In particular, we would like to thank the Chief Executive Officer at The Toronto Zoo for taking the time to meet with us, then agreeing to participate in the study and approving the release of the required data. The Senior Director of Marketing, Communications and Partnerships at the Toronto Zoo is also deserving of our gratitude for the time and effort spent organizing the supply and distribution of the visitation data. Furthermore, we would like to thank the former Vice-Principal of Research at the University of Toronto Scarborough for contacting the Toronto Zoo on our behalf, which was certainly instrumental in the working relationship we now have with the zoo.

References

Abegg, B., Konig, U., Burki, R., & Elsasser, H. (1997). Climate impact assessment. *Tourismus Die Erde, 128*, 105- 116.

Aylen, J., Albertson, K., & Cavan, G. (2014). The impact of weather and climate on tourism demand: the case of Chester Zoo. *Climatic Change, 127*, 183-197. http://dx.doi.org/10.1007%2Fs10584-014-1261-6

Barbosa, A. (2009). The role of zoos and aquariums in research into the effects of climate change on animal health. *International Zoo Yearbook, 43*, 131-135. http://dx.doi.org/10.1111/j.1748-1090.2008.00073.x

Becken, S. (2013). A review of tourism and climate change as an evolving knowledge domain. *Tourism Management Perspectives, 6*, 53-62. http://dx.doi.org/10.1016/j.tmp.2012.11.006

Butler, R. (1998). Seasonality in tourism: Issues and implications. *The Tourist Review, 53*, 18-24. http://dx.doi.org/10.1108/eb058278

Canadian Tourism Commission (CTC). (2015). Tourism as Canada's engine for growth: CTC 2014 annual report. Vancouver, Canada: CTC. Retrieved http://en.destinationcanada.com/sites/default/files/pdf/Corporate_reports /2014_annual_report_en_may5.pdf

Dawson, J., Scott, D., & McBoyle, G. (2009). Climate change analogue analysis of ski tourism in north-eastern USA. *Climate Research, 39*, 1-9. http://dx.doi.org/10.3354/cr00793

de Freitas, C. (2003). Tourism climatology: evaluating environmental information for decision making and business planning in the recreation and tourism sector. *International Journal of Biometeorology, 4,* 45-54. http://dx.doi.org/10.1007/s00484-003-0177-z

de Freitas, C., Scott, D., & McBoyle, B. (2008). A second generation climate index for tourism (CIT): Specification and verification. *International Journal of Biometeorology, 52,* 399-407. http://dx.doi.org/10.1007/s00484-007-0134-3

Fenech, A. (2009). *Rapid assessment of the impacts of climate change: an integrated approach to understanding climate change in the Halton region of Ontario, Canada* (Unpublished doctoral dissertation). University of Toronto, Toronto, Canada.

Fenech, A., Comer, N., & Gough, W. (2007). Selecting a climate model for understanding future scenarios of climate change. In A. Fenech & J. MacLellan J. (Eds) *Linking Climate Models to Policy and Decision-making.* Toronto, Canada: Environment Canada.

Fisichelli, N., Schuurman, G., Monahan, W., & Ziesler, P. (2015). Protected area tourism in a changing climate: will visitation at US national parks warm up or overheat? *PLOS ONE, 10*(6), 1-13. http://dx.doi.org/10.1371/journal.pone.0128226

Gössling, S., Scott, D., Hall, C. M., Ceron, J. P., & Dubois, G. (2012). Consumer behaviour and demand response of tourists to climate change. *Annals of Tourism Research, 39,* 36-58. http://dx.doi.org/10.1016/j.annals.2011.11.002

Gough, W. A., & Rosanov, Y. (2001). Aspects of Toronto's climate: heat island and lake breeze. *Canadian Meteorology & Oceanography Society Bulletin, 29,* 67-71.

Hewer, M., Scott, D., & Gough, W. A. (2014). Tourism climatology for camping: a case study of two Ontario parks (Canada). Theoretical & Applied Climatology. http://dx.doi.org/10.1007/s00704-014-1228-6

Intergovernmental Panel on Climate Change (IPCC). (2007). Climate change 2007: The physical science basis. In S. Solomon, D. Qin, M. Manning, Z. Chen, M. Marquis, K. Averyt, M. Tignor & H. Miller (Eds.) *Contribution of working group I to the fourth assessment report of the Intergovernmental Panel on Climate Change.* Cambridge, UK and New York, USA: Cambridge University Press.

Intergovernmental Panel on Climate Change (IPCC). (2010). Good practice guidance paper on assessing and combining multi model climate projections. Boulder, Colorado, USA: National Center for Atmospheric Research.

Intergovernmental Panel on Climate Change (IPCC). (2013). Summary for policymakers. Climate Change 2013: The Physical Science Basis. In T. Stocker, D. Qin, G. Plattner, M. Tignor, S. Allen, J. Boschung, A. Nauels, Y. Xia, V. Bex & P. Midgley (Eds.) *Contribution of working group I to the fifth assessment report of the Intergovernmental Panel on Climate Change.* Cambridge, UK and New York, USA: Cambridge University Press.

Jones, B., & Scott, D. (2006a). Climate change, seasonality and visitation to Canada's national parks. *Journal of Park & Recreation Administration, 24,* 42-62.

Jones, B., & Scott, D. (2006b). Implications of climate change for visitation to Ontario's provincial parks. *Leisure, 30,* 233-261.

Junhold, J., & Oberwemmer, F. (2011). How are the animal keeping and conservation philosophy of zoos affected by climate change? *International Zoo Yearbook, 45,* 99-107. http://dx.doi.org/10.1111/j.1748-1090.2010.00130.x

Kaján, E., & Saarinen, J. (2013). Tourism, climate change and adaptation: a review. *Current Issues in Tourism, 16,* 167-195. http://dx.doi.org/10.1080/13683500.2013.774323

Loomis, J., & Richardson, R. (2006). An external validity test of intended behaviour: comparing revealed preference and intended visitation in response to climate change. *Journal of Environmental Planning & Management, 49,* 621-630. http://dx.doi.org/10.1080/09640560600747562

Mohsin, T., & Gough, W.A. (2012). Characterization and estimation of urban heat island at Toronto: Impact of the choice of rural sites. *Theoretical & Applied Climatology, 108,* 105-117. http://dx.doi.org/10.1007/s00704-011-0516-7

Moreno, A., & Amelung, B. (2009). Climate change and tourist comfort on Europe's beaches in summer: A reassessment. *Coastal Management, 37,* 550-568. http://dx.doi.org/10.1080/08920750903054997

Morgan, R., Gatell, E., Junyent, R., Micallef, A., Ozhan, E., & Williams, A. (2000). An improved user-based beach climate index. *Journal of Coastal Conservation, 6*, 41-50. http://dx.doi.org/10.1007/BF02730466

Njoroge, J. M. (2015). Climate change and tourism adaptation: Literature review. *Tourism & Hospitality Management, 21*, 95-108.

Ontario Ministry of Tourism, Culture and Sport (OMTCS). (2015). Tourism research: Quick facts 2012. Retrieved from http://www.mtc.gov.on.ca/en/research/quick_facts/facts.shtml

Pang, F., McKercher, B., & Prideaux, B. (2013). Climate change and tourism: an overview. *Asia Pacific Journal of Tourism Research, 18*, 4-20. http://dx.doi.org/10.1080/10941665.2012.688509

Pearce-Kelly, P., Khela, S., Ferri, C., & Field, D. (2013). Climate-change impact considerations for freshwater-fish conservation, with special reference to the aquarium and zoo community. *International Zoo Yearbook, 47*, 81-92. http://dx.doi.org/10.1111/izy.12016

Pielke, R. A., & Wilby, R. L. (2012). Regional climate downscaling – what's the point? *Eos, 93*, 52-53. http://dx.doi.org/10.1029/2012EO050008

Rosselló-Nadal, J. (2014). How to evaluate the effects of climate change on tourism. *Tourism Management, 42*, 334-340. http://dx.doi.org/10.1016/j.tourman.2013.11.006

Rutty, M., & Scott, D. (2010). Will the Mediterranean become "too hot" for tourism? A reassessment. *Tourism & Hospitality Planning & Development, 7*, 267-281. http://dx.doi.org/10.1080/1479053X.2010.502386

Rutty, M., & Scott, D. (2013). Differential climate preferences of international beach tourists. *Climate Research, 57*, 256-269. http://dx.doi.org/10.3354/cr01183

Scott, D, Gössling, S., & de Freitas, C. (2008). Climate preferences for tourism: Evidence from Canada, New Zealand and Sweden. *Climate Research, 38*, 61-73. http://dx.doi.org/10.3354/cr00774

Scott, D., & Jones, B. (2006). The impact of climate change on golf participation in the Greater Toronto Area (GTA): A case study. *Journal of Leisure Research, 38,* 363-380.

Scott, D., & Jones, B. (2007). A regional comparison of the implications of climate change on the golf industry in Canada. *The Canadian Geographer, 51,* 219-232. http://dx.doi.org/10.1111/j.1541-0064.2007.00175.x

Scott, D., Gössling, S., & Hall, C. M. (2012). International tourism and climate change. *WIREs Climate Change, 3,* 213-232. http://dx.doi.org/10.1002/wcc.165

Scott, D., Jones, B., & Konopek, J. (2007). Implications of climate and environmental change for nature-based tourism in the Canadian rocky mountains: A case study of Waterton Lakes National Park. *Tourism Management, 28*, 570-579. http://dx.doi.org/10.1016/j.tourman.2006.04.020

Scott, D., McBoyle, G., & Mills, B. (2003). Climate change and the skiing industry in southern Ontario (Canada): Exploring the importance of snowmaking as a technical adaptation. *Climate Research, 23,* 171-181. http://dx.doi.org/doi:10.3354/cr023171

Scott, D., McBoyle, G., & Minogue, A. (2007). The implications of climate change for the Québec ski industry. *Global Environmental Change, 17*, 181-190. http://dx.doi.org/10.1016/j.gloenvcha.2006.05.004

Scott, D., McBoyle, G., & Schwartzentruber, M. (2004). Climate change and distribution of climatic resources for tourism in North America. *Climate Research, 27*, 105-117. http://dx.doi.org/doi:10.3354/cr027105

Scott, D., McBoyle, G., Minogue, A., & Mills, B. (2006). Change and the sustainability of ski-based tourism in eastern North America: A reassessment. *Journal of Sustainable Tourism, 14*, 376-398. http://dx.doi.org/10.2167/jost550.0

Steiger, R. (2011). The impact of snow scarcity on ski tourism: an analysis of the record warm season 2006/2007 in Tyrol (Austria). *Tourism Review, 66*, 4-13. http://dx.doi.org/10.1108/16605371111175285

Toronto Zoo. (2014). Saving and protecting species at home and abroad: Toronto Zoo annual report 2013. Toronto, Canada: Toronto Zoo. Retrieved http://www.torontozoo.com/pdfs/Toronto%20Zoo_2013_Annual_Report.pdf

Tourism Toronto. (2014). Elevating the "new Toronto": Tourism Toronto annual report 2013. Toronto, Canada: Tourism Toronto. Retrieved http://www.seetorontonow.com/getattachment/26d64e70-372a-4282-91a1-15fcc0b4b7f7/AnnualReport2013_email.pdf.aspx

United Nations World Tourism Organization (UNWTO). (2015). UNWTO annual report 2014. Madrid, Spain:

UNWTO. Retrieved http://dtxtq4w60xqpw.cloudfront.net/sites/all/files/pdf/unwto_annual_report_2014.pdf

Wilby, R. L., & Dawson, C. W. (2012). The statistical downscaling model (SDSM): Insights from one decade of application. *International Journal of Climatology, 33*, 1707-1719. http://dx.doi.org/10.1002/joc.3544

Wilby, R. L., Dawson, C. W., & Barrow, E. M. (2001). SDSM – a decision support tool for the assessment of regional climate change impacts. *Environmental & Modelling Software, 17*, 145-157. http://dx.doi.org/10.1016/S1364-8152(01)00060-3

Williams, P., Dossa, K., & Hunt, J. (1997). The influence of weather context on winter resort evaluations by visitors. *Journal of Travel Research, 36*, 29-36. http://dx.doi.org/10.1177/004728759703600205

Yukic, T. (1970). *Fundamentals of Recreation: 2nd Edition*. New York, USA: Harper & Row; 1970.

11

Interaction of Strength and Stress in High, Steep Rock Slopes

John V. Smith[1]

[1] School of Engineering, Royal Melbourne Institute of Technology University, Victoria, Australia

Correspondence: John V. Smith, School of Engineering, Royal Melbourne Institute of Technology University, Victoria, GPO Box 2476 Melbourne, Victoria 3001, Australia. E-mail: johnv.smith@rmit.edu.au

The research is financed by Royal Melbourne Institute of Technology University.

Abstract

The strength of rock mass and the stress in a slope are each complex fields of investigation. They are also intimately related as increasing confining stress makes a rock mass stronger and the strength of a rock mass can limit the magnitude of stress. Whereas these interactions are comparatively well understood for soils, principally through the advances of laboratory soil mechanics, the scale of rock masses, principally the presence of discontinuity surfaces, limits the capacity for laboratory investigation. The interaction of strength and stress in rock slopes is most evident in high, steep slopes where stress is typically greater. The slope angle and failure mechanisms occurring in the rock slope can reveal the ways that strength and stress interact to produce the observed morphology. McKay Bluff, near Nelson, South Island, New Zealand, is a high, steep rock slope affected by marine coastal erosion at its base. Finite element modeling illustrates sensitivities in determination of the stress magnitude in the slope. Engineering geology methods demonstrate the difficulty in precise determination of the rock mass strength. The ranges of these parameters are compared to find a compatible range for the interacting factors. The stress in a range of other high, steep slope types is reviewed and the implications for geomorphic analysis are discussed.

Keywords: rock slope, stress, rock mass strength, Nelson, New Zealand

1. Introduction

In a review of rock coast geomorphology Naylor, Stephenson, and Trenhaile (2010) noted "modeling would also benefit from collaboration with engineering geologists, who are making considerable progress in our ability to predict how rock masses are likely to erode." There are also opportunities for engineering geologists to learn from geomorphological approaches to the processes that control the height and steepness of natural slopes. The steepness of rocky coasts, for example, varies such that "near-vertical profiles occur where marine processes are dominant, whereas subaerial erosion tends to smooth and lower the cliff slope" (Kennedy & Dickson, 2007). The shape of naturally formed slopes in rock represents the complex interplay of the strength of rock and the low or high stress conditions to which it is subjected. At a coast this interplay can be driven by erosional undercutting of the slope toe leading to the slope forming its steepest stable slope.

1.1 Strength of Rock

The strength of intact rock samples can be measured in the laboratory. The strength is known to vary according to sample size and according to the confining stress during testing (Paterson & Wong, 2005). The strength of a rock specimen can be measured in confined (such as triaxial) or unconfined (such as uniaxial) conditions. Uniaxial compressive strength (UCS) refers to a standard sample size, shape and standard test conditions – including zero confining stress (ISRM, 1987).

In the field, measuring the strength of rock is a more complex problem as it is characterized by the interaction of intact rock strength and the spacing, orientation, length and frictional properties of discontinuities (together known as rock mass, Wyllie & Mah, 2004). Field mapping of these features can be used to assess the rock mass strength using methods such as the Geological Strength Index (GSI) within the Hoek-Brown rock mass strength model (Hoek & Brown, 1997; Marinos, Marinos, & Hoek, 2005). The Hoek-Brown model treats discontinuities as being effectively isotropic in distribution and orientation and therefore explicitly excludes the role of problematic structural orientations that are susceptible to kinematic failures such as planar and wedge block sliding (Hoek & Brown, 1997; Marinos et al., 2005).

Many rock slope stability problems are controlled by the kinematics of structures, mainly defects and weaknesses such as joints, bedding, foliation and faults (Wyllie & Mah, 2004; Youssef, 2012). These defects are typically oriented in particular directions such that the rock mass is anisotropic. Therefore the rock mass strength methods alone are insufficient and it is also necessary to analyse stability of structural blocks at the rock slope surface.

The mechanisms of deformation involving surfaces of weakness that allow blocks of rock to separate and fall or slide from a slope can be highly complex (e.g. Alejano, Gómez Márquez, & Martínez Alegría, 2010). Where deformation mechanisms can be inferred accurately, the stability of a slope can be analysed directly by a range of numerical methods (e.g. Stead, Eberhardt, & Coggan, 2006). The important role of discontinuities in the physical erosion of a rock mass has been applied in studies of coastal rock platforms (Naylor & Stephenson, 2010) and the size of coastal boulders (Stephenson & Naylor, 2011). The role of large geological structures such as faults, has been investigated for large landslides (Delgado et al., 2011) and the development of rocky coastal features, cliffs and embankments (Gómez Pujol, Vicente, García Tortosa, Alfaro, Estévez, López Sánchez, & Mallorquí, 2013). In particular, rock mass strength and kinematic slope instability has been shown to be highly influenced by the presence of major faults (Korup, 2004; Brideau, Yan, & Stead, 2009).

1.2 Stress in Slopes

The stability of a slope is typically analysed by assessing the equilibrium of forces in the slope. The balance of driving and resisting forces on an existing or newly formed failure surface leads to a limit-equilibrium analysis (Wyllie & Mah, 2004). This approach is not adequate for high, steep slopes as a clearly defined failure surface does not typically exist or develop, making the assessment of the stability of slopes in strong rocks more problematic than for soils and weak rocks (Wyllie & Mah, 2004).

Stress analysis provides additional important information needed to assess slope stability in high, steep slopes. Stress in a slope can be modeled by use of finite element analysis however, the results are typically highly sensitive to the choice of input parameters. In particular, elastic models do not limit the magnitude of stress and unrealistically high stresses can be found in such models. Where plastic materials are used the stress is limited by the strength of the rock and therefore the derived stress is highly sensitive to the selected strength model (Kinakin & Stead, 2005).

In this paper, the geomorphic and geological factors controlling strength and stress in high, steep rock slopes is illustrated using simple finite element modeling of a relevant field example. The main sensitivities are highlighted to provide general understanding of stress levels in such slopes. The role of stress in geomorphic processes is further illustrated for range of landform profile shapes.

2. Field Example McKay Bluff, New Zealand

2.1 Regional Setting

McKay Bluff is a high coastal cliff located approximately 10 km northeast of the city of Nelson on the South Island of New Zealand (Figure 1). Rocks eroded from the cliffs of McKay Bluff are considered to have contributed to the distinctive boulder bank that continues for approximately 13 km from the southern end of the bluff and forms a barrier separating the estuary, Nelson Haven, from Tasman Bay (Figure 1). Studies of the formation of the Nelson Boulder Bank have been published by Dickinson and Woolfe (1997), Johnston (2001), Hartstein and Dickinson (2001) and the geological background of the location is well documented in those works. An analysis of the wave energy in Tasman Bay, to which the cliffs of McKay Bluff are exposed, is provided by Hartstein and Dickinson (2006).

2.2 Rock Slope Characteristics

The slope described in this study comprises approximately 1000 m of Cable Granodiorite exposed along the coast (Figure 1). The mineral proportions and density of granodiorite are well defined and the average density of granodiorite has been taken as 2.73 g/cm3 (AusIMM, 1995).

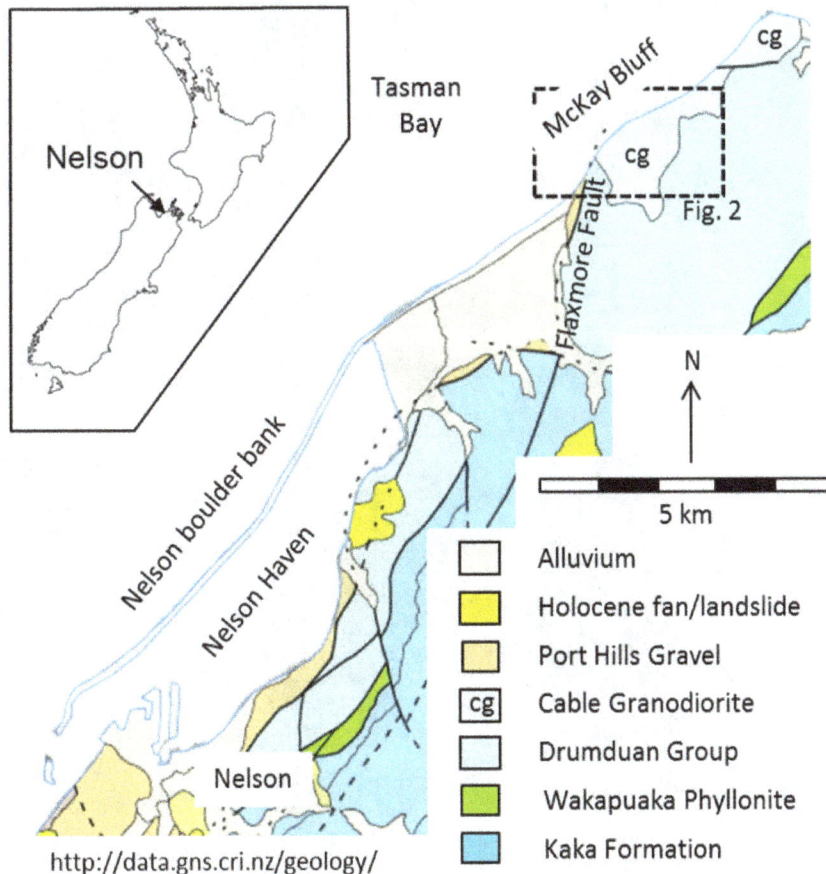

Figure 1. Location and geology map of McKay Bluff and the Nelson boulder bank, South Island New Zealand

Field data were collected from rock outcrops at five locations along the base of the slope (Figure 2, Table 1). At the northeastern end of this transect the coast makes a gentle convex bend to the east such that photographs along the coast can show a true profile of the slope (Figure 2 & 3).

McKay Bluff rises to a height of over 400 m above sea level and has a slope angle approximating 50° at its base and approximating 45° for a height of over 300 m (Figure 3). The base of the cliff is constantly being undercut by marine erosion and it is assumed the observed slope represents equilibrium between the active subaerial slope processes and the geological conditions in the slope. McKay Bluff provides an opportunity to study a high, steep slope and to investigate the mechanisms of slope instability.

2.3 Intact Rock Strength

The intact rock strength was measured using a Schmidt hammer which records the rebound of an impacting weight. The instrument used was a Proquest Rock Schmidt with an impact energy of 2.207 Nm. The instrument was used according to the International Society for Rock Mechanics (ISRM) recommended method which involves taking an average of 20 readings to derive each strength value (ISRM, 1987). Schmidt hammer rebound values (R) can be correlated with uniaxial compressive strength (UCS) and the elastic modulus (Katz, Reches, & Roegiers, 2000). A total of seven strength measurements were collected at the five locations (Table 2). The correlated UCS and elastic modulus of the rock are given in Table 2. The data on which the correlation is based does not support extrapolation above 300 MPa (Katz et al., 2000).

2.4 Rock Mass Characteristics

The joint spacing was measured at each of the five locations by counting the number of joints intersecting reference lines. At each location three, 2 meter long, reference lines oriented vertically and in two mutually perpendicular horizontal directions were used for joint spacing measurement. These data and the calculated volumetric joint count (Palmstrom, 2005) and block volume are provided in Table 1.

Figure 2. Topographic map satellite image and synthetic aerial view of the slopes at McKay Bluff (viewed toward the east). Google Earth (2008). Locations 1-5 referred to in the text are marked

Figure 3. Photograph of the slopes at McKay Bluff (viewed toward the north)

Table 1. Field locations (using hand-held GPS) and joint spacing data

Location	South	East	Average joints/m	Volumetric joint count	Block volume dm^3
1	41.18015	173.36762	4.3	8.7	41.5
2	41.17644	173.37077	4.5	9.0	37.0
3	41.17527	173.37172	4.8	9.7	29.9
4	41.17435	173.37229	4.0	8.0	52.7
5	41.17345	173.37367	3.5	7.0	78.7

Table 2. Field data on intact rock strength (Schmidt Hammer) following ISRM (1987) recommended method of 20 readings per UCS value. UCS and elastic modulus correlations are from Katz et al. (2000)

Location	Rebound number (R)	R standard deviation	UCS (MPa)	Elastic Modulus GPa	Comment
1	66	5.8	184	55	Red oxidized blocky rock
1	66.5	4.8	190	56	Green blocky rock
2	65.5	3.1	178	53	
3	69	7.3	225	63	
4	76.5	2.3	372	86	Fresh, wave-worn outcrop
5	60.5	5.6	127	42	
5	74.5	4.1	325	79	

The joints were observed to be slightly wavy on a large-scale and smooth to slightly rough on small scale observations, supported by measurement of the joint roughness coefficient using a brush-gauge profiling tool (Barton, 2013). The joint faces were slightly weathered with some being coated with a hard epidote alteration product less than 1 mm thick. At each location selected joint orientations were measured using a magnetic compass and corrected for magnetic declination of +22 degrees. A total of 60 joint measurements were made.

2.5 Rock Mass Strength

The Hoek-Brown rock mass strength criterion is a methodology used for determining the strength of a rock mass based mainly on field observations (Hoek & Brown, 1997; Marinos et al., 2005). The Hoek-Brown rock mass strength criterion is based on the uniaxial compressive strength (UCS), the intact rock property value (m_i) and the Geological Strength Index (GSI). It is rare for a rock mass to have a unique strength value and it is common to consider a range of values (Hoek & Brown, 1997). In particular, the GSI value was intended to be assessed visually and not considered in divisions below 5 points (Hoek & Brown, 1997; Marinos et al., 2005). Visual field assessment, joint spacing measurements and stereographic analysis of the joint orientations identified the rock mass as 'blocky' in the GSI framework ("very well interlocked undisturbed rock mass consisting of cubical blocks formed by three orthogonal discontinuity sets") with 'fair' joint conditions ("smooth moderately weathered or altered surfaces"). Other authors have provided quantitative guides for GSI (e.g. Cai, Kaiser, Uno, Tasaka, & Minami, 2004). Using the spacing and block volume measurements (Table 1) and a joint condition rating of fair a representative range of GSI from 45 to 55 was obtained. These parameters were applied to derive a range of rock mass strength values (RocLab, version 1.033, RocScience, Table 3, Figure 4).

The rock mass strength parameter used in this study is the unconfined rock mass strength which is applicable to the stress conditions near the slope face. Other strength parameters are used to assess the strength deeper into the slope such as the global rock mass strength which is higher and includes the effect of confining stress within the slope (Hoek, 2005).

Table 3. Representative rock mass strength parameters derived using the Hoek-Brown criterion (RocLab, V 1.033, RocScience)

Parameter	Low	High
RocLab Inputs		
Uniaxial compressive strength of intact rock, UCS (MPa)	125	250
Hoek-Brown parameter, m_i	29	29
Elasticity of intact rock, E_i (GPa)	43.1	66.4
Unit weight (kN/m³)	26	26
Geological Strength Index, GSI	45	55
Poissons Ratio	0.3	0.3
Peak m_b	4.067	5.813
Peak s	0.0022	0.0067
Residual m_b	1	1
Residual s	0.001	0.001
RockLab Outputs		
Uniaxial rock mass compressive strength(MPa)	5.6	20.1
Rock mass modulus (GPa)	9.63	27.13

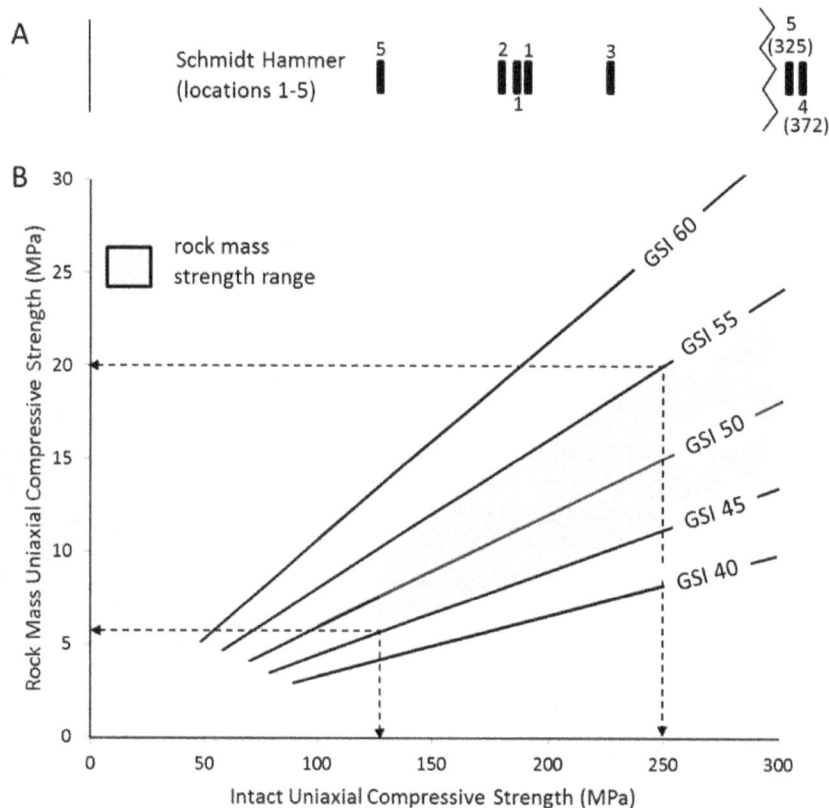

Figure 4. (A) Intact uniaxial compressive strength determined in the field by Schmidt hammer (scale at base of B. (B) Graph of rock mass uniaxial compressive strength (Rocscience RocLab V 1.033) for selected values of Geological Strength Index (GSI) versus uniaxial compressive strength range

2.6 Rock Slope Kinematics

The rocky outcrops on the slope face are of variable steepness and are locally affected by small-scale (<10 m) blocky failure zones. Some of these rocky outcrops are affected by wedge sliding failure formed by intersecting discontinuity planes (Figure 5). Blocky failure zones are also formed by the release of blocks on planes dipping steeply into the face (Figure 6). These failures can be described as toppling wedges as the steeply dipping planes allow the release of blocks from the slope.

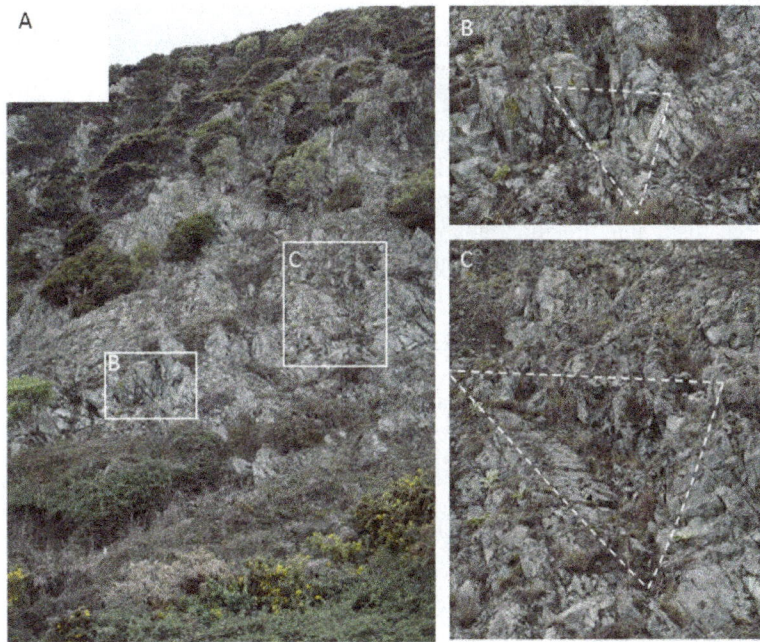

Figure 5. (A) Rocky outcrop approximately 10 m above beach level. (B and C) Sliding wedges (field of view 3 m across)

Figure 6. (A) Debris fan and rocky outcrop (viewed toward south). (B) Multiple toppling wedges (field of view 6 m across)

The active nature of these erosion mechanisms can be observed by the relatively fresh appearance of some wedge scars on the rock face and the presence of open cracks observed on the rock face. Such structural failures can occur in conditions of stress below or above the rock mass strength. If such structural failure processes are highly active it is possible that they will reduce the slope steepness to the extent that the stress in the slope is well below the rock mass strength. The cause of such structural instability must be assessed carefully, it has been proposed that some structural failure assumed to be directly gravity driven could be stress-induced (Smith 2014).

2.7 Rock Slope Stress

Stress is a three-dimensional tensor which is difficult to measure but deceptively easy to model. Stress in a slope is the distribution of forces, primarily due to gravity, through the topography. The fundamental controls on stress include density of the rock, height of the slope and steepness of the slope. Other factors also need to be considered (Table 4). Due to tectonic processes additional forces, typically horizontal, contribute to the total stress pattern. The World Stress Map (Heidbach, Tingay, Barth, Reinecker, Kurfeß, & Müller, 2008) provides a compendium of stress measurements from a range of sources. The nearest high quality reading to the site is an earthquake focal mechanism interpretation approximately 100 km southwest of the site. At that site the maximum stress is horizontal compression oriented west-northwest. This maximum compression direction is corroborated by other data from around the region (Heidbach et al., 2008).

Another influence is that stress cannot exceed the strength of the rock and the rock fails or yields where stress reaches equivalence with the strength. Where local failure or yielding occurs, stress is redistributed to surrounding parts of the rock. The presence of water also plays an important role in slope stability as the pore pressure impacts on rock and soil strength.

A finite element modeling software package has been used to provide a range of stress analyses in slopes and to assess sensitivities (Phase2, V. 8.020, RocScience). Finite element modeling (FEM) divides a two-dimensional area or three-dimensional volume into a mesh of polyhedral elements. Three-dimensional FEM is commonly available, however, two-dimensional FEM can give results that are comparable to three-dimensional analysis in situations where the geometry remains relatively consistent in the out-of-plane direction and the rockmass can be considered homogeneous and isotropic.

The FEM package solves the stress magnitude and orientation for each element providing a continuous stress model with a resolution dependent on the fineness of the mesh. Greater resolution can be achieved at the expense of longer computing times. The mesh size therefore represents a smoothing effect which must be taken into account when maximum or minimum stress values are quoted. The FEM can be interrogated to provide precise values but these values will vary depending on the mesh type and size.

A series of FEM stress analyses were conducted (Phase 2 V. 8.020, Rocscience) to illustrate difficulties in establishing meaningful results. The first and second models (Table 5) illustrate the sensitivity to the proximity of the area of interest in the model (in this case the slope and its base in particular) to the model boundary. Models 2, 3 and 4 (Table 5) show that where an angular corner exists in a model, increasing the resolution of the analytical mesh (i.e. reducing the element size) can result in great differences in maximum stress values.

Models 5 and 6 (Table 5) show the application of an inferred three-dimensional stress distribution with a maximum compression across the slope, an intermediate compression along the slope and the minimum stress being vertical (thrust faulting pattern). It can be seen that the stress ratio greatly influences the maximum stress compared to the equivalent isotropic model (Model 2 & 3, Table 5). Models 5 and 6 (Table 5) also illustrate the sensitivity to increasing resolution of the model (Figure 7A & B).

Models 7, 8 and 9 (Table 5) show the effect of smoothing the angular base of the slope with reference to isotropic stress conditions (Model 7) and an inferred stress condition range at the site (Models 8 & 9). Models 10, 11 and 12 (Table 5) have an increased radius of curvature at the base compared to the previous models. These models show a drop in stress at the base of the slope. The high stresses are still present but they are now separated from the surface by a low stress zone (Figure 7C). The low stress zone is caused by the mis-match between the applied stresses and the curvature of the slope. The phenomenon is well known in stress around tunnels in which good matching of curvature to stress conditions (e.g. circular in isotropic conditions or elliptical in non-isotropic stress conditions) is known to provide an even distribution of stress (Brady & Brown, 1993, Ren, Zuo, Xie, & Smith, 2014).

Table 4. Factors considered in assessing stress in slopes

Factor	Field data	Modeling	McKay Bluff, NZ
Slope height	Surveys	Model geometry	Approx 400 m total, main steep slope 200-300 m high.
Slope steepness	Surveys	Model geometry	Approx. 50° flattening progressively upward
Rock density	Laboratory measurement, assessment from mineral content, affected by void space	Material input	Granodiorite, mineral proportions and density well defined. Minimal reduction by fractures and voids
3D stress state including tectonic forces	Inferred from earthquakes and other field data	Adjust ratio of vertical and horizontal stress components. 2D models can incorporate through adjusting out-of-plane stress	Thrust tectonic regime. Nearest published stress data 100 km southwest. Horizontal compressive stress component across slope is maximum. Vertical stress component is minimum (thrust fault regime)
Yielding of rock	Rock mass strength assessment using engineering geology methods	Elastic models have no yielding and therefore can develop false high stress. Plastic models include yield but residual strength values poorly known	Elastic stress for the slope geometry exceeds rock mass strength. Further assessed through plastic strength models.
Structural failures	Sliding and toppling of blocks which can occur below rock mass strength	These mechanisms are modeled by limit-equilibrium and kinematic methods	Sliding and toppling wedge blocks observed
Water	Observations (e.g. springs), water table measurements	Water table surface or transient flow models	Wet, drained conditions
External loads	Sediments, engineering works, earthquakes	Model as overlying material or directly as applied distributed forces. Earthquakes modeled pseudo-statically as horizontal acceleration	Coastal sediment – load is very low compared to stresses in rock
Stress concentration	Local shape changes such as angularity or curvature of slope base. Local yielding to self-modify shape may occur.	Can impact on 'maximum' stress values observed at model surface. Extreme high magnitude can occur at corners. Unexpected low values can occur on curves	Slope base covered by sediment. No evidence of stress-induced concavity at base
Model boundary effects and mesh resolution	NA	Features of interest in model should be adequately distant from boundaries (e.g. 2 times slope height). Stress values measured can be highly sensitive to the model resolution. Coarse mesh models smooth stress concentration effects	Examples described in text

Models 1 to 12 (Table 5) are all based on elastic material properties. The deficiency of elastic models is that they can report stresses which are far higher than the strength of the rock mass. For an elastic model it is necessary and useful to observe the locations where the stress has exceeded the material strength. Once plastic properties are incorporated in the model these selected properties create a feedback to the stress conditions and it is possible to misinterpret whether the material has limited the stress or the stress is coincidently equivalent to the strength.

Models 13 and 14 (Table 5) apply plastic strength behaviour based on a low and high strength condition, respectively (Table 3). The stress conditions are equivalent to model 8 (Table 5) and the base of the slope of the three models is shown for comparison in Figure 8. Not surprisingly, the material has locally yielded causing significant changes to the stress patterns (Figure 8 B & C). The limit to the zone of intense yielding corresponds to the unconfined compressive strength of the materials (5.6 MPa & 20 MPa, respectively). This correspondence is also shown on Figure 4. This example illustrates the important point that for plastic materials the strength and stress are very intimately related.

The local stress reduction (Figure 8 B & C) indicates that local yielding or failure has occurred near the face of the slope. The yielded part of the rock mass no longer carries stress and consequently, the maximum stress magnitude at the surface (as listed in Table 5) is located above and/or below the yielded portion of the slope face.

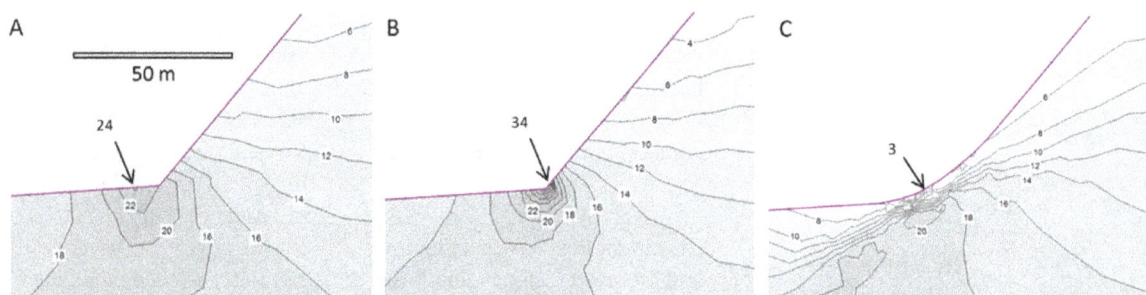

Figure 7. Finite element model (Phase 2 V. 8.020, Rocscience) illustrating differences in stress analysis for (A) an angular base with element size of approximately 5 m, (B) an angular base with element size of approximately 1.5 m and (C) a smooth shape at the base of a slope (radius of curvature approximately 50 m). The FEM output shows shaded contours of the magnitude of the maximum principal compressive stress with selected contours labelled (MPa)

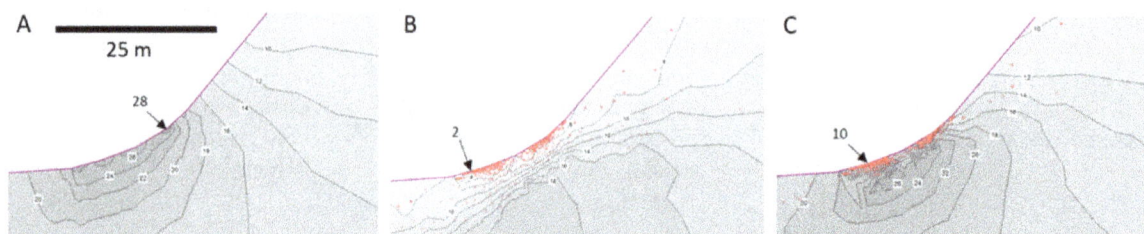

Figure 8. Finite element model (Phase 2 V. 8.020, Rocscience) illustrating differences in stress analysis for radius of curvature approximately 25 m. (A) Elastic material, (B) plastic material with low strength properties and (C) plastic material with low strength properties. Yielded elements shown as red symbols. See Table 3 for low and high strength properties. The FEM output shows shaded contours of the magnitude of the maximum principal compressive stress with selected contours labelled (MPa)

Table 5. Stress in a numerical model slope 400m high approximately 50° slope, 2D model with out-of-plane stress adjustments (V=vertical stress, Hp=horizontal stress parallel to slope, Ht=horizontal stress transverse to slope). (Phase2, V. 8.020, RocScience)

No.	Distance to model boundary relative to slope height	Approx. Element size at base (m)	Shape at base (radius of curvature, m)	3D stress V:Hp:Ht	Material type	Max. surface stress (MPa)	Comment
1	1	5	angular	isotropic	elastic	5	Poor model dimensions
2	>2	5	angular	isotropic	elastic	9	"Maximum stress"
3	>2	1.5	angular	isotropic	elastic	16	related to mesh density
4	>2	0.5	angular	isotropic	elastic	28	(i.e. resolution)
5	>2	5	angular	1:1.5:2.5	elastic	24	(Figure 7A)
6	>2	1.5	angular	1:1.5:2.5	elastic	34	(Figure 7B)
7	>2	1.5	25	isotropic	elastic	11	
8	>2	1.5	25	1:1:2	elastic	28	Figure 8A
9	>2	1.5	25	1:1.5:2.5	elastic	30	
10	>2	1.5	50	isotropic	elastic	11	
11	>2	1.5	50	1:1:2	elastic	7	Near-surface unloading at base due to shape. Minimum surface stress 4 MPa, 19 MPa at 3m below surface
12	>2	1.5	50	1:1.5:2.5	elastic	6	Near-surface unloading at base due to shape. Minimum surface stress 3 MPa, 20 MPa at 5m below surface (Figure 7C)
13	>2	1.5	25	1:1:2	plastic	8	Low strength yielding at approx. 6 MPa. Minimum surface stress at base 2 MPa (Figure 8B)
14	>2	1.5	25	1:1:2	plastic	20	High strength yielding at approx. 20 MPa Minimum surface stress at base 10 MPa (Figure 8C)

3. Discussion

Based on the field investigations and finite element modeling, it has been possible to bracket the range of strength of the rock mass. It is likely that the rock mass does not have a unique strength but has parts which cover this range. It has also been possible to derive expected stress conditions although the values are highly sensitive to curvature of the slope base (concealed by debris) and the three-dimensional stress regime (inferred only from published records a significant distance away). By considering low and high strength values in the inferred strength range it has been possible to assess the possible extent of yielding where stress is highest (base of the slope). In this way, the likely strength and stress conditions in the slope, while not necessarily 'determined', have been bracketed to a satisfactory level of detail for the aims of this study – which was to illustrate the interaction of the parameters. Clearly, the resources committed to data collection and modeling can be great and depend on the expected benefits of more precise estimates.

More generally, Kennedy and Dickson (2007) showed steep slope profile shapes representative of the relative efficiency of marine and sub-aerial erosion processes (Figure 9). It is understood that the profiles presented by Kennedy and Dickson (2007) are illustrative only. However, they form a useful framework for demonstrating the range of stress distributions which may occur in such slopes. In general, the magnitude of stress developed in a slope is proportional to the height and steepness of the slope (Table 6). The stress is also sensitive to the density of the rock mass and the three-dimensional stress conditions. Three of the illustrated profiles have an angular base which results in extremely high stress values in a high resolution elastic FEM model. In nature, it is expected that high stress would cause yielding of the material that would allow erosion of the slope to a less angular shape. Therefore determining the true maximum state of stress in a slope requires very detailed surveying of the slope shape, particularly at the base. However, the base of a slope is often concealed by debris making such surveying difficult. As an approximation to the maximum stress in slopes of the height and steepness shown in Figure 9, the results from the FEM model with coarse mesh at the toe (approximately 5 m across) can be considered as an initial guide. The models used elastic material properties and therefore the stress values are independent of the material strength. The stress values derived in these simple models can be compared to the strength of material to determine what form of strength-stress interaction is to be expected. If the elastic stress values are well below the rock mass strength then it is likely that stress-induced failures will not be significant. It would also be likely that the active erosional processes would be directly gravity driven and possibly structural in nature, such as block sliding or toppling. If the stress values are similar to, or above, the rock mass strength then plastic strength models should be applied to determine the zones of yielding. Zones of yielding in a model can be compared to the shape of the landform to validate the strength, as the yielded material is likely to have been eroded – leading to a change of shape and therefore a further change in stress.

Table 6. Stress in numerical model slopes (based on representative types of Kennedy and Dickson 2007). Analysis elements approx. 5m at base, stress isotropic, elastic material, unit weight 29 kN/m3. (Phase2, V. 8.020, RocScience)

Type	Height (m)	Max. slope angle (degrees)	Shape at base	Approximate max. stress (MPa)
Vertical	200	86	angular	8
Vertical	400	86	angular	16
Extremely steep	200	79	angular	7
Extremely steep	400	79	angular	14
Very steep	200	71	angular	5
Very steep	400	71	angular	10
Steep	200	60	smooth	3
Steep	400	60	smooth	6

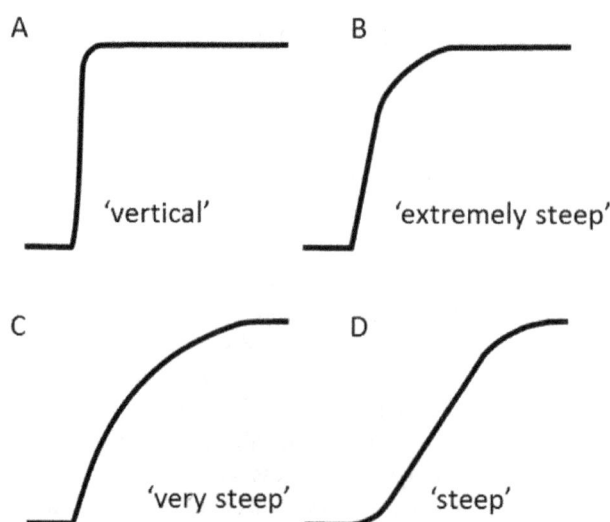

Figure 9. Coastal slope profile types with relative contribution of marine versus sub-aerial erosion decreasing from A to D (Kennedy and Dickson 2007). Short-hand names used for each type in this paper are in inverted commas

4. Conclusions

The stress in a slope is directly related to the height and steepness of the slope. The force of gravity is distributed through the rock mass as stress which increases in magnitude with the height and steepness of the landform profile. Where the strength of the rock is exceeded by the stress, the rock locally yields. This yielding or failure limits the magnitude of the stress that can develop in the slope. Also, local yielding weakens the rock and promotes reduction of slope angle and slope height through erosion and therefore reduces the stress in the slope. It is expected that the landform profiles of actively eroding natural coastal slopes represent the interplay of stress (controlled by height, steepness, rock density and limited by yield strength) and strength (controlled by the strength of the rock material and by the character and orientation of discontinuities in the rock mass). The stress in a slope can be estimated by finite element modeling however, the results are highly sensitive to a range of factors. The strength of a rock mass can be estimated using engineering geological methods but is expected to be a range rather than a unique value.

High, steep coastal cliffs at McKay Bluff near Nelson on the South Island of New Zealand have stress levels comparable to the range of strength of the rock mass. The lower end of strength estimates is likely to result in local yielding of the rock mass. The higher end estimates exceed the stress in the slope and slope failure mechanisms in these materials would be gravity-controlled structural sliding and toppling rather than stress-induced yielding.

Acknowledgments

The research was supported by RMIT University.

References

Alejano, L. R., Gómez Márquez, I., & Martínez Alegría, R. (2010). Analysis of a complex toppling-circular slope failure. *Engineering Geology, 114*, 93-104.

AusIMM. (1995). *Field geologists manual*. Australasian Institute of Mining and Metallurgy.

Barton, N. (2013). Shear strength criteria for rock, rock joints, rockfill and rock masses: Problems and some solutions. *Journal of Rock Mechanics and Geotechnical Engineering, 5*, 249–261.

Brady, B. H. G., & Brown, E. T. (1993). *Rock mechanics for underground mining*, Chapman and Hall.

Brideau, M. A., Yan, M., & Stead, D. (2009). The role of tectonic damage and brittle rock fracture in the development of large rock slope failures. *Geomorphology, 103*, 30–49.

Cai, M., Kaiser, P. K., Uno, H., Tasaka, Y., & Minami, M. (2004). Estimation of rock mass deformation modulus and strength of jointed hard rock masses using the GSI system. *International Journal of Rock Mechanics & Mining Sciences, 41*, 3–19.

Delgado, J., Vicente, F., García Tortosa, F., Alfaro, P., Estévez, A., López Sánchez, J. M., & Mallorquí, J. J. (2011). A deep seated compound rotational rock slide and rock spread in SE Spain: Structural control and DInSAR monitoring. *Geomorphology, 129*, 252-262.

Dickinson, W. W., & Woolfe, K. J. (1997). An in-situ transgressive barrier model for the Nelson Boulder Bank New Zealand. *Journal of Coastal Research, 13*, 937-952.

Gómez Pujol, L., Gelabert, B., Fornós, J. J., Pardo Pascual, J. E., Rosselló, V. M., Segura, F. S., & Onac, B. P. (2013). Structural control on the presence and character of calas: Observations from Balearic Islands limestone rock coast macroforms. *Geomorphology, 194*, 1-15.

Google Earth 6.0. (2014). Retrieved August 31, 2014, from http://www.google.com/earth/index.html

Hartstein, N. D., & Dickinson, W. W. (2001). Gravel Barrier Migration and Overstepping in Cable Bay Nelson New Zealand. *Journal of Coastal Research, 34*, 256-266.

Hartstein, N. D., & Dickinson, W. W. (2006). Wave energy and clast transport in eastern Tasman Bay New Zealand. *Earth Surface Processes and Landforms, 31*, 703-714.

Heidbach, O., Tingay, M., Barth, A., Reinecker, J., Kurfeß, D., & Müller, B. (2008). The World Stress Map database release. http://dx.doi.org/10.1594/GFZ.WSM.Rel2008, 2008

Hoek, E. (2005). Uniaxial compressive strength versus Global strength in the Hoek-Brown criterion Evert Hoek Vancouver. Retrieved March 30, 2005, from http://wwwrocsciencecom/library/rocnews/Spring2005htm accessed 17/7/2013

Hoek, E., & Brown, E. T. (1997). Practical Estimates of Rock Mass Strength. *International Journal of Rock*

Mechanics and Mining Sciences, 34, 1165-1186.

ISRM. (1987). Suggested methods for determining hardness and abrasiveness of rocks. *International Journal of Rock Mechanics and Mining Sciences & Geomechical Abstracts, 15*, 89-97.

Johnston, M. R. (2001). Nelson Boulder Bank, New Zealand. *New Zealand Journal of Geology and Geophysics, 44*, 79-88.

Katz, O., Reches, Z., & Roegiers, J. C. (2000). Evaluation of mechanical rock properties using a Schmidt Hammer. *International Journal of Rock Mechanics and Mining Sciences, 37*, 723-728.

Kinakin, D., & Stead, D. (2005). Analysis of the distributions of stress in natural ridge forms: implications for the deformation mechanisms of rock slopes and the formation of sackung. *Geomorphology, 65*, 85-100.

Korup, O. (2004). Geomorphic implications of fault zone weakening: slope instability along the Alpine Fault South Westland to Fiordland. *New Zealand Journal of Geology and Geophysics, 47*, 257-267.

Marinos, V., Marinos, P., & Hoek, E. (2005). The geological strength index: applications and limitations. *Bulletin of Engineering Geology and Environment, 64*, 55–65.

Naylor, L. A., & Stephenson, W. J. (2010). On the role of discontinuities in mediating shore platform erosion. *Geomorphology, 114*, 89-100.

Naylor, L. A., Stephenson, W. J., & Trenhaile, A. S. (2010). Rock coast geomorphology: recent advances and future research directions. *Geomorphology, 114*, 3-11.

Palmstrom, A. (2005). Measurements of and correlations between block size and rock quality designation (RQD). *Tunnelling and Underground Space Technology, 20*, 362–377.

Paterson, M. S., & Wong, T. F. (2005). Experimental rock deformation: the brittle field. *Springer, 314.*

Ren, G., Zuo, Z. H., Xie, Y. M., & Smith, J. V. (2014). Underground excavation shape optimization considering material nonlinearities. *Computers and Geotechnics, 58*, 81-87.

Smith, J. V. (2014). A new approach to kinematic analysis of stress-induced structural slope instability. *Engineering Geology, 394.* http://dx.doi.org/ 10.1016/ j.enggeo.2014.12.015

Stead, D., Eberhardt, E., & Coggan, J. S. (2006). Developments in the characterization of complex rock slope deformation and failure using numerical modeling techniques. *Engineering Geology, 83*, 217-235.

Stephenson, W. J., & Naylor, L. A. (2011). Geological controls on boulder production in a rock coast setting: insights from South Wales UK. *Marine Geology, 283*, 12-24.

Wyllie, D. C., & Mah, C. (2004). *Rock slope engineering.* Taylor & Francis.

Youssef, A. M., Maerz, N. H., & Al Otaibi, A. A. (2012). Stability of rock slopes along Raidah escarpment road, Asir Area, Kingdom of Saudi Arabia. *Journal of Geography and Geology, 4*(2), 48.

12

Shallow Subsurface Evidence for Postglacial Holocene Lakes at Ivanpah Dry Lake: An Alternative Energy Development Site in the Central Mojave Desert, California, USA

Douglas B. Sims[1] & W. Geoffrey Spaulding[2]

[1] Department of Physical Sciences, College of Southern Nevada, North Las Vegas, USA

[2] Terra Antiqua Research, Henderson, USA

Correspondence: Douglas B. Sims, Department of Physical Sciences, College of Southern Nevada, North Las Vegas, Nevada 89030, USA. E-mail: douglas.sims@csn.edu

A portion of this research (students) was funded by the NASA-CoP Grant (NNX14Q94A).

Abstract

Dry lakes, or playas, in the American Southwest have become important locations for alternative energy solar electric generation systems. Many of these playas have not been investigated for past environmental conditions prior to their development, causing alterations to the shallow sediment profile and altering subsurface chemical and physical data important to recent geologic and hydrologic histories of these areas. Recent studies of these features commonly have focused on surficial data with little or no subsurface geologic data to delineate wet and dry cycles of playa lakes. Many of the playa margins are covered with a surficial layer of alluvial and/or aeolian sands varying in thickness, which could interfere with correct interpretations of data gathered only on the surface material on these lands. Over the past 13,000 years Ivanpah Dry Lake has had at least three episodes of lake highstands up to ~10m in response to climate changes, based on data in this study with the oldest episode dates to ~13,000 cal yrs BP. This study has a link to other sites as many other playas are slated for energy development and could endure subsurface data alteration during site reworking and occupation, resulting in a loss in their collective historical geologic and hydrologic records if more complete studies of important sites are not carried out before developments begin.

Keywords: Ivanpah, pluvial lakes, playa, Lake Mojave, Holocene, Pleistocene, lacustrine, California

1. Introduction

Desert dry lakes and their margins in the American Southwest and elsewhere are considered potential sites for the development of alternative energy projects to supplant current energy sources that adversely affect climate, and climate change awareness in action policies by potentially affected institutions. Many of these areas are located within regions where there has been little focus into historical environmental conditions. In response to increased alternative energy needs, there are strategies in place to create solar energy zones that encroach onto and could disturb buried deposits across the deserts of the American Southwest, including Ivanpah Dry Lake (Fig. 1) ("Lake Ivanpah"). These developments are impacting the very sedimentological records that could provide information on the historical environmental conditions, and potential future environmental impacts. In other words, it seems that the vast magnitude of paleohydrologic changes that can occur in these basins is under-appreciated when facilities performance over decades are considered.

Understanding long-term environmental change in the arid Southwest rests in part on paleohydrologic studies, mostly of the Quaternary Period, pluvial lakes fed by rivers originating from beyond the desert (e.g. Blunt and Negrini, 2015; Enzel et al., 2003; Litwin et al., 1999; Nishizawa, 2010; Negrini et al., 2006; Wells et al., 2003), and paleospring deposits that reflect more local histories of recharge within the Great Basin and Mojave deserts (Quade et al., 1995; 2003). Such records of hydrologic responses to precipitation allow for greater delineation of the effects of changing climate. The effects of these changes have been profound on both surface processes and biotic communities in the region (e.g., Bull 1991; Spaulding 1990).

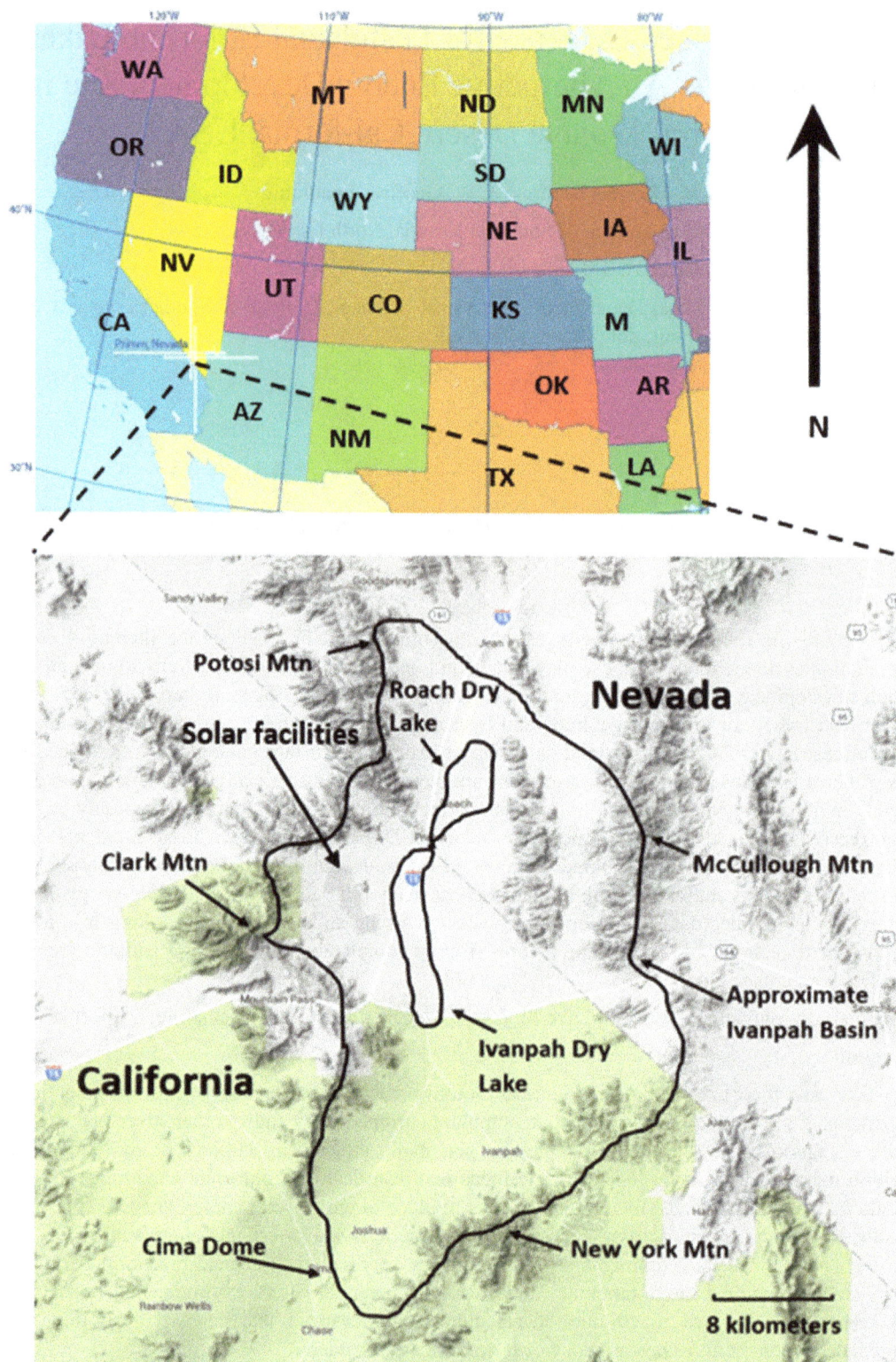

Figure 1. Location of Lake Ivanpah, California and approximate Ivanpah hydrologic basin

Desertification of the American Southwest followed the retreat of the continental ice sheets in the Northern Hemisphere, as an increase in thermal regimes accompanied a northward shift in circulation patterns to those characteristic of the present arid Southwest (Thompson et al., 1993). There have been studies in this region of late Quaternary Period precipitation and its effects on groundwater, climate change, and pluvial lake systems

(Quade et al., 1995; Pigati et al., 2011; Waring, 1920; Wells et al., 2003). Other research in this area evaluated Pleistocene Camelid tracks and aeolian distributions as they relate to climate change at Mesquite Lake, California, a dry lake system located 16.5 km northwest of Primm, Nevada (Shapiro, 2016; Whitney et al., 2015). Finally, environmental variability within this region, and more notably, Ivanpah Valley, is complex.

To better appreciate the environmental changes inherent to Lake Ivanpah, it is important to recognize that episodes of higher lake stands and increased paleospring discharge, occurred not only during the latest Pleistocene, but also during the succeeding Holocene (e.g. Wells et al., 2003; Noble et al., 2016; Quade et al., 1995; Pigati et al., 2011). The Pleistocene to Holocene transitional period experienced reversals in the desiccation of ancient lakes and extinct spring systems ("paleolakes" and "paleosprings"). So, while it has been reported that the "final" desiccation of paleolakes and ancient springs occurred between ca. 9,300 and 7,800 calibrated years BP (cal yrs BP), essentially concordant with the extirpation of mesic vegetation in current deserts (Wells et al., 2003; Pigati et al., 2011; Spaulding and Graumlich, 1986), there were important reversals during the succeeding postglacial. According to Noble et al. (2016), pluvial events of sufficient magnitude to fill some basins for decades or centuries occurred during the middle and late Holocene. Refilling of basins along the Mojave River reportedly occurred at least twice: ca. 3,600 yrs BP and again during the Little Ice Age, ca. 600 to 200 yrs BP (Enzel et al., 1992; Tchakerian and Lancaster, 2002). Other studies documented that the late Holocene Period witnessed a rebirth of paleospring systems throughout this region (e.g. Mehringer and Warren, 1976; Jones et al., 1999).

There has been comparatively little research on dry lakes with drainage basins lying entirely within the Mojave Desert. Unlike lakes Owens and Searles, on the Owens River that heads in the eastern Sierra Nevada, or Lakes Manix and Mojave on the Mojave River that heads in the Transverse Ranges, isolated dry lake basins show few geomorphological signs of highstands (i.e. wave-cut benches), fluvial deposits and/or terraces (Mifflin and Wheat, 1979; Reheis and Redwine, 2008). Comparing the apparent absence of pluvial lakes in the southern Great Basin to their relative abundance and great size in the north, Mifflin and Wheat (1979) developed a model relating the combined effects of an increased thermal regime in the south, lower elevations, and smaller mountains to insufficient runoff to support persistent lakes, even during glacial conditions. As a result, little research has focused on the paleohydrology of internal dry-lakes of the southern Great Basin. Studies have shown that the mere absence of surface evidence does not preclude the possibility of ancient lakes. Rather, studies require interpretation from the examination of aeolian dunes, infrared photographs, subsurface conditions, grain size, and sediment chemistry-proxy data to identify buried remnants of ancient lake beds (Sack, 1995; 2001).

2. Environmental Setting

Three solar electric energy generation projects are currently being built, or operational, in the Ivanpah Valley in the last five years. In order of their proximity (farthest from) to the current playa margin are: Stateline Solar Farm Development on the northeastern margin of Lake Ivanpah, on the toe of the Clark Mountain alluvial fan in California; Silver State and Silver State South Solar Energy Centers, east of Roach Dry Lake, northern portion of the Lucy Gray alluvial fan toe in Nevada; and Ivanpah Solar Electric Generating System, located further up the Clark Mountain alluvial fan. The Ivanpah facility is currently the world's largest concentrated solar thermal power station in operation while the other two facilities are large solar photovoltaic systems (CH2MHILL, 2009).

2.1 Hydrologic Setting

Although the Ivanpah basin is not part of the Mojave River drainage (see Enzel et al., 2003), tributaries to the lower Mojave River lie only ~25 km to the west in Shadow Valley. Approximately 20 km to the east on the other side of the McCullough Mountains, the Paiute Valley drains into the Colorado River. The Ivanpah basin incorporates a drainage network extending from the Teutonia Peaks area on Cima Dome, 25 km to the southwest in California, to Potosi Mountain, about 40 km to the east-northeast in the Spring Mountains of Nevada (Fig. 1). The watershed also includes the north flank of the New York Mountains and Mid Hills to the south and southwest. The Ivanpah Range, Mescal Range, and Clark Mountain provide runoff from the west, and, the Clark Mountain Range and southern-most Spring Mountains provide runoff from the north.

The Ivanpah basin also includes Roach Dry Lake, immediately to the northeast in Nevada, separated from Ivanpah playa only by the toe of the alluvial fan extending west from the Lucy Gray Mountains. Because this permeable sill is the only separation between the two playas, Roach Dry Lake and its drainage are part of the Ivanpah hydrologic system. Other studies of the hydrological processes of Lake Ivanpah have determined that Ivanpah Valley is entirely internally drained (House et al., 2010). Lake Ivanpah's surface drainage in California is approximately 1,128 km^2; its catchment in Nevada is approximately 831 km^2, for a total drainage basin area of 1,959 km^2 (756 mi^2). The surface area of Ivanpah playa is about 53 km^2 while Roach playa is approximately 15 km^2.

2.2 Topographic Setting

Ivanpah playa has a sinusoidal shape that is the result of two large alluvial fans intruding on its margins from the northeast and from the west; i.e. Lucy Gray and Wheaton Wash alluvial fans (Figs. 1-3). Both alluvial fans create a convex-shaped shoreline at their terminal edges. The arcuate margin of Lucy Gray Fan terminal edge projects west and extends for ~5 km south of the sill it creates, terminating at the playa in the north. Approximately 10 km south, on the opposed side of the playa, the toe of the Wheaton Wash Fan forms an eastward projecting convex shoreline. Fans are composed of moderately sorted, well stratified sands and gravels of pebble and boulder sizes with minor amounts of aeolian silts and sands at depth, with an estimated age of deposits of ~2.6 million years according to House et al. (2010). Both the Lucy Gray and Wheaton Wash Fan are lacking active fluvial processes, with recent flows confined to ephemeral systems nearer the source of materials (House et al., 2006; 2010).

The relief of the topography encompassing a basin's drainage is a critical factor in determining runoff; i.e. the higher mountains cause greater amounts of orographically induced precipitation (Mifflin and Wheat, 1979). Studies of Mojave River runoff for instance, show that water contained in Lake Mojave originated chiefly from surrounding mountains outside the Mojave Desert (Enzel et al., 2003). With the Mojave River in mind, the Ivanpah drainage basin includes some of tallest mountain ranges in the central Mojave Desert (i.e. New York Mountains [~ 2,296 m], Clark Mountain [~2,417 m], Potosi Mountain in the southern Spring Mountains [~2,595m], and ~2,142 m for McCullough Mountain) and therefore, runoff that entered the Ivanpah basin originates from the surrounding mountains.

The barren part of Ivanpah Basin's elongate and sinusoidal playa occupies only the northern 55% of the 19.3 km–long basin floor. Generally not more than 3 km wide, the playa extends south about 11 km from the vicinity of Primm, Nevada, to an ill-defined southern limit at about the 795 m contour. The barren northern and central portions of the playa are mapped as essentially level at ~794 m amsl (USGS, 1985). The playa surface continues farther south, but is mantled by an increasingly dense cover of saltscrub hummocks to the south. These hummocks, usually not exceeding a meter in height, consist of grey-brown to pale brown (10YR7/4), silty fine sand anchored by saltscrub vegetation, and typified by Allscale (*Atriplex polycarpa*), a species of saltbush. South of the Nipton Road, blow-outs that have stripped back the mantling saltscrub hummocks reveal a mosaic of playa surface and carbonate-rich paleospring outcrops armored by nodular tufa concretions (Robinson et al., 1999). Saltscrub rims the entire playa except for its north end, and it gives way upslope to a creosote bush (*Larrea tridentata*) - Burrobush *(Ambrosia dumosa)* scrub transition zone.

Moving south, the floor of Ivanpah playa rises in elevation well before its southern end is reached. Because the surface of Ivanpah playa is not level, evidence of shallow water in the south indicates deeper water to the north. Previous research near Nipton Road identified oxidized lake muds in yardangs at 800.1 m amsl, only ~0.9 m above the adjacent playa but >6 m above the elevation of the pan in the north (Robinson et al., 1999). Their investigation found bedded, sandy alluvium yielding abruptly to a charcoal-rich, dark-grey lake mud at ~1 m depth. The top of this lake mud was surveyed at 799.9 m amsl, again ~6 m above the playa floor in the northern Ivanpah basin, yielding an age of 9,560 ± 70 ^{14}C yrs BP (Robinson et al., 1999). Assuming inundation is needed for playa-mud deposition, and that no substantial tectonic warping of the basin floor has occurred, their study indicated that a lake of at least moderate depth occupied the northern Ivanpah basin.

There is no abrupt southern boundary to Ivanpah playa. Rather, it is characterized by an increasingly dense cover of saltscrub hummocks and tongues of gravelly alluvium, the latter from the broad delta-fan expanding onto the basin floor from the south at the mouth of Cima Wash. The name Cima Wash was applied to the system of anastomosing sandy washes that extends ~25 km south and west from the southern playa margin to the southeast flank of Cima Dome. As this broad drainage runs down the southwestern Ivanpah Valley, it is joined by washes draining the north flank of the New York Mountains and Mid-Hills. Remote imagery and field assessment places the distal toe of the Cima Wash delta-fan, and the southern limit of Ivanpah playa, ~1 km south of Nipton Road.

2.3 Ivanpah Playa

The location of Lake Ivanpah is a topographic depression at the base of the Ivanpah, Mescal, and Clark Mountain ranges, just west of the Nevada state line. The study area is hot and dry with temperatures fluctuating between ~1°C in winter (January/December) to >41°C in summer (July/August). Precipitation averages <11.5 cm annually near the basin bottom and higher in the surrounding mountain ranges (Robinson et al., 1999).

Geographically, this relatively high basin (playa surface is ~ 794 m above mean sea level [amsl]) is located within the Mojave Desert. Roach Dry Lake is connected to the Ivanpah playa at the border of Nevada and California (Primm, NV) and is part of the Ivanpah Catchment Basin. It lies at ~794 m amsl and is separated from Ivanpah by a narrow sill produced by the merger of the Lucy Gray alluvial fan and a limestone ridge extending south from the Spring Mountains Range (Carr and Pinkston, 1987; Garside et al., 2009). The Ivanpah catchment in California is approximately 1,128 km^2 with the Nevada portion being ~831 km^2 (totaling ~1,959 km^2) with Ivanpah playa covering ~53 km^2 and Roach playa ~15 km^2.

Lacustrine sediments have been described as having distinct lithological units with defined beds (laminated) of reworked material containing ubiquitous amounts of carbonates, much greater than sediment organic carbon, with large amounts of soluble salts (Eugster and Kelts, 1983). Other authors have defined lacustrine sediments as originating from incoming sediments due to wetting events, depositing sediments in an aqueous environment which in turn becomes a playa surface with aridity (Kliem et al., 2012; May et al., 2015). Thus, the superficial portion of a lacustrine unit will appear as a classic playa surface, the upper most portion of a lacustrine unit according to Kliem et al., (2012).

2.4 Ancient Shorelines and Lacustrine Features

Identifying, correlating and mapping shoreline features in an area where the climate has been relatively stable since the end of the Pleistocene can be challenging (Sack, 1995). Shorelines usually leave demarcations ("highstands") of their farthest advance and later, may be buried or obliterated by successional advances but leaving visible geomorphic evidence (Adrian et al., 1999). Ivanpah has no such obvious preserved visible lake highstands as describe by other authors (i.e. Negrini et al., 2006; Enzel et al., 1992; Wells et al., 2003). Studies have however, documented the reworking and/or burial of shorelines and their associated lacustrine sediments by active alluvial fans and evidence of fluvial and aeolian processes that obliterate and bury geomorphic evidence making them difficult to locate surficially with aerial photographs, topographic maps and ground reconnaissance activities (Sack, 1995).

There was surficial geological mapping of portions of this basin that identified no obvious ancient shorelines (House et al., 2010; 2006; Ramelli et al., 2006; Schmidt et al., 2006). These studies however did map "playa fringe deposits" to elevations >10m above the current playa margin. Visual evidence in the form of shorelines is only one factor in pluvial lake demarcation: soils, sedimentologic, sediment chemistry, subsurface stratigraphy, and geomorphic evidence also can be used to identify pluvial lakes lacking obvious shorelines (Sack, 1995; 2001; Schnurrenberger et al., 2003). Current efforts do not contradict prior mapping efforts (House et al., 2010, 2006; Ramelli et al., 2006; Schmidt et al., 2006), but are complementary to them. However, because the possibility of a pluvial lake having occupied the basin is absent from their interpretive geological framework (House et al., 2010, 2006; Ramelli et al., 2006; Schmidt et al., 2006.), and because their work was restricted to the Nevada portion of the basin, those data sets have limited transfer value to this effort.

Within the Ivanpah basin nonetheless, there are yardangs, loose and un-weathered surface sand, strath surfaces (Fig. 2A), hummocks (Fig. 2B), both modern (Fig. 2C) and ancient (Fig. 2D) beach surfaces, north-south trending low lying aeolian dunes (Fig. 2E) with discontinuous sand sheets, and exposed lacustrine sediments just off the perimeter of the playa margin. These features are subtle evidence of preserved landforms that could easily be overlooked by analysis of aerial photography and topographic map review, even supported with some ground reconnaissance. This research does not map geomorphic evidence on the surface, but rather, evaluates subsurface conditions at one location on the eastern side of Lake Ivanpah.

3. Objectives

The purpose of this research is to demonstrate that a large (depth >5 m) body of water episodically occupied the basin in lieu of a pluvial lake, and expanded well beyond the edge of the current playa as supported predominantly by data gathered from subsurface sediments. To do this, subsurface conditions on the playa margins were investigated using sedimentary stratigraphy and chemical proxies to identify subsurface anomalies potentially related to lake transgressions over alluvial fan toes. These chemical proxies (metals, SO_4^{2-}, Cl^-, $CaCO_3$, pH, Mg/Ca ratios, radiocarbon dating (^{14}C), $\delta^{13}C$, and $\delta^{18}O$) were used to locate a suitable location for further excavation so that subsurface sedimentological, geomorphic, and lacustrine event evidence could establish chronology, and be used to describe the magnitude of environmental change within the Ivanpah basin. Data was comparable with records of lake-level changes from other paleolakes, particularly those that received most of their runoff from mountains both inside and outside the Mojave Desert.

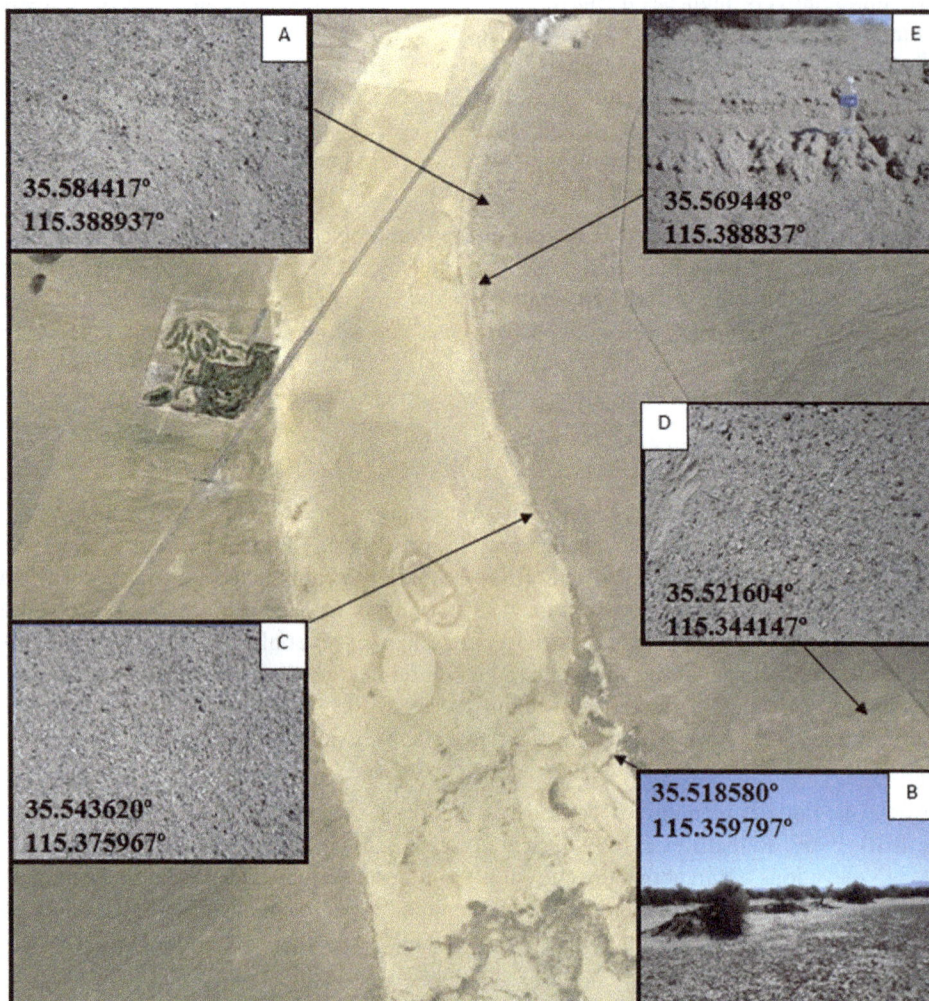

Figure 2. Subtle geomorphic evidence across Lake Ivanpah

4. Methods and Materials

4.1 Field Sampling Methods

Methods employed are similar to Rodrigues-Filho and Muller (1999) and adapted for this region and environment. Their study involved coring to a depth of 12 m with mechanical assistance for sediment characterization. This study was however, limited to hand-driven cores by the BLM due to the culturally sensitive environment of the eastern side of Ivanpah. With this limitation, coring was used only for the identification of subsurface anomalies and to determine the optimal location for further excavation for sediment characterization.

The eastern side of Ivanpah was chosen for this study because the western side of the lake has been heavily impacted by large solar energy projects (Fig. 3A). The first phase consisted of the 12 shallow cores to ~50 cm depth (see supplementary data for geolocations) starting near the center of the dry lake and moving eastward onto the Lucy Gray Mountains bajada (Fig. 3B). Core samples were collected with a hand driven portable AMS hollowstem auger (2.54 cm x 60 cm) kit with a split spoon insert and a high-density polyethylene (**HDPE**) liner to collect a continuous 50 cm profile. Liners were removed, capped and placed in a cooler for transport to the laboratory. After cores were analyzed and an anomaly was identified (chemical proxy), one location (LGL 5) was excavated by hand to 1.96 m bgs to identify and document substrata features where chemical proxies (the anomaly) were identified. LGL 5 substrata were described and catalogued so that the processes that emplaced the fluvial and aeolian sediments could be linked by their depositional development and chemical proxies detected.

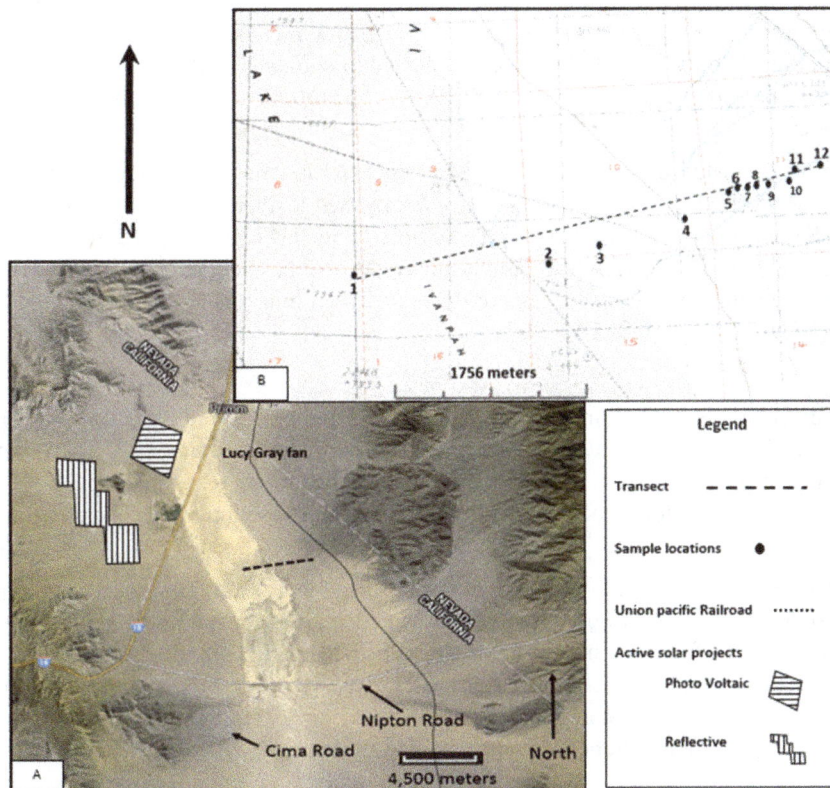

Figure 3. Aerial overview of Lake Ivanpah and features

Note: (A); Lucy Gray Transect (LGL) and sample locations (B). Lat./Long. for locations (1-12) are available in suplamentary data.

Samples were collected between May 2014 and December 2015 from lands administered by the United State Department of Interior, Bureau of Land Management (BLM), Needles, California, Field Office. The approach was designed to assess chemical proxies such as trace metals (As, Ba, Cr, Cu, Li, Mn, Ni, Pb, Ti, V) and crustal metals (Al, Ca, Fe, K, Mg, Zn) and wet chemistry (Cl$^-$, SO$_4^{2-}$, pH, CaCO$_3$), Mg/Ca ratios, C^{14}, δ^{13}C and δ^{13}O in sediments across a ~4.5 km transect.

Google Earth was not utilized for surface elevations within the Ivanpah basin because of an error observed. To plot sampling locations and the single excavation site accurately, each sampling site was recorded with a Trimble Survey R10 GNSS rover and base station receiver, GLONASS L1/L2 signal, Trimble HD-GNSS™ processing engine, Trimble SurePoint™ electronic bubble and traceable tilt values with a Trimble 360 satellite tracking. The Trimble xFill™ is designed to work in radio and cellular black spots with an accuracy of ± 0.1 m with all locations tied to United States Geological Survey benchmarks.

4.2 Laboratory Procedure

Each core sample was dissected into 10 cm sections to provide separate data locations at depth. Core dissection was performed by scoring around the HDPE insert, then breaking at the scoring location so that each 50 cm core could be processed as 10 cm sections (0-10, 10-20, 20-30, 30-40, 40-50), resulting in 60 separate samples along the LGL transect. Sediment samples were removed from each section, placed in clean HDPE sample containers, marked with location and depth, and transported to the laboratory.

4.3 Metals Analysis

Researchers Rodrigues-Filho and Muller (2001) have demonstrated that chemical weathering in saturated sediments causes the accumulation of less mobile metals and soluble constituents in a horizontally bedded fashion. Horizontally distributed metals in lake bed sediments are described by Rodrigues-Filho and Muller (1999) as a reliable indicator of previous lake bed features. Therefore, trace elements (As, Ba, Cr. Cu, Li, Mn, Ni, Pb, Ti, V) and crustal metals (Al, Ca, Fe, K, Mg, Zn) in sediments were analyzed using United States Environmental Protection Agency (USEPA) Solid Waste 846 (SW-846) protocols (USEPA, 1997) were performed at the College of Southern Nevada. These protocols included quality control measures to validate data integrity. Sediment samples (~1 gram) were digested by USEPA method 3050B (hot aqua-regia digestion), diluted to 50 ml after digestion with deionized water, allowed to cool, settle, then filtered to remove particulate matter. Analysis was performed according to USEPA method 6010B; inductively coupled plasma-optical emission spectrometer.

Instrument calibration consisted of six points across a concentration range, including a blank, with a linear regression (R^2) of >0.995 (USEPA, 1997). Analytical integrity was further verified by using a USEPA certified reference sample (RTC Corporation: CRM022-020, Sample 5, lot D522) purchased from a USEPA certified reference laboratory. The certified reference sample measures within specified windows (±15%) and cannot exceed specified requirements. Additionally, samples were analyzed in triplicate with a required USEPA relative percent difference (RPD) of <20% (USEPA, 1997).

4.4 Wet Chemistry

Analysis for wet chemistry were performed at the College of Southern Nevada. Sediment pH is an important measurement for assessing potential availability of beneficial nutrients in sediment. USEPA method 9045 was applied using a H^+ ion-selective glass electrode in a saturated paste to measure sediment pH (USEPA, 1997). The pH meter was calibrated by selection of pH standards that bracketed the sediment pH and measured to within 0.1 standard units.

Sulfate (SO_4^{2-}) was determined by USEPA method 9038 (USEPA, 1997). Ten (10) g of a sample was placed into a 100 mL HDPE bottle, 50 mL of 24°C distilled water was added and agitated for 60 minutes to dissolve SO_4^{2-} ions, then allowed to settle, followed by filtration through Whatman^{TM} 42 filter paper. The SO_4^{2-} ion was precipitated (with barium-chloride [$BaCl_2$] to barium-sulfate ($BaSO_4$) forming a suspension under controlled conditions to determine the concentration using a nephelometer and a calibration curve for sulfate (USEPA, 1997). Sample weight and solution volume were accounted in the data.

Sediment chloride was determined potentiometrically using a chloride ion (Cl^-) selective electrode (ISE) in conjunction with a chloride combination ISE on a pH meter with an expanded millivolt scale. Each sample was determined by measuring ~10 g into a 100 ml HDPE bottle, adding 50 ml of 24°C distilled water and agitated for 60 minutes to dissolve Cl^- ions. Once the sediment was agitated, the solution was filtered through a Whatman^{TM} 42 filter paper. Each sample was compared to a calibration curve with final concentration determined by accounting for sample weight and solution volume.

Calcium carbonate ($CaCO_3$) was measured by gravimetric techniques outlined by Bauer et al. (1972). One (1) g of sediment was measured into a tared 25 mL beaker to the nearest 0.001 g. Ten (10) mL of 5N hydrochloric acid (HCl) were added to the beaker covered with a watch glass to prevent splashing. The sample was swirled twice at 10 minute intervals for 30 minutes to allow the HCl and sediment reaction to dissolve all available carbonates. Samples were placed into an oven for 12 hours at 105 °C, removed and allowed to cool before the weight was recorded on both the beaker and the sediment. The difference between the original beaker and sediment weight and the final weight, after the HCl addition, was recorded as mg kg^{-1} $CaCO_3$.

4.5 Mg/Ca ratios, ^{14}C, $\delta^{13}C$ and $\delta^{18}O$

Mg/Ca ratios were used to examine possible links between past environments and associated temperatures. Magnesium is incorporated into calcite deposited in sediments upon evaporation of water. With the incorporation of Mg as an impurity in calcite (i.e. endothermic), more is incorporated into growing crystals at higher temperatures (Morse et al., 1997). Mg is a divalent cation that substitutes for Ca during the development of biogenic calcium-carbonate (Barker et al., 2005; Barker et al., 2003). When Mg is incorporated into calcite, its incorporation is influenced by the temperature of the surrounding water; increasing with rising temperatures (Ries, 2004; Barker et al., 2005).

Korber et al. (2007), Morse et al. (1997), and Sun et al. (2015) suggest Mg/Ca ratios >3.5 represent modern freshwater. Higher atmospheric temperatures will produce higher Mg/Ca ratios from the aquatic chemical system, whereas lower atmospheric temperatures will produce lower Mg/Ca ratios from the same aquatic chemical system (Barker et al., 2005; Cleroux et al., 2008; Regenberg et al., 2009). For this study, ratios < 2.0 represent a cooler period with decreased evaporation; ratios > 3.5 represent higher temperatures with increased evaporation.

All ^{14}C samples were performed by Beta Analytic Inc. in Miami, Florida and the University of Arizona Accelerator Mass Spectrometry Laboratory in Tucson, Arizona. Beta Analytic Inc. and the University of Arizona Accelerator Mass Spectrometry Laboratory data are conveyed as Conventional Radiocarbon Age (CRA) reported in ^{14}C and cal yrs BP (2 σ age range). Samples for ^{14}C dating were composed of one bulk sample (charcoal present) and four grab samples (detrital carbon). The bulk sample was ~10 kg collected from within a subsurface beach zone (LU3) layer that was floated for visible charcoal and sent to the laboratory for analysis. The four grab samples came from single locations of ~300 g of sediment collected at each location from within the excavation (LGL5), and sent for analysis.

Laboratory procedures for ^{14}C analysis first visually inspected samples for size, homogeneity, debris, inclusions, clasts, grain size, organic constituents and potential contaminants. Their first step was to disperse sediment samples in de-ionized water, homogenized through stirring and sonication, then sieved through a 180 μm sieve with materials passing through the sieve used for the analysis. The next step in preparation was soaking samples in 2.5N HCl at 90°C for a minimum of 1.5 hours to ensure removal of carbonates (inorganics) so that a measure of organic carbon could be achieved. Following the HCl at 90°C, serial rinses with 70°C de-ionized water were performed until neutrality was reached. Any debris (i.e. micro-rootlets) was discarded during rinsing. Samples were then dried in an oven at 100°C for 12-24 hours and then concentrated HCl was applied to a representative sub-sample under magnification to confirm the absence of carbonates. If carbonate was still present, the laboratory would repeat the HCL processes to remove any inorganic carbonates from the sample. As a final point, although the use of ^{14}C dating does present possible erroneous results, though slight, it is the standard used by researchers for bulk dating of detrital carbon in dry lake sediments, even if no visual carbon is present (i.e. Enzel et al., 2003; Negrini et al., 2006; Quade et al., 2003; Wells et al., 2003).

The University of Arizona Environmental Isotope Laboratory, Department of Geosciences analyzed sediment samples for δ ^{13}C and δ^{18}O in mineral carbonates. Values for δ ^{13}C are important for their understanding of past environments; an increase in prime productivity causes a corresponding rise in δ ^{13}C as supplementary ^{12}C is locked up in plants. When organic carbon is buried, more ^{12}C is locked out of the sediment system than is found in background systems. Samples analyzed for δ^{18}O ratios were used to indirectly determine the temperature of the surrounding water at the time the sediment was deposited. When δ^{18}O ratios vary slightly, it is a response to temperature of the surrounding water. Other factors related to δ^{18}O are salinity and volume of water in a given area. Studies have used this technique on gastropods, endocarps, and other carbonate matrices; this study used δ^{18}O and δ ^{13}C to detect chemical anomalies.

4.6 Quality Control/Quality Assurance

Quality assurance/quality control protocols followed USEPA methods (USEPA, 1997). All analyses required replicated certified reference material purchased from a USEPA certified supplier. Instrument detection limits (IDL) were calculated by multiplying the standard deviation (SD) of a fortified reagent blank by three times. Statistical data were calculated excluding values that were below IDL values. Sample quality control was processed per protocols as outlined in USEPA SW-846 procedures (USEPA, 1997). Finally, all samples were analyzed in triplicate following USEPA requirements for relative percent deviation (RPD) of <20% to be acceptable (USEPA, 1997).

4.7 Stratigraphy and Chronologic Context

Chemical proxy data ("the anomaly") were compared to this local stratigraphy as well as broader geomorphic features identified along the LGL transect for selection of a point for further excavation. To document the stratigraphy contributing to the anomalies in initial chemical proxy data, a larger (1.5 m x 1.5 m wide) test pit was excavated at LGL 5 (Fig. 3B). Strata were exposed in this excavation on two vertical walls with north-south and east-west axes to a depth of 1.96 m, sketched and photographed. Samples for isotopic analyses were collected from different strata, taking care to avoid sampling across stratigraphic discontinuities. Sediment from identified layers were sieved for gradation and recorded. Water depth is expressed in m, and changes with difference in depth (Δ_w) from the current Ivanpah playa floor of ~794 m amsl. For this study, Rehies et al. (2012) definition was applied for "shallow" lake depths of less than 5 m ($\Delta_w{\leq}5$ m), "moderate" depth to Δ_w of 5-8 m, and "deep" lake conditions of $\Delta_w{>}8$ m. For the purpose of this paper, the end of the Pleistocene is placed at 12,000 yrs BP, the early to middle Holocene boundary at 7,800 yrs BP, and the beginning of the late Holocene at 4,000 yrs BP.

5. Results

5.1 Sediment Chemical Proxy Anomaly

Coring was used only for the identification of subsurface anomalies so that a location could be identified for further excavation. Elevation change at Ivanpah is ~12.7 m from LGL 1 (794.4 m), at the playa base surface, to 807.1 m amsl at LGL 12: the Lucy Grey bajada creosote bush scrub transition zone (Fig 2b). Table 1 shows a distinct anomaly in data with Al, Fe, K, and V following a uniform distribution below the surface across LGL 1 through 12, as did other constituents such as metals, SO_4^{2-}, Cl⁻, pH, $CaCO_3$ (see supplementary tables). Aluminum, Fe, K, and V are higher within the horizontally positioned ("bedded ") anomaly and bracketed by lower concentrations above and below the max concentrations (Table 1; highlighted in grey and lighter grey) based on ±1σ. Data shows Al, Fe, K and V have similar distributions at approximately -20 to -40 cm bgs at LGL 1 through LGL 6, surfacing at LGL 7 (799.7 m), descending to 20-40 cm bgs by LGL 8, surfacing again at LGL 11 (804.2 m) and descending to 20-40 cm bgs at LGL 12. Largely, this anomaly is positioned bgs with a uniform depth except at LGL 7 and LGL 11, where it surfaces.

Crustal (e.g. Al, Fe, Ca) and trace metals (e.g. Cr, Cu, Pb, V) ranged between 3,500 and 46,010 mg kg⁻¹ Al; 5,150 and 29,780 mg kg⁻¹ Fe; and 8.7 and 83 mg kg⁻¹ V across the LGL transect (see supplementary tables). These findings are not unusual as oxidizing environments increase the presence of FeOOH in horizontally bedded sediments (Stumm and Morgan, 1996; Sanchi et al., 2015), providing adsorptive capacity for charged species dissolved in surface and percolating waters. Data presented in Table 1, including supplementary tables, shows Al, Fe, Zn and other metals (e.g. As, Ba, Ca, Cr. Cu, Li, K, Mg, Mn, Ni, Pb, Ti, V) tend to follow this horizontal bedding ("an anomaly") at depth (~20-40 cm) from LGL 1 out ~4.5 km east to LGL 12.

Sediment pH also measured along the LGL transect ranged between 6.81 and 8.50 standard units (SU) with a distribution comparable to metals and other chemical proxies (Table 2). Core sample pH, although not significantly different across the LGL transect, shows the anomaly at depth and followed metal accumulations presented in Table 1, a common association know between pH and metals (Tardy and Roquin, 1992; Watanabe et al., 2001). These findings are not unusual as sediment pH will support similar chemistries in horizontal beds where there are analogous redox conditions (Tardy and Roquin, 1992).

Data for $\delta^{13}C$ and $\delta^{18}O$ were also used to identify patterns below the surface with two locations (LGL 3 and 5; Fig. 3B) along the LGL transect selected for measurements. Table 3 displays $\delta^{13}C$ and $\delta^{18}O$ with comparable alignments as presented in Tables 1 and 2. LGL 3 is located within the playa (794.5 m amsl) base surface, whereas LGL 5 is located on the Lucy Grey bajada (796.4 m amsl) ~1.9 m above LGL 3. Data for $\delta^{13}C$ and $\delta^{18}O$ at LGL 3 and 5 have a similar orientation, although with minor differences; -2.20, -2.56 $\delta^{13}C$ and -4.71, -5.10 ‰ $\delta^{18}O$, respectively.

Table 1. Selected chemical proxies for locations 1 - 12 of the LGL transect with Max ±1σ ranges

	Location	1	2	3	4	5	6	7	8	9	10	11	12
	Elevation	794.4	794.7	794.5	796.3	796.4	797.4	799.7	800.6	801.9	803.2	804.2	807.1
	Depth												
Al	10	11,600	12,500	16,500	18,750	28,860	27,850	33,160	4,015	36,920	6,700	44,580	31,200
	20	20,550	21,000	17,000	16,600	36,290	41,920	44,760	7,200	38,670	3,500	43,010	45,920
	30	20,600	20,300	24,800	19,500	49,670	40,010	44,750	9,350	41070	7,650	41,920	46,010
	40	20,900	23,500	24,100	17,100	39,370	32,550	34,760	8,200	40,960	7,360	36,410	33,010
	50	20,400	21,050	20,800	16,200	32,220	35,600	26,130	5,800	36,000	6,080	31,440	32,000
	STDV	4035	4188	3863	1426	7990	5674	8028	2081	2301	1657	5443	7638
	Max +1σ	24935	27688	28663	20926	57660	47594	52788	11431	43371	9307	50023	53648
	Max -1σ	16865	19312	20937	18074	31380	36246	36732	7269	38769	5993	39137	38372
Fe	10	17200	15560	15700	9050	23340	21810	25920	5150	24800	8100	29000	23400
	20	17350	16265	16805	13350	25020	27250	27190	8305	25800	8950	29100	28630
	30	16550	24550	26400	24800	29780	26600	27450	18250	26210	9250	28700	29110
	40	27650	27800	26900	25900	26690	21880	24080	29750	25830	7800	25600	23000
	50	27500	26350	26500	17750	25050	24680	21760	16800	24270	7900	23300	22800
	STDV	5782	5774	5684	7251	2434	2554	2377	9632	812	657	2591	3190
	Max +1σ	33432	33574	32584	33151	32214	29804	29827	39382	27022	9907	31691	32300
	Max -1σ	21868	22026	21216	18649	24256	24696	25073	20118	25398	8593	26509	25920
K	10	22270	27820	26790	10150	8815	8280	9581	1920	9144	9250	9513	8065
	20	22405	30330	29300	17705	9734	9576	9611	2614	9753	9575	9404	9439
	30	29480	25040	28000	21395	9620	9600	9784	2831	9461	16200	9564	9217
	40	32890	34050	28650	24980	9532	8635	9341	2214	9312	10800	9093	8900
	50	30380	31670	27390	20150	9070	8556	7913	2402	9448	8800	1280	8850
	STDV	4863	3479	993	5539	393	616	762	352	224	3041	3633	522
	Max +1σ	37753	37529	30293	30519	10127	10216	10546	3183	9977	19241	13197	9961

Max -1σ		28027	30571	28307	19441	9341	8984	9022	2479	9529	13159	5931	8917
	10	46	39	39	16	55	46	65	8.7	51	12	79	57
V	20	44	37	37	28	57	65	71	14	69	15	83	78
	30	48	42	41	34	77	61	71	19	67	16	82	78
	40	48	44	43	41	67	49	62	18	61	11	68	69
	50	44	45	40	31	61	43	58	13	63	14	57	66
STDV		2.0	3.4	2.2	9.2	8.9	9.7	5.7	4.1	7.0	2.1	11.1	8.8
Max +1σ		50	48	45	50	86	75	77	23	76	18	94	87
Max -1σ		46	42	41	32	68	55	65	15	62	14	72	69

Note: Dark shading is maximum value in vertical profile; lighter shading is ±1σ of maximum value detected in vertical profile; elevation in meter; depth in cm below surface; concentrations in mg kg^{-1}; σ is standard deviation. Additional data available in supplementary tables.

Table 2. Selected pH values found in the Lucy Grey Line

	LGL Locations			
Depth	3	5	7	11
10	7.21	7.39	7.62	7.31
20	7.41	7.30	7.63	7.27
30	7.54	7.62	7.27	7.19
40	8.05	7.15	7.16	7.22
50	7.96	7.81	7.34	7.54

Note: pH values are in S.U.; depth in cm below surface; shaded values match Table 1

The lowest $\delta^{13}C$ and $\delta^{18}O$ values detected at LGL 3 (-2.56 ‰, -5.42 ‰, respectively) were at 0-10 cm bgs. LGL 5 $\delta^{13}C$ and $\delta^{18}O$ revealed a comparable pattern to LGL 3; $\delta^{13}C$ ranged between 1.39 ‰ and -1.33 ‰ with 1.39 ‰ at 50 cm bgs and -1.33 ‰ at 0-10cm bgs. This alignment shows the highest $\delta^{13}C$ at 40-50 cm bgs with the lowest values at 0-10 cm bgs. Distribution of $\delta^{13}C$ and $\delta^{18}O$ align with other chemical proxies with a distinct change (Δ) in values based on depth. The $\Delta_\delta^{18}O$ is similar to $\Delta_\delta^{13}C$; highest at 40-50 cm bgs with the lowest values detected at 0-10 cm bgs. Data for both $\delta^{13}C$ and $\delta^{18}O$ illustrate a similar anomaly below the surface as seen with other constituents. The observed pattern is consistent when the local system is under similar hydrologic balances caused by environmental changes (e.g. Watanabe et al., 2001, Regenberg et al., 2009). Data also show the highest $\delta^{13}C$ value (1.39 ‰) at LGL 5 at 40-50 cm bgs with the lowest value (-1.33 ‰) at 0-10 cm bgs. At shallower depths (0-10 cm), $\Delta_\delta^{13}C$ decreases, likely the result of increased temperatures over the past 1.0 ka yrs BP. Moreover, $\delta^{13}C$ data indicate that the deeper levels (40-50 cm bgs) were likely long-standing, water-rich environments. LGL 5 $\delta^{18}O$ values were highest (-2.85 and -5.30 ‰, respectively) at 40-50 cm and 20-30 cm bgs with the lowest (-5.66 and -5.42 ‰, respectively) at 0-10 cm bgs. Data show $\delta^{13}C$ and $\delta^{18}O$ within the LGL5 anomaly are consistent with mid-Holocene. The $\delta^{13}C$ data ranged from -0.11 to 0.03‰ and $\delta^{18}O$ data ranged from -3.56 to-2.85‰.

Table 3. $\delta^{18}O$ and $\delta^{13}C$ ratios for LGL 3 and 5

Location	depth	$\delta^{13}C$	STDV	$\delta^{18}O$	STDV
	10	-2.56	0.036	-5.42	0.022
	20	-2.21	0.012	-5.10	0.037
3	30	-2.48	0.020	-4.71	0.062
	40	-2.20	0.017	-5.34	0.021
	50	-2.30	0.005	-5.30	0.034
	0	-1.33	0.024	-5.66	0.029
	20	-1.33	0.025	-4.92	0.035
5	30	-0.11	0.031	-3.56	0.045
	40	0.03	0.011	-4.23	0.018
	50	1.39	0.017	-2.85	0.072

Note: Depth: cm below surface, values are in ‰; Laboratory ID: AA K883

Metal ratios (Mg/Ca) shown in Table 4 have comparable alignments to other sediment chemistry presented in Tables 1-3; following the detected anomaly. Data show 81% of Mg/Ca ratios are <3.5 and 21% are >3.5; values set by other authors to indicate higher temperatures and increased evaporation (Barker et al., 2005; Cleroux et al., 2008; Regenberg et al., 2009). Only 5% of ratios are <2.0; a value symptomatic of cooler temperatures with decreased evaporation (Barker et al., 2005; Regenberg et al., 2009). Data for LGL 3: 30-50 cm; LGL 5: 20-40 cm; and LGL 7 and 11: 0-20 cm, show elevated Mg/Ca ratios within the anomaly, indicative of long-standing water. Table 4 displays a comparable pattern as shown in other data tables; lowest Mg/Ca ratios are situated within the identified chemical-proxies. These findings show that Ivanpah had periods of moist-humid conditions followed by aridity caused by environmental fluctuations.

Table 4. Mg/Ca ratios of the LGL transect

Depth*	Locations											
	1	2	3	4	5	6	7	8	9	10	11	12
10	3.83	3.64	4.03	2.07	2.89	2.45	2.44	2.54	2.32	2.08	2.85	1.97
20	3.89	3.94	3.86	2.83	2.48	2.99	2.36	2.30	2.35	2.03	2.95	2.29
30	3.94	4.32	3.19	2.73	3.38	3.12	2.31	2.08	2.82	2.24	2.62	2.31
40	3.16	4.21	3.60	1.86	3.29	2.61	2.29	2.83	2.80	2.09	2.61	2.41
50	3.72	4.38	3.60	1.82	2.99	2.79	2.31	2.72	2.99	2.13	2.49	2.47

Note: * cm below surface

5.2 Geomorphic and Stratigraphic Features at LGL-Test Pit 5 Discussion

Chemical proxies (shallow coring) led to a 1.5 x 1.5 m wide by ~2 m depth excavation at LGL 5 (796.4 m amsl; 35.523617°, -115.341933°) to explore the anomaly identified at 20-40 cm bgs (Table 1, Fig. 4A). Research shows the presence of possible lacustrine units below the surface; these units are designated as Lake Unit (LU) 0, 1, 2 and 3 with LU3 subdivided into LU3a, b, c, and d. Five samples were collected from LGL 5 and sent for ^{14}C analysis and are presented within their respective sections below. The excavation sidewalls displayed a mixture of compositions dominated by sand, silt, and clay, with some clastic materials at shallow depths, unconformities indicative of three possible lake beds, and two apparent playa surfaces exhibiting mud cracks.

The surface (unit ALØ) of LGL5 is composed of a relatively thin (~10 cm bgs) layer of light brown (Color hue 7.5YR 6/4), very fine to fine sandy loam with granules, pebbles and small gravels at the base (Fig. 4F). ALØ is dominated by an upward sequence of fine-earth textures and a non-laminated matrix, aeolian in origin, becoming coarser at depth, grading to pea-sized pebbles, clasts, and small gravels situated on a boundary surface (LU3d) at depth (-10cm). ALØ represents a late Holocene-to-modern deposit situated on an older unit (LU3).

LU3 is a relatively thick unit (LU3a, b, c, d) ranging from ~10 to 60 cm bgs with a base elevation of 795.8 m amsl when adjusted for ground elevation. Materials within LU3d (~10 to 20cm bgs) is an aridisol composed of light brown to reddish brown, sandy loam (color hue 7.5YR 6/4) with subrounded to subangular texture, showing classic oxidation (Fig. 4B; Retallack, 1988). This unit contains 19.4% pea-sized gravels (0.5-1.0 cm), 72.9% sands, and 8.0 % silts (Table 5), consolidated by clays with carbonate strings forming a weak vesicular surface that appears erosionally truncated. The surface of LU3d displays a disconformity (truncated) representing a rapid regressive phase overlain by AlØ and leaving no subsequent visual evidence. The lack of well-developed soils in LU3d is consistent with a younger age for the most recent transgressional phase situated on a base of gravels and sands of LU3a, b, and c. Lastly, LU3d deposition is difficult to determine, the structure reduces vertical infiltration, making it problematic to date owing to very high inorganic carbon, though it reflects the cumulative effects of pedogenic processes with time.

LU3a, b, and c consist of 27.5% gravels, 71.4% sands and 1.1% silts, sedimentologically characteristic of a fluvial deposit beach facies containing a poorly sorted matrix (Table 5; Figs. 4). The unit grades upward from sub-rounded basal gravels (~2-5 cm) and clasts (~2-5 cm) with minor sand in LU3a (Fig. 4C); pea-sized beach gravels (~0.7-2 cm) in LU3b (Fig. 4F); and sand/loamy sand in LU3c (Fig. F) yielding a composite (LU3a, b, c) age of 1,050-1,025 cal yrs BP (Table 6). All subunits within LU3 (a-c) are light brown with a color hue 7.5YR 6/3. The uppermost portion of LU3c is positioned at 796.2 m amsl when adjusted for current surface elevation. Evidence shows that a shoreline advanced eastward from changing lake volumes, water deepened at LGL 5, depositing sandy loam with pea-sized beach gravels within unit LU3a-c forming a classic transgressional sequence of poorly sorted, lacustrine gravel beach facies (Bacon et al., 2006). Structurally, LU3 (a-c) is ~30 cm thick fluvial deposit of sands, pea-sized gravels and clasts, all positioned on an older lacustrine unit boundary surface (LU2).

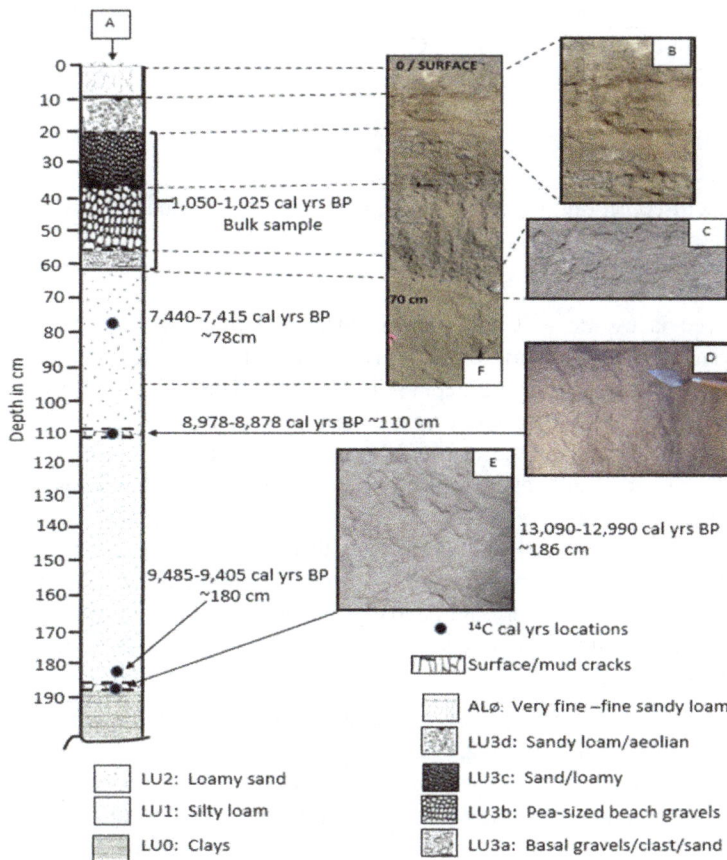

Figure 4. LGL 5 stratigraphic cross section from transect

Unit LU2 is positioned at 795.8 m amsl when adjusted for elevation and consists of <0.1 % gravels, 89.1% sands and 10.4% silts and <0.5% clays (Table 5). This unit consist of a light brown loamy sand (color hue 7.5YR 6/4) containing carbonate strings within the upper few (~3) cm of a thicker component ranging from ~63 cm to ~109 cm bgs yeilding a date of 7,440-7,415 cal yrs BP at ~78 cm bgs (Table 6). Under magnification (4x), bedding in LU2 consists of loamy sand-sized grains arranged in parallel lines (partings and intervening) with fine to very fine sands erosionally truncated at ~63 cm bgs (Fig. 4F) exhibiting minor surface cracking. LU2 sediments are very well sorted and interbedded, with very thin (i.e. 1-3 cm thick) lamination throughout, interpreted as beach sand overlain by a coarser fluvial unit LU3 as described by McKee and Weir (1953). LU2 is likely the result of a low energy suspension event overlain LU1.

Table 5. Grain size, color hue of LGL5 sediments

Unit ID	~Depth (cm)	Color hue	Gravel %	Sand %	Silt %	Clay%
ALØ	0-10	7.5YR 6/4	1.0	83.4	15.5	<0.1
LU3d	10-21	7.5YR 6/4	19.4	72.9	8.0	<0.1
LU3c	21-36	7.5YR 6/3	20.8	74.1	5.1	<0.1
LU3b	36-55	7.5YR 6/3	23.7	74.8	1.8	<0.1
LU3a	55-63	7.5YR 6/3	37.6	62.3	0.1	<0.1
LU2	63-109	7.5YR 6/4	<0.1	89.3	10.5	<0.1
LU1	109-186	7.5YR 7/3	<0.1	25.7	74.2	<0.1
LU0	186 -unknown	7.5YR 6/4	<0.1	44.2	53.1	2.6

Note: LU0 was explored only to 196 cm below surface

Table 6. Radiocarbon dates from LGL5 sediments

Lab ID	Unit ID	Depth[1] (cm)	Materials dated	$\delta^{13}C$ (‰)	[14]C age[2] (yrs BP)	cal yrs BP[3] (2σ age range)
Beta432398	LU3(a-c)	-30 to -61	Bulk C_{org}	-11.5	$1,060 \pm 30$	1,050-1,025
Beta434682	LU2	- 78	Grab MC_{org}	-17.5	$6,500 \pm 31$	7,440-7,415
AA107397	LU1	-110	Grab MC_{org}	-18.1	$7,940 \pm 31$	8,978-8,878
Beta434906	LU1	- 180	Grab MC_{org}	-21.6	$8,400 \pm 30$	9,485-9,405
Beta436547	LU0	- 186	Grab MC_{org}	-17.0	$11,160 \pm 40$	13,090-12,990

Note: C: carbon, M: microscopic, org: organic, [1] cm below surface; [2] reported in conventional radiocarbon years; [3] reported in calibrated years BP with 95% probability

LU1 is a well preserved lacustrine unit commencement at ~110 cm bgs, extending to ~186 cm bgs, and appearing to have been deposited by low energy suspension. The superficial portion of this unit dated to 8,978-8,878 cal yrs BP at ~110 cm bgs and at depth (~180 cm bgs) dated 9,485-9,405 cal yrs BP. The surface of LU1 (~110 cm) is desiccated, showing signs of cracking ("mud-cracks") with a crust rich in swelling clays exhibiting evaporite-staining of the surface (~2 cm), indicative of an established desert playa (Fig. 4D). The sediment matrix consists of very thin laminated pink silt loam (color hue 7.5YR 7/3) containing <0.4 % gravels, 24.2% sands, 74.6% silts, and 0.8% clays (Table 5). In contrast to other units, LU1 appears more sorted with overlapping of edges ("imbricated"), and sand-size grains that are marginally rounded. The appearance of this unit is likely the effects of a repeated influx of water followed by episodic evaporation resulting in rough-crusted surfaces similar to those described by Adams and Sada (2014) and Tollerud and Fantle (2014). LU1 is dissimilar to units LU2, LU3 and ALØ; sediment matrix is very thin laminated silt loam, indicative of a lacustrine water facies overlain a deeper unit (LU0).

The lowest unit (LU0) encountered was a preserved lacustrine feature of light brown clay (color hue 7.5YR 6/4) beginning at ~186 cm bgs, dating to 13,090-12,990 cal yrs BP at 186 cm bgs. It is important to note that this unit was not explored beyond ~196 cm bgs owing to limited excavation techniques and safety. Sediments consisted of very thin laminated, fine-grained, imbricated clay-rich matrix composed of <0.1 % gravels, 44.2% sands, 53.1% silts and 2.6% clays (Table 5). The surface of LU0 contained surface-cracking ("mud-cracks") with evaporite-staining (~2.5 cm), composed of well-sorted, firm, illuviation of argillic clays that were bedded, very plastic, and with particle sizes ranging between 1 mm and 2 μm (Fig. 4E). This unit is progressively thicker with well-developed grains, ped-cutans, and blocky to prismatic structure. LU0 is clearly the result of chemical weathering of local silicate-bearing materials that produced primary clay deposits. LU0 is in stark contrast to other units; the matrix is composed of firm, very thin, laminated clays indicative of a long-standing, lacustrine, deep-water facies. Finally, the age of LU0 (13,090-12,990 cal yrs BP at ~186 cm bgs) places the surface firmly within the Pleistocene - Holocene transitional period.

6. Discussion

Major environmental events inferred from the sedimentological evidence at Ivanpah were compared to those of other studies from localities within or bordering the Mojave Desert. Documented events are likely associated with other recognized climatic shifts in the region, for instance, Wells et al. (2003) and others describe climatic shifts (drying events) within the region at ca. 9.2, 7.6, 8.2, 3.2 ka yrs BP with several minor shifts (wetting events) occurring ca. 2.5, 2.0, 1.5, 1.0 and 0.8 ka yrs BP, respectively (Cook et al., 2007; Harvey et al., 1999; Holmquist et al., 2016). Sediments from the LGL 5 excavation yielded data documenting multiple sequences with LU3d characterized as an aridisol sandy loam showing minor pedogenic processes (Figs. 4F/B). This unit's (LU3d) deposition and age is problematic though, evidence suggests it is younger than LU3a-c (1,050-1,025 cal yrs BP). Within this time frame, LU3d is probably linked to the wetting period (~800 to 600 cal yrs BP) described by Mensing et al. (2008) (Fig. 4B). They documented that after the wetting period a drier, arid climate settled into this region. This drier period is likely responsible for ALØ; an aeolian sandy-silt bed deposited on the north-northeastern lateral of the Ivanpah valley.

Unit LU3a-c is a classic, near-shore littoral deposit of interbedded gravel and sand beach facies grading upward from basal to pea-sized gravels containing sands and minor silts as described by Bacon et al. (2006). LU3 (a-c) age of 1,050-1,025 cal yrs BP (Figs. 4F/B/C) places this unit within Holmquist et al. (2016) and Cook et al. (2007) period of warming over the Western United States, promoting increased monsoonal activities ~1,000 yrs BP. Other authors also describe sporadic wetting periods around the same time that caused monsoonal events (spring to winter) beginning ~1,900 and ending ~1,100 cal yrs BP (Negrini et al., 2006). Holmquist et al. (2016)

further describes a period ending ~1,280 cal yrs BP when this region became significantly wetter; perhaps providing ideal stormy lake conditions to establish the beach zone of LU3$_{(a-c)}$ that overlays the unit of LU2.

The surface unit LU2 is erosionally truncated with very minor desiccation cracking dating to 7,440-7,415 cal yrs BP at ~78 cm bgs (Fig. 4F). Researchers Tollerud and Fantle (2014) studied similar environments documenting minor desiccation of playa surface and the stages at which mud-cracks form (Tollerud and Fantle, 2014). They also describe the subsurface of similar mud-cracked units as very thin laminated sediments resulting from a low energy suspension environment. Their efforts described surfaces similar to the minor desiccation of LU2 where the evolution of the mud-cracks promotes a smooth, evaporite-stained, strong surface with minor cracking of ~3-5 cm long 0.5-1.5 cm thick, as seen with the surface of LU2.

Seismological evidence for LU2, a unit ~46 cm thick, is indicative of a period of drying that occurred after a wetting phase (Fig. 4F). This unit is composed of classic fine, very thin laminated sediments commencing at ~3-4 cm below the desiccated surface (~63 cm bgs) to a depth of ~109 cm bgs. It is dissimilar to LU3$_{(a-d)}$ as it contains substantially more sands and silts, with no gravels. Evidence indicates that it is the result of a classic low energy depositional lacustrine environment similar to that described by Tollerud and Fantle (2014). Murchison and Mulvey (2000) and Patrickson et al. (2010) documented a drying events between 8,200 and 7,650 yrs BP causing recessional phases of the Great Salt Lake located northeast of Ivanpah. Furthermore, Blunt and Negrini et al. (2015) recognized events west-northwest of Ivanpah where a Tulare Lake shoreline (ca. 8,166-7,984 and 8,857-9,008 cal yrs BP) began a recessional phase due to aridity as well. It is probable that LU2 is the result of those documented drying events and subsequent desiccation of LU2.

Below LU2 is LU1, a unit composed of very fine, very thin laminated sediments commencing at ~109 cm to a depth of ~185 cm bgs. Of importance at LU1 is the surface mud-cracks, which are ~3-10+ cm long by a ~2-5 cm thick; they are smooth, contain minor evaporite-staining, and have a strong surface (Fig. 4D). Below the surface cracking are very thin laminated sediments suggestive of a classic low energy depositional lacustrine environment as described by others (Negrini et al., 2006; Tollerud and Fantle, 2014). The surface (~109 cm bgs) of LU1 yielded an age of 8,978-8,878 cal yrs BP with a lower depth (~180 cm) dating to 9,485-9,405 cal yrs BP (Fig. 4A).

LU1 links with documented regional shifts that affected the southwestern United Stated (e.g. Great Salt and Tulare lakes). Harvey et al. (1999) documented that during the early Holocene (ca. 9,500 -8,600 cal yrs BP), the jet stream moved north, and allowing monsoonal activity to enter the American Southwest inundating geographic basins. Such a shift in environmental conditions would have been the reason for the existence of LU1: a thick lacustrine facies. Following this event (Harvey et al. 1999), a drying episode described by Murchison and Mulvey (2000) likely led to the desiccation of the surface of LU1. Data suggests that the surface of LU1 is the product of the early Holocene lake termination event described by Murchison and Mulvey (2000).

The deepest unit, LU0 (13,090–12,990 cal yrs BP), represents a period of a long-standing water environment during the Pleistocene – Holocene transition (ca. 14,600–8,000 yrs BP) as described by Antinao and McDonald (2013). Moreover, at Tulare Lake, a period of increased wetting was documented (ca. 12,605 to 12,971 cal yrs BP), promoting perennial lakes within the region (Bacon et al., 2006, Kirby et al., 2015), possibly driving the wetting events at Lake Ivanpah. Furthermore, authors Harvey et al. (1999) describe an arid phase (ca. 13,000-10,500 yrs BP) following the perennial lake period, likely related to Bacon et al. (2006) and Kirby et al. (2015) work, resulting in the termination of pluvial lakes in this region. These drying events are probably the cause of the recessional phases identified at LU0 dating to 13,090 – 12,990 cal yrs BP (Fig. 4E). Evidence clearly puts unit LU0 within the early Pleistocene – Holocene transition; however, it is also feasible that LU0 is the result of the end of a late pluvial lake phase. Because of the limitation on the excavation imposed by the BLM, it is difficult to extrapolate much more from LU0, as additional work including deeper excavations would be required to fully describe this unit.

7. Summary and Conclusions

Dry lake systems located within the Mojave Desert are potentially good sites for alternative energy such as solar energy collection however, they have not been sufficiently studied for late Pleistocene - early Holocene lacustrine environments. Development could adversely impact valuable sedimentological resources contained within the shallow subsurface. This study used shallow sediment chemical proxies and characterization to assess subsurface conditions for buried lakebed features at Lake Ivanpah. The number and depth of shallow cores, and uses of trenching, to describe shallow sedimentary facies for this paper were adequate for limited delineation of wet and dry periods associated with lacustrine lake occupations, but does need more comprehensive data collection (especially subsurface data) and interpretation of this important period of desertification in recent geologic history.

Results at Lake Ivanpah illustrated that transitional lakes existed in the basin for two Holocene and one Pleistocene-Holocene transitional periods. Although there are no signs of visual wave-cutting at Ivanpah, chemical proxies and sediment characterization of site LGL5 showed a buried shoreline ("beach zone") and two lacustrine features. Furthermore, chemical proxies also identified two possible truncated shorelines at sites LGL7 and 11. With this in mind, evidence shows lake high-stands likely occurred at water depths of ~1.8, ~5.3 and ~9.8 m, respectively, in the Ivanpah valley. Lacustrine units identified beneath the surface at LGL5 were dated to ca. 1,050-1,025 (LU3a-c), 7,440-7,415 (LU2), and 9,485-9,405 (LU1) cal yrs BP with the deepest lake unit (LU0) dating to ca. 13,090-12,990 cal yrs BP. This evidence clearly shows that Lake Ivanpah has a significantly more complex history than just a single continuous lake episode of mid- to late-Holocene in age but rather, multiple episodes of lake stands, the oldest being a Pleistocene-Holocene (or older) transitional lake.

It is likely that other lacustrine units exist beneath the Ivanpah playa. The lack of visual evidence is most likely the result of an abrupt regional climatic shift <1,000 yrs BP, resulting in shoreline truncation and obliteration. Understanding environmental change in these basins is needed to better recognize climate changes during the late Pleistocene-early Holocene and regional desertification. Further studies of the surface of the Ivanpah and Roach dry lakes and surrounding landscapes is required to better characterize the overall basin. Finally, a more comprehensive surface and subsurface study is required to describe past conditions and depth of lacustrine environments at Ivanpah before extensive alternative energy developments disrupt shallow sedimentary evidence necessary for future research.

Acknowledgments

The authors would like to thank the personnel of the United States Department of interior, Bureau of Land management, Needles Office for their cooperation during this research. We would like to thank the numerous students from the College of Southern Nevada, Department of Physical Sciences, for their time in the field, laboratory and data interpretation. The NASA-CoP Grant (NNX14Q94A) provided funding to students at the College of Southern Nevada. We would also like to thank the reviewers whose time and expertise added great value to this paper and how it is presented.

References

Adams, K., & Sada, D. W. (2014). Surface water hydrology and geomorphic characterization of playa lake systems: Implications for monitoring the effects of climate change. *Journal of Hydrology, 510*, 92-102. http://dx.doi.org/10.1016/j.jhydrol.2013.12.018

Adrian, M. H., Wigand, P. E., & Wells, S. G. (1999). Response of Alluvial Fan Systems to the Late Pleistocene to Holocene Climatic Transition: Contrasts between the margins of pluvial lakes Lahontan and Mojave, Nevada and California, USA. *Catena, 36*(4), 255-281. http://dx.doi.org/10.1016/S0341-8162(99)00049-1

Antinao, J. L., & McDonald, E. V. (2013). An enhanced role for the Tropical Pacific on the humid Pleistocene-Holocene transition in southwestern North America. *Quaternary Science Reviews, 78*, 319-341. http://dx.doi.org/10.1016/j.quascirev.2013.03.019

Bacon, S. N., Burke, R. M., Pezzopane, S. K., & Jayko, A. S. (2006). Last glacial maximum and Holocene lake levels of Owens Lake, eastern California, USA. *Quaternary Science Review, 25*, 1264-1282. http://dx.doi.org/10.1016/j.quascirev.2005.10.014

Barker, S., Cacho, I., Benway, H., & Tachikawa, K. (2005). Planktonic foraminiferal Mg/Ca as a proxy for past oceanic temperatures: a methodological overview and data compilation for the Last Glacial Maximum. *Quaternary Science Review, 24*, 821–834. http://dx.doi.org/10.1016/j.quascirev.2004.07.016

Barker, S., Greaves, M., & Elderfield, H. (2003). A study of cleaning procedures used for foraminiferal Mg/Ca paleothermometry. *Geochemistry,* Geophysics, *Geosystems,* *4*(9), 1-20. http://dx.doi.org/10.1029/2003GC000559

Bauer, H. P., Beckett, P. H. T., & Bie, S. W. (1972). A rapid gravimetric method for estimating calcium carbonate in soils. *Plant and Soil, 37*, 689-690. http://dx.doi.org/10.1007/BF00264189

Blunt, A. B., & Negrini, R. M. (2015). Lake levels for the past 19,000 years from the TL05-4 cores, Tulare Lake, California, USA: geophysical and geochemical proxies. *Quaternary International, 387*, 122-130. http://dx.doi.org/10.1016/j.quaint.2015.07.001

Bull, W. B. (1991). Geomorphic responses to climatic change. Oxford University Press, Oxford, U. K.

Carr, M. D., & Pinkston, J. C. (1987). Geologic map of the Goodsprings District, southern Spring Mountains, Clark County, Nevada, Miscellaneous Field Studies Map - U. S. Geological Survey, MF-1514.

CH2MHILL (2009). Biological assessment for the Ivanpah solar electric generating system: Ivanpah (SEGS) project. Retrieved from http://www.blm.gov/style/medialib/blm/ca/pdf/needles/lands_solar.Par.30845.File.dat/ ISEGS_Biological_Assessment_Dec09.pdf

Cléroux, C., Cortijo, E., Anand, P., Labeyrie, L., Bassinot, F., Caillon, N., & Duplessy, J. C. (2008). Mg/Ca and Sr/Ca ratios in planktonic foraminifera: Proxies for upper water column temperature reconstruction. *Paleoceanography*, *23*(3), 1-16. http://dx.doi.org/10.1029/2007PA001505

Cook, E. R., Seager, R., Cane, M. A., & Stahle, D. W. (2007). North American drought: reconstructions, causes, and consequences. *Earth Sci. Rev., 81*, 93-134. http://dx.doi.org/10.1175/JCLI4042.1

Enzel, Y., Brown, W. J., Anderson, R. Y., McFadden, L. D., & Wells, S. G. (1992). Short-duration Holocene lakes in the Mojave River drainage basin, southern California. *Quaternary Research, 38*, 60-73. http://dx.doi.org/10.1016/0033-5894(92)90030-M

Enzel, Y., Wells, S. Lancaster, G., N. (2003). Late Pleistocene lakes along the Mojave River, southeast California. In Paleoenvironments and Paleohydrology of the Mojave and Great Basin Deserts (Y. Enzel, S. G. Wells, N. Lancaster, Eds.). *Geological Society of American Special Paper, 368*, 61-77. http://dx.doi.org/10.1130/0-8137-2368-X.1

Eugster, H. P., & Kelts, K. (1983). Lacustrine chemical sediments. In: Goudie AS. Pye K. (Eds) Chemical sediments and geomorphology. Academic Press, London.

Garside, L. G., House, P. K., Burchfiel, B. C., & Rowland, S. M. (2009). Preliminary geologic map of the Jean quadrangle, Clark County, Nevada. Nevada Bureau of Mines and Geology Open-file Map 09-5, scale 1:24,000.

Harvey, A. M., Wigand, P. E., & Wells, S. G. (1999). Response of Alluvial Fan Systems to the Late Pleistocene to Holocene Climatic Transition: Contrasts between the Margins of Pluvial Lakes Lahontan and Mojave, Nevada and California, USA. *Catena, 36*(4), 255-281. http://dx.doi.org/10.1016/S0341-8162(99)00049-1

Holmquist, J. R., Booth, R. K., & MacDonald, G. M. (2016). Boreal peatland water table depth and carbon accumulation during the Holocene thermal maximum, Roman Warm period, and Medieval Climate anomaly. Palaeogeography, Palaeoclimatology, Palaeoecology, 444, 15-27. http://dx.doi.org/10.1016/j.palaeo.2015.11.035

House, P. K., Buck, B. J., & Ramelli, A. R. (2010). Geologic assessment of piedmont and playa flood hazards in the Ivanpah Valley area, Clark County, Nevada (online version): Nevada Bureau of Mines and Geology Report 53.

House, P. K., Ramelli, A. R., & Buck, B. J. (2006). Surficial geologic map of the Ivanpah Valley area, Clark County, Nevada: Nevada Bureau of Mines and Geology Open-File Report 06-8, scale 1:50,000.

Jones, T. L., Brown, G. M., Raab, L. M., McVickar, J. L., Spaulding, W. G., Kennett, D. J., York, A. L., Walker, P. L. (1999). Environmental Imperatives Reconsidered: Demographic Crises in Western North America During The Medieval Climatic Anomaly. *Current Anthropology, 40*(2), 137-170. http://dx.doi.org/10.1086/200002

Kirby, M. E., Knell, E. J., Anderson, W. T., Lachniet, M. S., Palermo, J., Eeg, H., ...Hiner, C.A. (2015). Evidence for isolation and pacific forcing of late glacial through Holocene climate in the Central Mojave Desert (Silver Lake, CA). *Quaternary Research, 84*, 174-186. http://dx.doi.org/10.1016/j.yqres.2015.07.003

Kliem, P., Enters, D., Hahn, A., Ohendorf, C., Lise-Pronovost, A., St-Onge, G.,... PASADO Science Team (2012). Lithology, radiocarbon chronology and sedimentological interpretation of the lacustrine record from Laguna Potrok Aike, southern Patagonia. *Quaternary Science Reviews, 71*, 54-69. http://dx.doi.org/10.1016/j.quascirev.2012.07.019

Kober, B., Schwalb, A., Schettler, G., & Wessels, M. (2007). Constraints on paleowater dissolved loads and on catchment weathering over the past 16 ka from [87]Sr/[86]Sr ratios and Ca/Mg/Sr chemistry of freshwater ostracode tests in sediments of Lake Constance, Central Europe. *Chemical Geology, 240*(3-4), 361-376. http://dx.doi.org/10.1016/j.chemgeo.2007.03.005

Litwin, R. J., Smoot, J. P., & Smith, G. I. (1999). Calibrating late Quaternary terrestrial climate signals: Radiometrically dated pollen evidence from the southern Sierra Nevada, USA. *Quaternary Science Reviews, 18*, 1151–71. http://dx.doi.org/10.1016/S0277-3791(98)00111-5

May, J. H., Barrett, A., Cohen, T. J., Jones, B. G., Price, D., & Gliganic, L. A. (2015). Late Quaternary evolution of a playa margin at Lake Frome, South Australia. *Journal of Arid Environments, 122*, 93-108. http://dx.doi.org/10.1016/j.jaridenv.2015.06.012

McKee, E. D., & Weir, A. (1953). Terminology for stratification and cross-stratification. Bull. Geol. Soc. Amer., 64, 381-390.

Mehringer, P. J., & Warren, C. N. (1976). Marsh, dune, and archaeological chronology, Ash Meadows, Amargosa Desert, Nevada. In Elston, R., Ed., "Holocene environmental change in the Great Basin." *Nevada Archaeological Survey Research Paper*, 6, 120-150.

Mensing, S., Smith, J., Norman, K. B., & Allan, M. (2008). Extended drought in the Great basin of western North America in the last two millennia reconstructed from pollen record. *Quaternary International, 188*(1), 79-89. http://dx.doi.org/10.1016/j.quaint.2007.06.009

Mifflin, M. D., & Wheat, M. M. (1979). Pluvial Lakes and Estimated Pluvial Climates of Nevada. *Nevada Bureau of Mines and Geology Bulletin* 94.

Morsel, J. W., Wang, Q., & Tsio, M. Y. (1997). Influences of temperature and Mg:Ca ratio on $CaCO_3$ precipitates from seawater. *Geology, 25*, 85-87. http://dx.doi.org/10.1130/0091-7613(1997)025<0085:IOTAMC>2.3.CO;2

Murchison, S. B., & Mulvey, W. E. (2000). Late Pleistocene and Holocene shoreline stratigraphy on Antelope Island. *Utah Geological Survey Misc. Pubulication, 1*, 77-83. http://dx.doi.org/10.1016/j.yqres.2009.12.006

Negrini, R. M., Wigand, P. E., Draucker, S., Gobalet, K., Gardner, J. K., Sutton, M. Q., & Yohe, R. M. (2006). The Rambla highstand shoreline and the Holocene lake-level history of Tulare Lake, California, USA. *Quaternary Science Reviews*, 25, 1599-1618. http://dx.doi.org/10.1016/j.quascirev.2005.11.014

Nishizawa, S. (2010). The Bonneville lake basin shoreline records of large lake and abrupt climate change events. Unpublished dissertation, University of Utah, Department of Geography.

Noble, P., Zimmerman, S., Ball, I., Adams, K., Maloney, J., Smith, S. (2016). Late Holocene subalpine lake sediments record a multi-proxy shift to increased aridity at 3.65 kyr BP, following a millennial - scale neopluvial interval in the Lake Tahoe watershed and western Great Basin, USA. *Geophysical Research Abstracts* 18.

Patrickson, S. J., Sack, D., Brunelle, A. R., & Moser, K. A. (2010). Late Pleistocene to early Holocene lake level and paleoclimate insights from Stansbury Island, Bonneville basin, Utah. *Quaternary Research, 73*, 237-246. http://dx.doi.org/10.1016/j.yqres.200 9.12.006

Pigati, J. S., Miller, D. M., Bright, J. E., Mahan, S. A., Nekola, J. C., & Paces, J. B. (2011). Chronology, sedimentology, and microfauna of groundwater discharge deposits in the central Mojave Desert, Valley Wells, California. *Geological Society of America Bulletin, 123*, 2224-2239.

Quade, J., Forester, R. M., & Whelan, J. F. (2003). Late Quaternary Paleoclimatic and Paleotemperature Change in southern Nevada. In Paleoenvironments and Paleohydrology of the Mojave and southern Great Basin Deserts (Y. Enzel, S. G. Wells, and N. Lancaster, Eds.). *Geological Society of American Special Paper, 368*, 165-188. Boulder, CO.

Quade, J., Mifflin, M. D., Pratt, W. L., McCoy, W., & Burckle, L. (1995). Fossil spring deposits in the southern Great Basin and their implications for changes in water-table levels near Yucca Mountain, Nevada, during Quaternary time. *Geological Society of America Bulletin, 107*, 213-230.

Ramelli, A. R., House, P. K., & Buck, B. J. (2006). Preliminary Surficial Geologic Map of the Ivanpah Valley Part of the Roach and desert 7.5' Quadrangles, Clark County, Nevada. Nevada Bureau of Mines and Geology. Open-file Report 611-F. Carson City, NV.

Regenberg, M., Steph, S., Nurnberg, D., Tiedemann, R., Garbe-Schonberg, D. (2009). Calibrating Mg/Ca ratios of multiple planktonic foraminiferal species with ^{18}O-calcification temperatures: Paleothermometry for the upper water column. *Earth and Planetary Science Letters, 278*, 324-336. http://dx.doi.org/10.1016/j.epsl.2008.12.019

Reheis, M., & Redwine, J. L. (2008). Lake Manix shoreline and Afton Canyon terraces: Implications for incicion of Afton Canyon. *Geological Society of America, 439*, 227-259. http://dx.doi.org/10.1130/2008.2439(10)

Reis, J. (2004). Effects of ambient Mg/Ca ratio on Mg fraction in calcareous marine invertebrates: A record of the oceanic Mg/Ca ratio over the Phanerozoic. *Geological Society of America, 32*(11), 981-984.

Retallack, G. J. (1988). Field recognition of paelosols. *Geological Society of America*, Special paper 216.

Robinson, M. C., Flint, S., & Spaulding, W. G. (1999). Molycorp, Inc. Mountain Pass Mine cultural resources investigations testing and evaluation report for CA-Sbr9387, CA-SBr-9388, and CA-SBr9389/H, Ivanpah Dry Lake, San Bernardino County, California. Applied Earth Works, Hemet, CA. Retrieved from https://www.scribd.com/doc/314014653/Robinson-et-al-1999-Ivanpah-CA-Report

Rodrigues-Filho, S., & Muller, G. (1999). A Holocene sedimentary record from Lake Silvana, SE Brazil: evidence for paleoclimatic changes from mineral, trace-metal, and pollen data. Volume 88 of Lecture notes in earth sciences. *Springer*, p. 96.

Rodrigues-Filho, S., Behling, H., Irion, G., & Muller, G. (2001). Evidence for Lake Formation as a Response to an Inferred Holocene Climatic Transition in Brazil. *Quaternary Research, 57*(1), 131–137. http://dx.doi.org/10.1006/qres.2001.2281

Sack, D. (1995). The shoreline preservation index as a relative-age dating tool for late Pleistocene shorelines: An example from the Bonneville basin, USA. *Earth Surf Process Landf, 20*, 363-377. http://dx.doi.org/10.1002/esp.3290200406

Sack, D. (2001). Shoreline and basin configuration techniques in paleolimnology, *in* Last, W.M., J.P. Smol (Eds), 2001. Tracking environmental change using lake sediments. Vol. 1: Basin analysis, coring and chronological techniques. Kluwer Academic Publishers, Dordrecht, The Netherlands.

Sanchi, L., Menot, G., & Bard, E. (2015). Environmental controls on paelo-pH at mid-latitudes: A case study from Central and Eastern Europe. *Paleogeography, Paleoclimatology, Paleoecology, 417*, 458-466. http://dx.doi.org/10.1016/j.palaeo.2014.10.007

Schmidt. K. M., & McMackin, M. (2006). Preliminary surficial geologic map of the Mesquite Lake 30' x 60' Quadrangle, California and Nevada. U.S. Geological Survey Open File Report 2006-1035. Denver, CO.

Schnurrenberger, D., Russell, J., & Kelts, K. (2003). Classification of lacustrine sediments based on sedimentary components. *J Paleolimnology, 29*, 141-154. http://dx.doi.org/10.1023/A:1023270324800

Shapiro, R. (2016). Camelid record of Mesquite Lake, California: Impact of earliest Holocene climate change. *The 30th Annual Desert Symposium*, p. 41.

Spaulding, W. G. (1990). Vegetational and climatic development of the Mojave Desert: The last glacial maximum to the present. In "Packrat middens: The last 40,000 years of biotic change" (J.L. Betancourt, T.R. Van Devender, and P.S. Martin, Eds.). University of Arizona Press, Tucson. pp. 166-199.

Spaulding, W. G., & Graumlich, L. J. (1986). The last pluvial climatic episodes in the deserts of southwestern North America. *Nature, 320*, 441-444.

Stumm, W., Morgan, J. J. (1996). Aquatic Chemistry Chemical Equilibria and Rates in Natural Waters. 3rd Ed., John Wiley & Sons, Inc., New York, p. 1022.

Sun, Z., Xu, G., Hao, T., Huang, Z., Fang, H., & Wang, G. (2015). Release of heavy metals from sediment bed under wave-induced liquefaction. *Marine Pollution Bulletin, 79*(1-2), 209-216. http://dx.doi.org/10.1016/j.marpolbul.2015.06.015

Tardy, Y., Roquin, C. (1992). Geochemistry and evolution of lateritic landscapes. In: C.R.M. Butt and H. Zeegers (Editors). Handbook of Exploration Geochemistry, 4. Regolith Exploration Geochemistry in Tropical and Subtropical Terrains. Elsevier, Amsterdam, pp. 407-443.

Tchakerian V. P., & Lancaster, N. (2002). Late Quaternary arid/humid cycles in the Mojave Desert and western Great Basin of North America. *Quaternary Science Reviews, 21*, 799–810. http://dx.doi.org/10.1016/S0277-3791(01)00128-7

Thompson, R. S., Whitlock, C., Bartlein, P. J., Harrison, S. P., & Spaulding, W. G. (1993). Climatic changes in the western United States since 18,000 yr B.P. In "Global climates since the last glacial maximum" (H. E. Wright, Jr., J. E. Kutzbach, T. Webb, III, W. F. Ruddiman, F. A. Street-Perrott, and P. J. Bartlein, Eds.), pp. 469-513. University of Minnesota Press, Minneapolis, MN.

Tollerud, H. J., & Fantle, M. S. (2014). The temporal variability of centimeter-scale surface roughness in a playa dust source: Synthetic aperture radar investigation of playa surface dynamics. *Remote*

Sensing of Environment, 154, 285-297. http://dx.doi.org/10.1016/j.rse.2014.08.009

US Geological Survey (USGS) (1985). Ivanpah Lake, Calif. – Nev. 7.5-minute Topographic Quadrangle. Provisional Edition. Lakewood, CO.

USEPA. (1997). SW-846, Test Methods for Evaluating Solids and Wastes – Physical/Chemical Methods, U.S. Environmental Protection Agency. December, CD-Rom.

Waring, G. A. (1920). Ground Water in Pahrump, Mesquite, and Ivanpah Valleys Nevada and California. Water–Supply Paper, 450-C. pp. 51-85.

Watanabe, T., Winter, A., & Oba, T. (2001). Seasonal changes in sea surface temperature and salinity during the Little Ice Age in the Caribbean Sea deduced from Mg/Ca and $^{18}O/^{16}O$ ratios in coral. *Marine Geology, 173*, 21-35. http://dx.doi.org/10.1016/S0025-3227(00)00166-3

Wells, S. G., Brown, W. J., Enzel, Y., Anderson, R.Y., & McFadden, L. D. (2003). Late Quaternary geology and paleohydrology of pluvial Lake Mojave, southern California. In Paleoenvironments and Paleohydrology of the Mojave and southern Great Basin Deserts (Y. Enzel, S. G. Wells, and N. Lancaster, eds.). *Geological Society of American Special Paper, 368*, 79-114.

Whitney, J. W., Breit, G. N., Buckingham, S. E., Reynolds, R. L., Bogle, R. C., Luo, L., Goldstein, H. L., & Vogel, J. M. (2015). Aeolian responses to climate variability during the past century on Mesquite Lake Playa, Mojave Desert. *Geomorph, 230*, 13-25. http://dx.doi.org/10.1016/j.aeolia.2016.09.001

Supplementary Data

Locations

	Depth	1	2	3	4	5	6	7	8	9	10	11	12
	10	11,600	12,500	16,500	18,750	28,860	27,850	33,160	4,015	36,920	6,700	44,580	31,200
	20	20,550	21,000	17,000	16,600	36,290	41,920	44,760	7,200	38,670	3,500	43,010	45,920
Al	30	20,600	20,300	24,800	19,500	49,670	40,010	44,750	9,350	41070	7,650	41,920	46,010
Units	40	20,900	23,500	24,100	17,100	39,370	32,550	34,760	8,200	40,960	7,360	36,410	33,010
mg	50	20,400	21,050	20,800	16,200	32,220	35,600	26,130	5,800	36,000	6,080	31,440	32,000
kg-1	Mean	18810	19670	20640	17630	37282	35586	36712	6913	38724	6258	39472	37628
	STDEV	4035	4188	3863	1426	7990	5674	8028	2081	2301	1657	5443	7638
	MAX	20900	23500	24800	19500	49670	41920	44760	9350	41070	7650	44580	46010
	MIN	11600	12500	16500	16200	28860	27850	26130	4015	36000	3500	31440	31200

	Depth	1	2	3	4	5	6	7	8	9	10	11	12
Fe	10	17200	15560	15700	9050	23340	21810	25920	5150	24800	8100	29000	23400
Units	20	17350	16265	16805	13350	25020	27250	27190	8305	25800	8950	29100	28630
mg kg-1	30	16550	24550	26400	24800	29780	26600	27450	18250	26210	9250	28700	29110
	40	27650	27800	26900	25900	26690	21880	24080	29750	25830	7800	25600	23000
	50	27500	26350	26500	17750	25050	24680	21760	16800	24270	7900	23300	22800
	Mean	21250	22105	22461	18170	25976	24444	25280	15651	25382	8400	27140	25388
	STDEV	5782	5774	5684	7251	2434	2554	2377	9632	812	657	2591	3190
	MAX	27650	27800	26900	25900	29780	27250	27450	29750	26210	9250	29100	29110
	MIN	16550	15560	15700	9050	23340	21810	21760	5150	24270	7800	23300	22800

	Depth	1	2	3	4	5	6	7	8	9	10	11	12
K	10	22270	27820	26790	10150	8815	8280	9581	1920	9144	9250	9513	8065
Units	20	22405	30330	29300	17705	9734	9576	9611	2614	9753	9575	9404	9439
mg kg-1	30	29480	25040	28000	21395	9620	9600	9784	2831	9461	16200	9564	9217
	40	32890	34050	28650	24980	9532	8635	9341	2214	9312	10800	9093	8900
	50	30380	31670	27390	20150	9070	8556	7913	2402	9448	8800	1280	8850
	Mean	27485	29782	28026	18876	9354	8929	9246	2396	9424	10925	7771	8894
	STDEV	4863	3479	993	5539	393	616	762	352	224	3041	3633	522
	MAX	32890	34050	29300	24980	9734	9600	9784	2831	9753	16200	9564	9439
	MIN	22270	25040	26790	10150	8815	8280	7913	1920	9144	8800	1280	8065

V Units mg kg-1	Depth	1	2	3	4	5	6	7	8	9	10	11	12
	10	46	39	39	16	55	46	65	8.7	51	12	79	57
	20	44	37	37	28	57	65	71	14	69	15	83	78
	30	48	42	41	34	77	61	71	19	67	16	82	78
	40	48	44	43	41	67	49	62	18	61	11	68	69
	50	44	45	40	31	61	43	58	13	63	14	57	66
	Mean	46	41	40	30	63	53	65	15	62	14	74	70
	STDEV	2	3.4	2.2	9.2	8.9	9.7	5.7	4.1	7	2.1	11.1	8.8
	MAX	48	45	43	41	77	65	71	19	69	16	83	78
	MIN	44	37	37	16	55	43	58	9	51	11	57	57

Mg Units mg kg-1	Depth	1	2	3	4	5	6	7	8	9	10	11	12
	10	6700	6700	6650	4480	8640	8299	9180	3519	9698	3952	10680	8814
	20	6710	6750	6600	6100	9790	11170	10510	4370	9535	4025	10480	10520
	30	6600	6675	6800	6600	12200	11050	10710	4380	10430	4143	9030	10630
	40	6850	8100	6900	6650	10790	9114	9178	4092	10370	3870	9100	9100
	50	7400	7700	6700	6100	9170	9490	7849	3700	9553	3658	8574	8900
	Mean	6852	7185	6730	5986	10118	9825	9485	4012	9917	3930	9573	9593
	STDEV	319	668	120	882	1412	1251	1164	391	446	182	944	903
	MAX	7400	8100	6900	6650	12200	11170	10710	4380	10430	4143	10680	10630
	MIN	6600	6675	6600	4480	8640	8299	7849	3519	9535	3658	8574	8814

Ca Units mg kg-1	Depth	1	2	3	4	5	6	7	8	9	10	11	12
	10	25700	24400	22800	4300	25050	20300	21300	8950	22940	7420	30430	17360
	20	26100	26650	25450	17550	24310	33480	24840	9000	22900	7950	30910	24070
	30	26050	28490	24450	16950	41210	34500	27850	9100	29500	9270	27320	24590
	40	21700	34100	22050	12370	35600	23750	23810	10330	29050	8100	25340	22100
	50	27600	33750	22150	11080	31390	26510	18120	9070	21120	7800	21360	22010
	Mean	25430	29478	23380	12450	31512	27708	23184	9290	25102	8108	27072	22026
	STDEV	2210	4312	1504	5352	7148	6153	3678	584	3883	697	3927	2852
	MAX	27600	34100	25450	17550	41210	34500	27850	10330	29500	9270	30910	24590
	MIN	21700	24400	22050	4300	24310	20300	18120	8950	21120	7420	21360	17360

Zn Units mg kg-1	Depth	1	2	3	4	5	6	7	8	9	10	11	12
	10	85	77	75	32	59	59	75	20	69	21	82	66
	20	85	80	81	61	73	82	83	24	70	23	82	84
	30	80	69	82	70	98	80	82	34	77	31	81	87
	40	92	89	83	76	83	62	67	32	76	28	72	80
	50	88	84	74	62	80	65	56	23	68	24	71	77
	Mean	86	80	79	60	79	70	73	27	72	25	78	79
	STDEV	4.4	7.5	4.2	16.9	14.3	10.6	11.3	6.1	4.2	4	5.6	8.1
	MAX	92	89	83	76	98	82	83	34	77	31	82	87
	MIN	80	69	74	32	59	59	56	20	68	21	71	66

As Units mg kg-1	Depth	1	2	3	4	5	6	7	8	9	10	11	12
	10	2.5	0.9	2	1.1	0.03	0.01	2.9	0.02	3.60E-05	0.9	0.06	0.06
	20	5.5	1.4	1.3	1.1	0.09	0.03	0.4	0.01	1.60E-05	2.8	0.01	0.01
	30	0.8	5.3	5.4	1.8	0.05	0.03	1.9	0.02	3.60E-05	0.9	0.03	0.04
	40	1.8	6.1	4.3	2.3	0.01	0.01	1.8	0.01	1.60E-05	1	0.01	0.01
	50	4.3	1.1	3.2	1.9	0.04	0.04	1.4	0.01	1.60E-05	0.9	0.01	0.01
	Mean	3	3	3	2	0	0	2	2	0	0	1	0
	STDEV	1.9	2.5	1.7	0.5	0	0	0.9	0.9	0	0	0.8	0
	MAX	6	6	5	2	0	0	3	3	0	0	3	0
	MIN	1	1	1	1	0	0	0	0	0	0	1	0

	Depth	1	2	3	4	5	6	7	8	9	10	11	12
Ba	10	255	352	246	217	310	353	275	133	210	102	315	273
Units	20	215	198	313	266	364	424	308	148	270	114	283	287
mg kg-1	30	208	241	312	269	432	418	306	160	275	123	270	291
	40	360	238	320	234	362	407	239	97	269	121	243	260
	50	350	205	235	204	357	331	260	81	234	117	187	250
	Mean	278	247	285	238	365	387	278	124	252	115	260	272
	STDEV	73	61.9	41.1	29	43.6	41.9	29.7	33.7	28.4	8.3	48.1	17.4
	MAX	360	352	320	269	432	424	308	160	275	123	315	291
	MIN	208	198	235	204	310	331	239	81	210	102	187	250

	Depth	1	2	3	4	5	6	7	8	9	10	11	12
Cr	10	29	25	22	12	36	31	40	6.2	30	9.9	46	33
Units	20	29	24	26	19	39	41	42	9.7	43	11	50	46
mg kg-1	30	31	21	25	22	47	38	42	14	32	12	41	42
	40	32	29	27	25	41	30	37	14	32	10	42	38
	50	27	27	23	22	40	37	29	8.6	31	9.8	32	33
	Mean	30	25	25	20	41	35	38	11	34	11	42	38
	STDEV	1.9	3	2.1	4.9	4	4.7	5.4	3.4	5.3	0.9	6.7	5.7
	MAX	32	29	27	25	47	41	42	14	43	12	50	46
	MIN	27	21	22	12	36	30	29	6	30	10	32	33

	Depth	1	2	3	4	5	6	7	8	9	10	11	12
Cu	10	37	33	35	14	28	25	31	7.3	29	10	37	26
Units	20	38	35	42	27	43	34	35	10	33	12	37	36
mg kg-1	30	39	36	40	30	39	30	35	13	32	13	33	37
	40	45	42	41	34	31	27	28	12	32	10	30	28
	50	42	24	39	30	32	30	29	18	29	10	24	25
	Mean	40	34	39	27	35	29	32	12	31	11	32	30
	STDEV	3.3	6.5	2.7	7.7	6.2	3.4	3.3	4	1.9	1.4	5.4	5.7
	MAX	45	42	42	34	43	34	35	18	33	13	37	37
	MIN	37	24	35	14	28	25	28	7	29	10	24	25

	Depth	1	2	3	4	5	6	7	8	9	10	11	12
Li	10	38	36	29	11	10	10	9.3	7.4	10	9.8	13	8.9
Units	20	41	40	32	23	21	17	11	9.5	12	10.5	12	12
mg kg-1	30	39	43	30	29	26	18	12	11	11	11	9.6	11
	40	45	43	33	30	26	12	10	10	9.7	9.9	10	9.5
	50	65	27	34	24	24	10	9.6	9.7	9.5	9.5	9.5	9.6
	Mean	46	38	32	23	21	13	10	10	10	10	11	10
	STDEV	11.2	6.7	2.1	7.6	6.7	3.8	1.1	1.3	1	0.6	1.6	1.3
	MAX	65	43	34	30	26	18	12	11	12	11	13	12
	MIN	38	27	29	11	10	10	9	7	10	10	10	9

	Depth	1	2	3	4	5	6	7	8	9	10	11	12
Mn	10	800	720	709	322	535	527	741	432	589	474	745	536
Units	20	810	718	703	610	585	650	723	502	613	468	748	740
mg kg-1	30	800	703	710	579	700	610	691	550	633	476	671	756
	40	883	820	779	706	588	571	609	474	603	402	576	700
	50	860	767	780	685	650	582	472	400	599	396	493	695
	Mean	831	746	736	580	612	588	647	472	607	443	647	685
	STDEV	38.4	48	39.6	153.6	64.1	45.8	110.2	58.7	16.7	40.5	110.8	87.5
	MAX	883	820	780	706	700	650	741	550	633	476	748	756
	MIN	800	703	703	322	535	527	472	400	589	396	493	536

	Depth	1	2	3	4	5	6	7	8	9	10	11	12
Ni	10	26	20	23	12	21	20	21	7.4	23	10	29	22
Units	20	24	20	23	20	24	27	27	10	26	11	29	30
mg kg-1	30	25	23	26	22	33	26	27	13	27	12	30	30
	40	28	25	25	25	27	21	22	12	27	11	25	27
	50	25	20	24	20	24	24	18	9.1	22	10	21	23
	Mean	26	22	24	20	26	24	23	10	25	11	27	26
	STDEV	1.5	2.3	1.3	4.8	4.5	3	3.9	2.2	2.3	0.8	3.8	3.8
	MAX	28	25	26	25	33	27	27	13	27	12	30	30
	MIN	24	20	23	12	21	20	18	7	22	10	21	22

	Depth	1	2	3	4	5	6	7	8	9	10	11	12
Pb	10	33	14	21	9.5	11	13	12	6.2	14	5.2	19	17
Units	20	24	13	20	10	14	16	17	7.9	18	6.1	17	18
mg kg-1	30	23	17	24	13	15	16	19	8.6	16	6.4	15	18
	40	32	17	25	15	13	12	10	5.5	14	5.1	13	16
	50	30	16	20	15	12	14	9.8	4.6	12	5.3	12	15
	Mean	28	15	22	13	13	14	14	7	15	6	15	17
	STDEV	4.6	1.8	2.3	2.6	1.6	1.8	4.2	1.7	2.3	0.6	2.9	1.3
	MAX	33	17	25	15	15	16	19	9	18	6	19	18
	MIN	23	13	20	10	11	12	10	5	12	5	12	15

	Depth	1	2	3	4	5	6	7	8	9	10	11	12
Ti	10	480	292	323	253	741	609	860	145	810	270	1100	780
Units	20	439	384	387	206	680	941	871	237	860	311	1130	1058
mg kg-1	30	455	636	470	386	1250	841	990	409	961	371	1008	1008
	40	480	591	416	331	1100	625	810	391	851	331	940	950
	50	473	526	389	276	921	610	648	267	800	307	740	860
	Mean	465	486	397	290	938	725	836	290	856	318	984	931
	STDEV	18	144.2	53.2	69.9	239.4	155.6	124	110.4	63.9	36.9	155.6	112.1
	MAX	480	636	470	386	1250	941	990	409	961	371	1130	1058
	MIN	439	292	323	206	680	609	648	145	800	270	740	780

	Depth	1	2	3	4	5	6	7	8	9	10	11	12
SO$_4^{2-}$	10	6878	1523	1520	687	3531	1189	1081	854	690	1854	1520	690
Units	20	9556	7046	9891	680	687	1844	1523	1687	1356	1520	1525	1390
mg kg-1	30	12066	29303	8531	12903	695	6870	1021	1189	352	520	680	1300
	40	30290	29500	12066	20250	519	1022	884	855	190	499	180	520
	50	26950	9054	9054	2010	2528	1691	851	519	180	488	181	510
	Mean	17148	15285	8212	7306	1592	2523	1072	1021	554	976	817	882
	STDEV	10697	13178	3977	8869	1361	2454	269	441	494	660	675	430
	MAX	30290	29500	12066	20250	3531	6870	1523	1687	1356	1854	1525	1390
	MIN	6878	1523	1520	680	519	1022	851	519	180	488	180	510

	Depth	1	2	3	4	5	6	7	8	9	10	11	12
Cl$^-$	10	88553	162170	301068	1280	530	870	1060	226	282	207	2081	2295
Units	20	41920	87950	172407	3018	1238	4925	730	2230	664	197	630	4858
mg kg-1	30	38374	158900	170078	2700	220	392	910	730	668	227	186	123
	40	46097	140400	53182	1980	520	1031	377	150	179	214	146	197
	50	50360	58900	117818	1690	280	640	550	110	160	211	143	210
	Mean	53061	121664	162911	2134	558	1572	725	689	391	211	637	1537
	STDEV	20343	45973	91245	716	405	1890	273	897	256	11	833	2071
	MAX	88553	162170	301068	3018	1238	4925	1060	2230	668	227	2081	4858
	MIN	38374	58900	53182	1280	220	392	377	110	160	197	143	123

	Depth	1	2	3	4	5	6	7	8	9	10	11	12
CaCO₃	10	172	153	148	132	177	482	490	141	474	120	549	501
Units	20	169	156	186	165	470	465	483	127	468	143	470	477
mg kg-1	30	124	105	149	151	410	475	420	153	470	120	453	470
	40	163	225	158	169	422	480	485	155	460	122	480	420
	50	223	289	141	155	478	469	474	133	470	123	463	360
	Mean	170	186	156	154	391	474	470	142	468	126	483	446
	STDEV	35.3	71.9	17.6	14.5	123.4	7.2	28.8	12.2	5.2	9.8	38.2	56.2
	MAX	223	289	186	169	478	482	490	155	474	143	549	501
	MIN	124	105	141	132	177	465	420	127	460	120	453	360

	Depth	1	2	3	4	5	6	7	8	9	10	11	12
pH	10	7.21	7.3	7.21	7.46	7.39	6.94	7.62	8.05	6.81	8.15	7.31	6.84
Units	20	7.41	7.32	7.41	7.66	7.3	6.84	7.63	8.02	7.35	8.29	7.37	6.89
S.U	30	7.54	7.23	7.54	7.86	7.62	7.33	7.27	8.25	6.97	8.5	7.47	7.23
	40	8.05	7.43	8.05	7.67	7.15	7.42	7.16	8.34	7.02	8.31	7.22	7.01
	50	7.96	7.47	7.96	7.68	7.81	7.62	7.34	8.25	7.64	8.39	7.54	6.97
	Mean	7.6	7.4	7.6	7.7	7.5	7.2	7.4	8.2	7.2	8.3	7.4	7
	STDEV	0.4	0.1	0.4	0.1	0.3	0.3	0.2	0.1	0.3	0.1	0.1	0.2

Supplementary Tables

Core Locations

Sample ID	Latitude	Longitude
1	35.518350°	-115.379900°
2	35.518400°	-115.359616°
3	35.519900°	-115.354383°
4	35.521400°	-115.345967°
5	35.523617°	-115.341933°
6	35.523900°	-115.340566°
7	35.524267°	-115.338217°
8	35.524367°	-115.337933°
9	35.524600°	-115.336166°
10	35.524483°	-115.333917°
11	35.525400°	-115.333417°
12	35.525933°	-115.330558°

13

Volcanic Deposits and Volcanic Hazard in Santo Domingo de Heredia, Costa Rica

Martín Rojas-Barrantes[1] & Mario Fernández-Arce[2]

[1] Dirección de Geología y Minas, MINAE, Costa Rica

[2] Escuela de Geografía - PREVENTEC, Universidad de Costa Rica, Costa Rica

Correspondence: Martín Rojas-Barrantes, Dirección de Geología y Minas, MINAE, Curridabat 5583-1000, San José, Costa Rica. E-mail: martinr@minae.go.cr

The research is financed by (Sponsoring information).

Abstract

The present research aims to investigate more precisely about the geology of the Eastern region of the Santo Domingo County. Santo Domingo is part of the structural plateau in the center of Costa Rica, which is located at the foot of the Cordillera Volcánica Central (CVF) [Central Volcanic Front] and is covered by volcanic deposits. On this plateau, called Central Valley, is the highest percentage of the population of the country and therefore, a large sector of the Costa Rican population is exposed to volcanic eruptions of the volcanoes in the CVF. For existing the national system for risk management and a law that demands actions to local authorities to prevent and mitigate disaster, it is necessary to identify the threats that exist in the cantons (counties) of Costa Rica. This will serve to take the prevention and mitigation actions necessary to reduce the impact of volcanic eruptions in the area of Santo Domingo.

The research method consisted of review and analysis of previous works through literature research, data collection and analysis of boreholes from records of water-supply wells and open pits, and field work to better know the geology of the area. The results indicate that there are deposits of powerful volcanic eruptions of pyroclastic fall deposits (volcanic ash and lapilli) that mostly form clayey soils and lahars deposits that practically covers the entire territory. Underlying these deposits there is a pyroclastic flow deposit (ignimbrite), followed by lapilli tephra (a layer of pumice of at least 2 meters thick) that mark a change in the volcanic activity. Such pyroclastic flow is overlaid by an igneous presumably sub-volcanic activity of andesites interlayered with ancient tuffs, with a considerable thickness of over 350 meters according with borehole data and the exposure recognition on Pará river study sites. According to site locations (P1 to P23) of volcaniclastic deposits, there is evidence of an important environmental impact caused by the last eruptions of the CVF volcanoes. The real and current volcanic threat to the population of the County is the fall-out of ash emitted from the Turrialba and Irazú volcanoes. From local observations along the Virilla and Pará rivers sections, there is no evidence of younger pyroclastic flows overlying the volcanic sequence.

Keywords: Costa Rica, County, lava flow, Pará River, volcaniclastic, pyroclastic flow, tuff, volcanic threat

1. Introduction

The Central volcanic front (CVF) of Costa Rica is part of the Central American volcanic front (CAVF) that extends from Mexico-Guatemala to central Costa Rica aligned parallel to the Middle American Trench, where Cocos and Caribbean plates interact originating an active volcanism. The CVF of Costa Rica is defined as the Northern domain geochemically distinctive with Galapagos-OIB signature, compare with the rest of the Central American volcanism (Gazel et al., 2009). The CVF of Costa Rica is constituted of massive composite shield volcanoes, Platanar, Poás, Barba, Irazú, and Turrialba, the largest in both area and volume of the CAVF. The CVF borders the low- relief of Central Valley, the most populated area of Costa Rica (Marshall, 2007). This low-relief is deeply incised by river canyons, cutting the underlying Quaternary volcanic sequence. The CVF volcanoes constitutes the greatest in terms of area and volume to any volcanoes in Central America, being Barva the largest that looks like expansive rise (van Wyk de Vries, Grosse & Alvarado, 2007). In this scenario, an important problem related to a complex geological active settling for geo-hazards arises, particularly volcanic hazard associated to active volcanism.

There is an agreement that dormant volcanoes are often the most dangerous ones. The mayor problem in volcanic hazard regarding reduction of volcanic risk is that most dangerous volcanoes and calderas are located in countries or regions densely populated without economic and scientific resources and political decision to adequately study and monitor them (Alvarado et al., 2007). It is essential for volcanic hazard maps detailed stratigraphic studies and distribution of deposits as well as historic and prehistoric records of eruptions to adequately determine patterns of activity, affected areas and the repose periods of a volcano. In this sense, accessibility and good exposure of outcrops is essential and conditional for this purpose (Alvarado et al., 2007). Hazard assessment in general has been improved according to intuitive criterion for those historically most active volcanoes, but should endure on those that have no historical activity like Barva, Orosí and Tenorio volcanoes in Costa Rica. This work provide new information for this important problem as a preliminary step research and look for more exposing sites where essential and useful information can be obtained.

In this paper, we analyze the most important outcrops for a specific area of the territory of the canton (County) Santo Domingo de Heredia, corresponding to the Pará River. We also include open pit and boreholes data for geologic correlation (Figure 1). It contains precise descriptions and stratigraphic columns for further analysis of volcanic hazard in the area. The current research focuses on the Eastern part of the County (Pará River) as part of a project to map the entire territory. In this context, this is an initial research advance for further work in the Virilla, Tibás and Bermudez rivers and surrounding areas to obtain the geological and hazard maps of the County. The current obtained data, is useful not only for a more precise recognition of volcaniclastic deposits regarding of the characteristics and spreading capacity, that poses a major threat in volcanic hazards but to correlate with other sites in the County for better understand of the thicknesses and extension of them. The study reveals that there are important outcrops along the Pará River not yet recognize in the literature, and constitute a valuable information as a starting point to integrate in further research of the interpretation of volcanic hazard not yet complete for the area. The importance to determine unknown sites or areas from outcrops not yet explored, in areas of quite difficult access is relevant for the purpose according with the objectives of the present work.

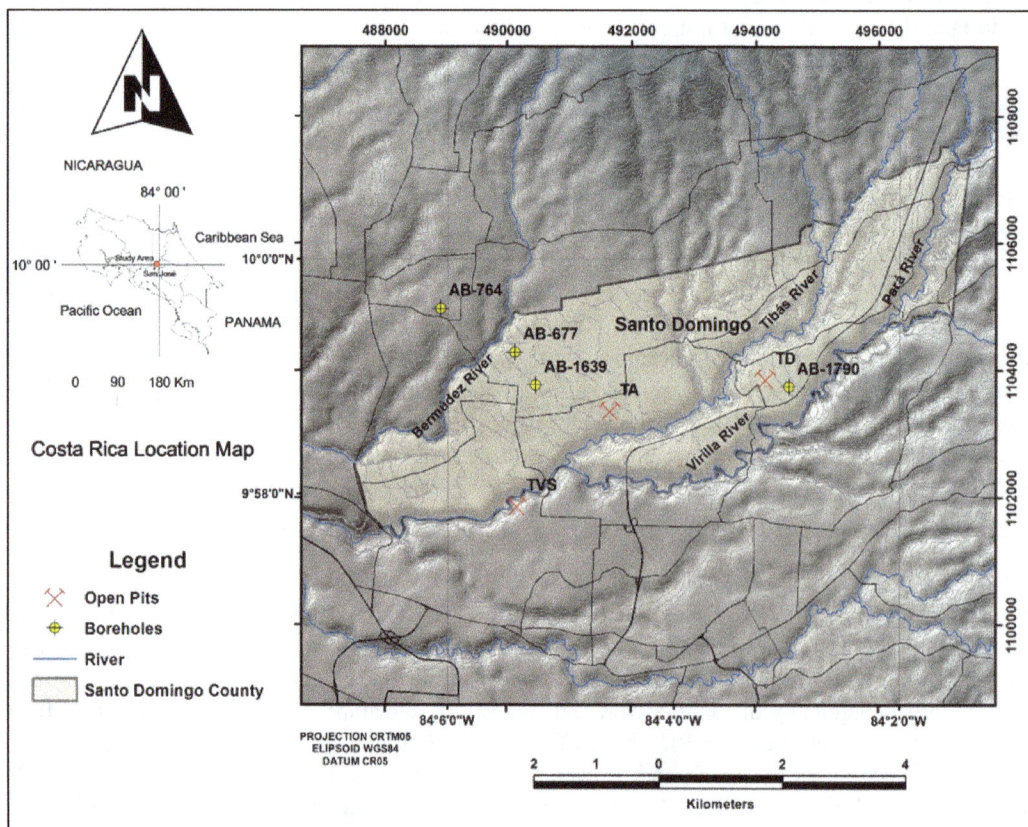

Figure 1. Location map of the territory of Santo Domingo. The area of the County is in pale yellow color. Well sites are represented as yellow circles and open pit mines are shown in red color

Although it is known that the structural plateau of Central Costa Rica is a deposit of volcanic materials, no detailed description of the local (Santo Domingo) volcanic products has been published yet. Such a task is of utmost importance to understand the threat, vulnerability, and volcanic risk for the inhabitants of Santo Domingo. Pérez, Alvarado & Gans (2006) pointed out that it is important to recognize events of great magnitude, like the Tiribí Tuff eruption that occurred 322 000 years ago in the geological record of the Barva volcano, which still poses a very serious threat for the 2.4 million inhabitants living at its foot and southern slope. Soto & Paniagua (1992) concluded that the southern flank of the Central Volcanic Range (Costa Rica), from San Ramon to Turrialba, is an area susceptible to suffer considerable damage from plinian eruptions, pyroclastic flows and mud flows originated in the Barva and Turrialba volcanoes. Prosser & Carr (1987) stated that information about the geological history of the volcanoes is essential to estimate the nature of the eruptions and are useful for the assessment of volcanic threat.

This work was carried out to better understand the volcanic threat for Santo Domingo and improve local disaster risk management. Through the knowledge of the threat, local committees in charge of risk management may have better arguments and criteria for organizing the communities around the subject. There is no doubt that such committees work more effectively if they know and are aware of their environment and threats. The risk management requires that the entire population know the threats to which they are exposed and become aware of the risk that they represent.

2. Method

The research included bibliographic review and field work to recognize volcanic outcrops in order to better understand the geological history of the territory and describe the type of volcanic deposits precisely. This work had three stages: literature review, collection of data from water-supply wells and open pit mines used in the past for extraction and exploitation of aggregates, and field work to identify types of volcanic deposits in the lower basin of the Pará River and its Eastern confluence with the Virilla River. At the onset, we review information from scientific articles, books, and geological reports existing in three public institutions [University of Costa Rica, National Service of Underground Water and Irrigation (SENARA) and the Direction of Geology and Mines (MINAE)]. The review of previous works consisted mainly in the analysis of different definitions and descriptions of rock units and how to correlate this information with the new data obtained. Then, the new data was integrated to make a new assessment of the volcanic stratigraphy as a preliminary stage for further analysis. Stratigraphic data from wells were identified and located in the National Service of Underground Water and Irrigation (SENARA) and in the Santo Domingo Municipality. In the same way, borehole data were validated and analyze according of location, type of deposits and thicknesses in order to correlate with the fieldwork and open pit data. Boreholes and surface deposits allow us to have underground control of the geology to make more precise correlation (Colima and Tiribí Formations). Open pit mines exploited in the past for aggregates were consulted through the mining records in the Direction of Geology and Mines of Costa Rica. The importance of this exposing rocks outcrops is the general overview from a slope cutting that allow to make clear identifications of the geology to construct a stratigraphic column of a small area. This new data is integrated with the fieldwork and borehole data to stablish a common correlation framework. Fieldwork studies consisted of visiting and studying different outcrops points along the Pará River of the study area to map surface deposits. Stratigraphic columns allowed were used to know more accurately the distribution and thickness of the volcanic deposits in the County. The new data analysis consisted in recognize types of volcaniclastic deposits associated to explosive volcanism. The recognition was useful to verify the type of eruption and the uppermost stratigraphy, and to validate the threat from recent eruptions from the CVF. Residents of Santo Domingo joined the investigation both in the provision of data and fieldwork studies.

3. Results

3.1 Geology of the Central Valley of Costa Rica

The geology of the Central Plateau of Costa Rica was studied by Williams (1952) who proposed three main units: (1) lava flows deposited on the early Tertiary topography when the volcanic activity began in the Cordillera Central which he named Lavas Intracañon; (2) Nue'e ardente deposits consisting of tuffs and ignimbrites and, (3) recent lava flows, later defined as Post-Avalanche Lavas (Fernández, 1989). Such units were later classified as formations and were named Colima, Tiribí and Barva (SENARA-BGS, 1985). This volcanic activity has been explained as effusive fissure eruptions (Protti, 1986; Kussmaul, 1988 in Alvarado & Gans, 2012), although some of them could be explained as volcanism from the medium to distal part of the Paleo-Volcanic Chain (Alvarado & Gans, 2012) and the Barva volcano whose activity began after the volcanic activity that generated the Aguacate Group.

The Colima Formation was defined by Fernández (1968) and studied by Echandi (1981). The latter divided it into three members: (1) <u>Belén</u>. This lower member contains the first lava layers deposited on Tertiary sediments and possibly on the Aguacate Complex in some sectors. The member appears in the river bed of the Virilla River and is well distributed along the Central Valley. Its composition are andesitic lavas with pyroxenes, with some transitions between andesites and basalts, interrupted by mantles of ash. It is estimated that this member has a thickness of up to 30 meters in the Virilla River canyon, (2) <u>Puente de Mulas</u>. It was formed after an interruption in the volcanic activity. This part of the formation consists of tuffaceous pyroclastic flow deposits both at the base and in its upper part as well as characteristic ignimbrites in the central part. Based on drilling data, its estimated thickness is up to 50 meters. The ignimbrites are dark gray or brown color and have abundant lapilli fragments in a welded matrix and a somewhat developed columnar joint structure, (3) <u>Linda Vista</u>. The composition of this member are andesitic and latiandesitic breccias with augite and glass (Echandi, 1981), with a dense central core. Borehole data reports a thickness of 270 meters for Linda Vista and at least seven andesitic lava flows and pyroclastic flows. They are associated to fissure emissions along fractures oriented NE - SW (Valverde, 2003).

The Tiribí Formation emerges in the North and Northwest of the Central Valley and consists of deposits of basal pumice up to 3 meters thick, followed by deposits of ash, lapilli and blocks with local intercalations of ignimbrites and andesitic lava flows. The formation has an average thickness of 45 meters, covers an area of 500 km^2 and is characterized by a flat surface and is quite tabular in form. Its deposits overlying the formations Colima, Pacacua, Peña Negra, Grifo Alto and La Cruz, and are overlain by alluvions, lahars and ash of the Barva Formation (Valverde, 2003). According to Echandi (1981), this formation is divided into the following members: La Caja, Nuestro Amo and Electriona. La Caja is formed of little welded tuffs, Nuestro Amo consists of nue'e ardent deposits and the components of Electriona are well-welded ignimbrites. The Tiribí Formation is younger than the Colima Formation, its age is Pleistocene according to Denyer & Arias (1991).

The Barva Formation materials are recent volcanic products of the Barva volcano. Four lava flows are recognized in it: San Rafael, San Antonio, Ciruelas and Cebadilla. Lithologically, the composition of these flows is andesitic-basaltic lavas rich in olivine, with thicknesses from 10 to 80 meters and intercalations of ash and occasional lapilli (Valverde, 2003).

3.2 Volcanic Deposits in Santo Domingo

Santo Domingo is located in the central part of the Central Valley of Costa Rica and therefore its soils are of volcanic origin. Below are details of the deposits of volcanic materials in open pit mines, wells and in an array of outcrops located along the Pará, Agrá, and Tibás Rivers (Table 1 and Figure 5).

Table 1. Boreholes and open pit data for the study area of Santo Domingo

Open Pit / Borehole	ID	Northern (m)	Eastern (m)	Elevation (m.	Thickness (m)	Depth (m)	
						Maximum	Elevation
Vargas Solera	TVS	1 101 890.9	490 163.1	1101.9	70.0	----	1050.0
Arizona	TA	1 103 369.2	491664.5	1181.5	70.0	----	1180.0
Dent	TD	1 103 866.5	494 164.7	1251.2	62.0	----	1250.0
AB-1639	AB-1639	1 103 770.4	490 465.1	1170.6	----	50.0	1120.6
AB-677	AB-677	1 104 280.7	490 135.6	1170.0	----	175.7	994.3
AB-764	AB-764	1 104 971.9	488 916.5	1176.7	----	146.3	1030.4
AB-1790	*AB-1790*	*1 103 746.2*	*494 544.6*	*1261.7*	----	*200.0*	*1061.7*

3.2.1 Geology of the Vargas Solera Open Pit (TVS)

The Vargas Solera open pit is located in the Santa Rosa district of Santo Domingo, on the right bank of the Virilla River, within the CRTM05 coordinates 1 101 870.9 North and 490 163.1 East (Table 1). Locally, there are volcanic deposits of the Colima and Tiribí Formations, with a minimum and maximum thickness of 40 to 55 meters (Calvo, 1998).

The basal unit is comprised of massive lavas of andesite composition up to 10 meters thick (Calvo, 1998; Obando, 2008) that emerge in the course of the Virilla River. According to Echandi (1981) this unit is correlated with the Belén Member of the Colima Formation.

The intermediate unit is a package of lavas and breccias of lava that overlay the basal unit and consists of three layers: a lower horizon of brecciated lava with vesicular texture, dark gray color (fresh) to reddish brown (altered); an intermediate horizon that corresponds to a layer of dark gray aphanitic andesitic lava with flow-laminated structure and jointed (fractures with or without clay mineral fill), and an upper layer similar to the lower one. According to Echandi (1981), this unit is correlated to the Linda Vista Member of the Colima Formation.

The upper unit consists of tuffs and ignimbrites of the Tiribí Tuff. It has a basal layer of non-consolidated pumice up to 2 meters thick, overlain by dark gray ignimbrite and tuff, with a minimum thickness at the site of 15 meters. According to Echandi (1981) this unit can be correlated with the Upper Member of the Tiribí Tuff. Over this unit is a clay soil of brown color of 5 meters thick (Figure 2).

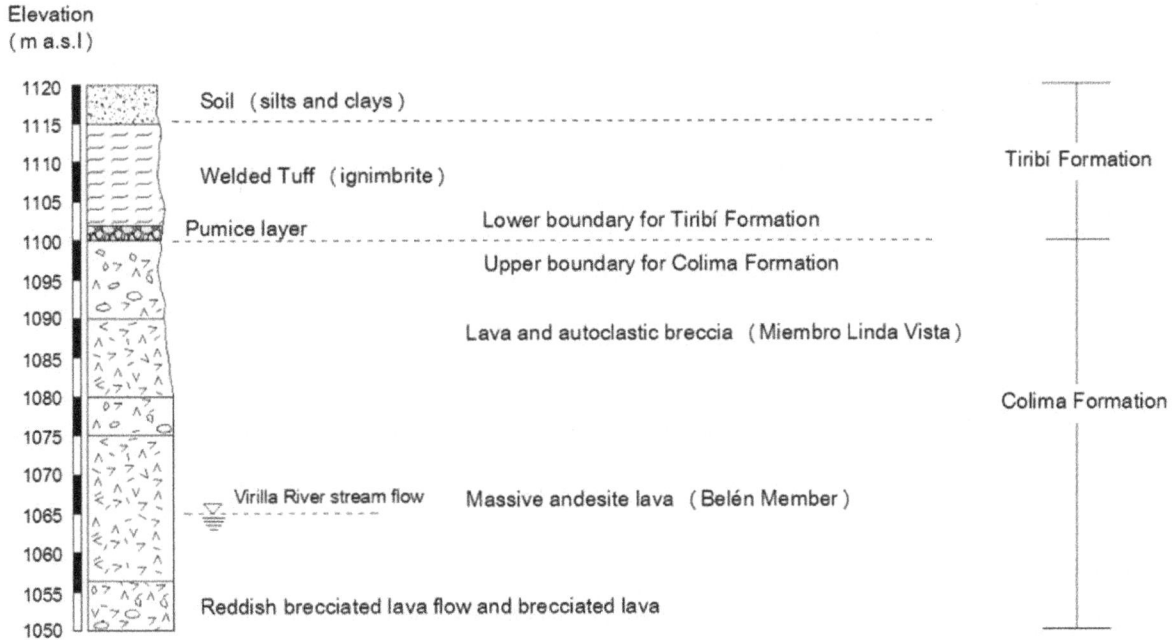

Source: Modify from Calvo, G., 1998 & Obando, J., 2008

Figure 2. Litho-stratigraphy of the Vargas Solera open pit (TVS)

3.2.2 Geology of the Arizona Open Pit (TA)

The Arizona pit was a site of extraction of volcanic material for construction aggregates. It is located in Santo Tomás of Santo Domingo, within the CRTM05 coordinates 1 103 369.2 North and 491 664.5 East (Table 1). The concession covers an area of 17.6 hectares (0.176 km^2).

On the site, there are lavas of the Colima Formation and tuffs from the La Caja Member of the Tiribí Formation (Figure 3). Two types of pyroxene bearing andesitic lavas from Colima are recognized: (1) gray porphyritic lava, like the scoria on the outside, with few joints, (2) light gray lava with fine crystals, dense, hard and compact, with vertical joints, outer section very fractured, with well-developed flow-banded structure (Fernandez, 1989). In between the two lava flows there is a very porous brecciated lava, with evidence of calcination and with a thickness of approximately 2 meters. Petrographically, the rocks are andesitic and latiandesitic lavas with glass and augite (Echandi, 1981).

The morphology of the stratigraphic sequence is irregular, characterized by undulations of mounds and lava fronts (Obando, 1990). The total thickness of the Colima member is 30 meters (Fernandez, 1989).

The upper stratigraphic sequence consists of ignimbrites, tuffs and pumice, which expose a regular horizontal to near horizontal distribution, which is reflected in the flat upper morphology. In this sequence, there is a thickness between 2-3 meters of white pumice of rhyolitic composition, with an approximate thickness of 6 meters of ignimbrite, a section of sound, gray tuff with ash, porous matrix, slightly compact, with a thickness of three meters and weathered tuff, similar to the previous, with a thickness of about 20 meters.

According to Marin & Goic (1999), the Electriona Member covers the whole area of concession of the Arizona pit. Within this member, the thickness of the volcanic materials in the pit is about 70 meters.

Elevation
(m a.s.l)

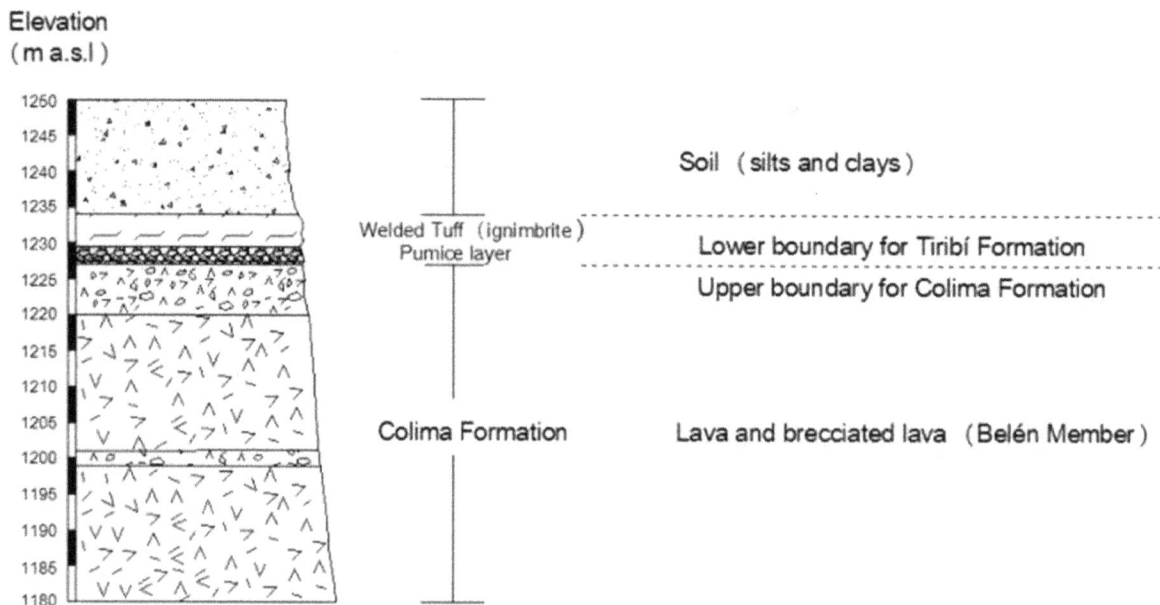

Source: Modify from Echandi, 1981; Fernández, 1989; Obando, 1990; y Marín y Goic, 1999)

Figure 3. Litho-stratigraphy of the Tajo Arizona open pit (TA)

3.2.3 Geology of the Dent Open Pit (TD)

The site is located in the San Miguel district, within the CRTM05 coordinates 1 103 866.5 North and 494 164.8 East. On the topographic sheet Abra, scale 1: 50 000, from the National Geographic Institute (IGN), the site is among the coordinates 218 000 - 219 000 North and 530 000 – 531 000 East. On the side of the road, next to the pit some caves which were used for years to extract pumice stone are noticed (Salazar, 1993). The Dent pit corresponds to one site for extraction of pumice stone which was open in the 40's of the last century. A mining record for this pit was requested from the Direction of Geology and Mines of Costa Rica in April of 1993 to exploit the material to be used as an aggregate for construction purposes. The concession was granted and the extraction began. The exploitation lasted until March 2004, when the mining record was filed and the extraction of material ended.

According to Salazar (1993), the geology of the site corresponds with the Linda Vista Member of the Colima Formation (Figure 4). Galleries of digging show a lenticular stratum of white to cream pumice, vesicular, granular, light, and friable, with alignments of oxides and hydroxides of iron (Salazar, 1993; Valverde, 1994). The roof of the galleries is a brecciated tuff of purple withe color. On the floor of the digging galleries, there is an altered volcanic rock that can be considered as volcanic bombs (Salazar, 1993). The thickness of the gallery walls does not exceed 2.5 meters (Salazar, 1993), which is confirmed by Valverde (1994) who reported a compact lens of 2 - 2.5 meters thick intercalated dark tuffaceous deposits and occasional small lenses of clays.

The exploitation and extraction method was through galleries with halls and pillars for supporting the excavation itself. The tunnels were built on the line of maximum gradient. The pillars are in the pumice layer. Some of these tunnels were excavated more than 70 years ago but are preserved in good condition (Valverde, 2003). During the exploitation, the extraction technique was manual.

A borehole close to the Dent pit shows 200 meters of volcanic deposits of lavas and autobreccias from the Lower Colima Formation, and lavas, brecciated lavas and tuffs from the Upper Colima Formation, before the deposition of the plinian pumice eruption that encloses the Dent pit in this site (Figure 4). There is no report of the pumice layer from borehole AB-1790 despite the short distance from TD pit. A correlative analysis from Dent and Vargas Solera pits and P13, suggest this layer should appear over the 1190 m.a.s. l, and in principle should be present in borehole AB-1790. However, this could be explained by the lack of precise direct information from field drilling data, besides being probably be related to the lack or reduced thickness of this layer *in situ*, or due to the drilling process (auger action) incorporated with mud drilling of the relative low consistency and density of this volcanic deposit.

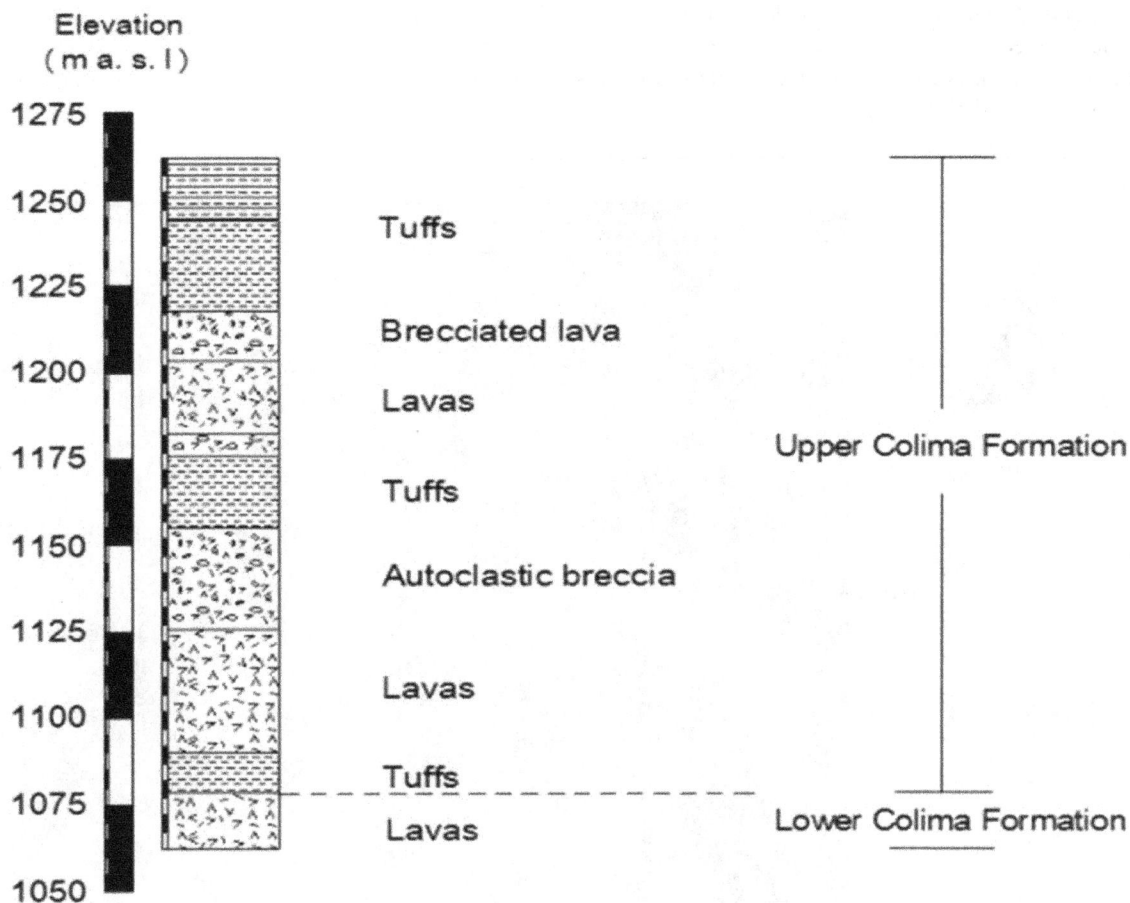

Figure 4. Litho-stratigraphy registered in borehole AB-1790

3.2.4 Geology of the Pará River

A series of volcanic outcrops from the upper Virilla River basin (P1 to P4), lower Pará River basin (P5 to P22) and Agrá River (P23) were subjected to analysis (Figures 5).

At (P1), downstream of the confluence of the Pará and Virilla rivers, emerges a deposit composed of angular clasts of lava in a sandy silty and quite indurated matrix. The mentioned deposit is located on the slope of the Virilla River valley and interpreted as a pyroclastic flow from Tiribí Formation. A similar deposit is found in P18 overlying the pumice layer and Colima Formation. Upstream of Virilla River (P2) and near the confluence of the Pará and Virilla rivers (P3 and P4) is an effusive volcanic deposit made of one or more blocky jointed lava flows with reddish alteration in between lava blocks or fractures (Figure 5). This material is interpreted as a lava flow front whose blocky structure is associated with mechanical fracturing of the igneous body during emplacement (autoclastic). Petrographic analysis in thin section, for one sample from P3 suggests a lava flow of andesite composition with intersertal porphyritic texture, composed of phenocrysts of felspar (plagioclase ≈ 10%), clinopyroxene <3-4%, orthopyroxene (hypersthene <1-2%), and opaques <2%, in a matrix of microlithic plagioclase, interstitial glass and opaques (Figures 6A and 6B).

At P5, upstream on the Pará River, there are deposits of explosive volcanism activity, of 1.3 meters thick overlaying the volcanic deposit of P4, in which we recognize three different eruptive events, separated by the development of a highly oxidized to hematite soil. The base of this sequence is a lapilli tephra layer, a compact and hardened pyroclastic fall deposit of cream to yellowish color, somehow correlated with the pumice layer. Over the lapilli tephra is a very hardened paleosol of brown to yellowish color and 0.3 m thick. There is a layer of 0.2 m thick cream color tuff overlying the paleosol. In the uppermost part of the sequence, there is 1 meter thick, gray and very compact indurated lithic tuff which was analyzed under a microscope in order to have some information about the type of eruption activity in the geologic record specifically in this site. It, clearly shows a pyroclastic fall deposit composed of an ash matrix partially altered to clay minerals, crystals of feldspars (plagioclase), clinopyroxene (augite), lithic fragments (lavas) that are apparently not juvenile and accessory lithics altered to sericite and secondary quartz, probably from hydrothermal alteration at the volcanic conduit (Figure 6C), pumice fragments with round shaped vesicles partially altered to clay mineral (Figure 6D) and the hydrous mineral biotite is present (Figure 6E).

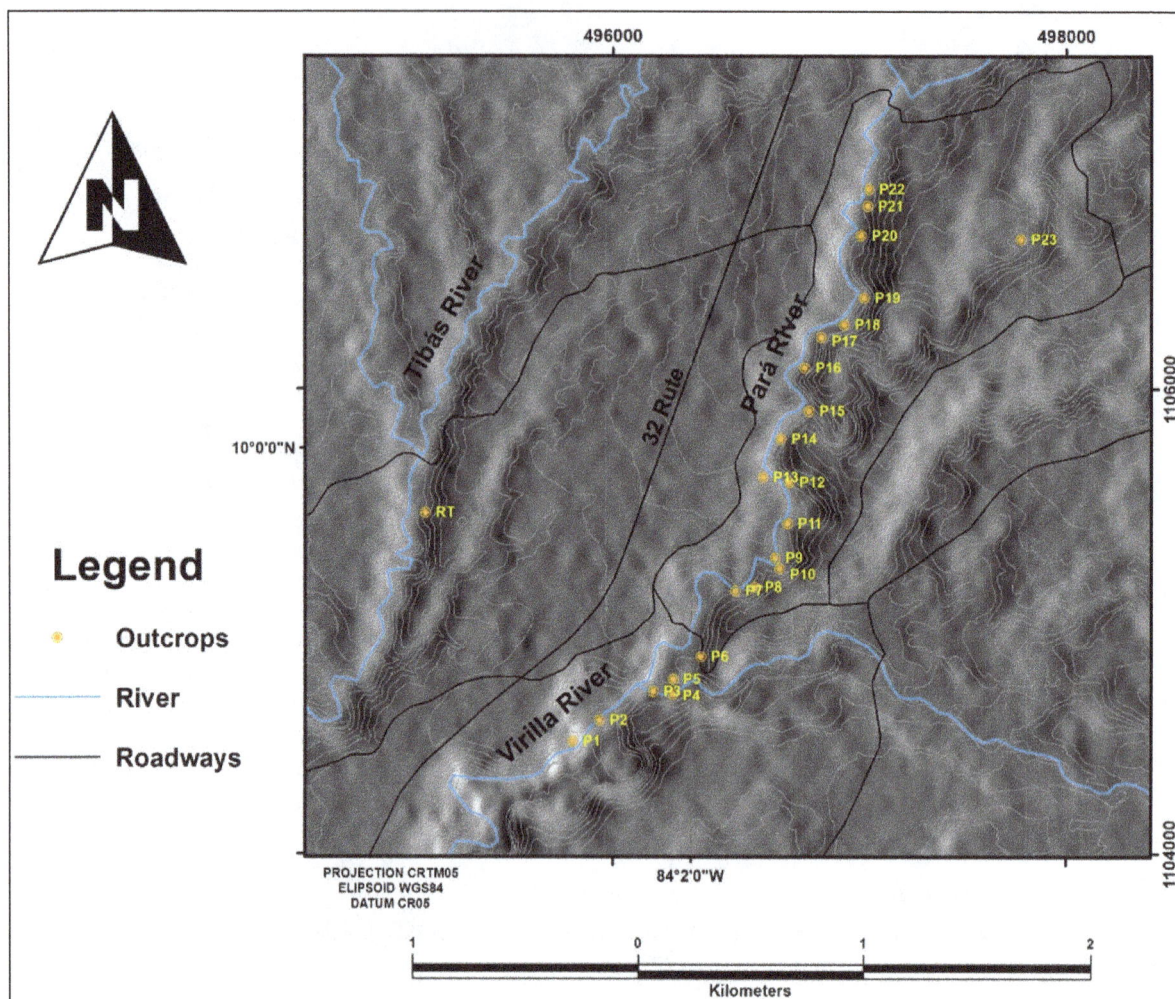

Figure 5. Location map for the outcrops subject to analysis on the Virilla River, Pará River and Agrá River

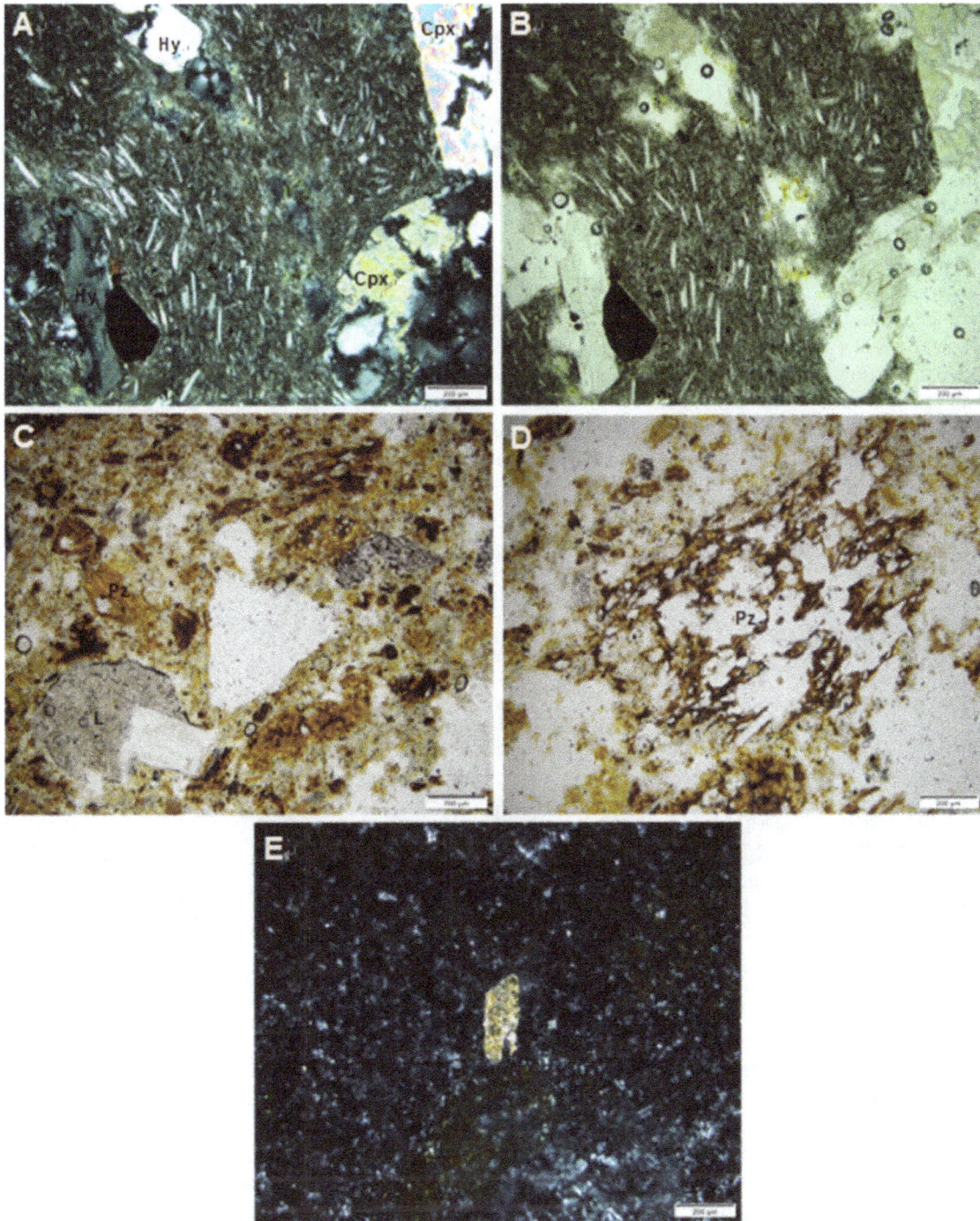

Figure 6. Photomicrographs for the samples in outcrops P3 and P5. (A) Intersertal phorpyritic texture in a groundmass enclosing plagioclase, clinopyroxene and hypersthene for sample in P3 (crossed nicols), classify as andesite for the Upper Colima Formation; (B) same as (A) in plane polarised light; (C) Pyroclastic fall deposit in plane polarised light showing details for the components of ash matrix, feldspar crystals, lithics fragments of lava and pumice fragments in sample from P5; (D) Details for pumice fragment showing no sign of devitrification and round shape vesicles in sample from P5 (plane polarized light); (E) Presence of hydrous minerals as biotite in sample from P5. Hy: hypersthene; Cpx: clinopyroxene; Pz: pumice; L: lithic fragments; Bi: biotite

Over the previous sequence (P6) there is a lahar, the youngest volcaniclastic deposit mapped in the area, and paleo-alluvions at its base which are interpreted as part of the current lahar during its deposition or as paleo-alluvions resulting from ancient fluvial action in the Pará River basin. The sequence from P7 to P19 along the river bank of Pará River, constitutes an exposure of lava flows varying from scoraceous black to massive gray porphyritic aphanitic textured of andesite, in a blocky-jointed to flow-laminated structure. These lava flows are associated with the Upper Colima Formation. Three remarkable characteristics are present in the sequence: (1) the considerable exposing thickness of lava flows along the Pará River canyon; (2) the presence in P13 of the plinian pumice layer overlain by a pyroclastic flow on the right stream slope of the canyon. Both lithologic units are correlated with the Tiribí Formation and similar in description to the Dent pit according to Salazar (1993). The pumice fragments are slightly altered to clay but they still evidence a fibrous silicic texture, with low content of primary minerals of magnetite and plagioclase. There is evidence of pumice stone extraction activities from the last century thorough digging galleries on this site (Figure 7); (3) a clear evidence of ancient brown-altered lithic-crystal tuffs (P18), overlain by flow-laminated andesite lava at the contact, to blocky-jointed and massive lava at top. This correlates with the description of interlayer tuffs from the Colima Formation in other places of the Central Valley according to Echandi (1981).   Upstream on the river (P19), there is a blocky-jointed to massive lava flow in contact with and overlying a volcanic breccia (in appearance a debris flow). This volcaniclastic deposit consists of an advanced altered brown color matrix and clay silty composition, containing lava clasts up to 50 cm, little altered and subangular in shape. At P19 the contact is clear between both lithologies. Upstream from P19 emerges the basal part of the previously mentioned lava flow, which looks like a brecciated deposit.

Figure 7. Digging galleries used in the last century to exploit pumice stone, at point (P13) river right of the Pará River canyon. The right picture shows the hardened compact and dense pyroclastic fall deposit of pumice

Upstream from P19, at sites P20, P21, P22 and P23 (Agrá River) secondary volcaniclastic deposits clearly emerge as debris flows/lahars, probably as different events that are differentiated by the degree of alteration of the matrix (in some cases much more hardened) and volcaniclastic components (lithics and lava clasts, scorias and pumice fragments, feldspar and pyroxene crystals) (Figure 5). These deposits include layers or lenses of hardened volcanic sands and pyroclastic material as part of the volcanic flow and deposition mechanism. In addition, alluvial lenses possibly incorporated into the flow of the lahar current were observed. In the Agrá River (P23) east from Pará River the same lahar is observed close to San Jerónimo (not shown).

From the sites subject to study (Pará River), there is no evidence of the Tiribí Tuff (ignimbrite), and that may be related to the spreading capacity and exposure which are not always continuous over the area, besides the soil and vegetation covering in the area that make continuous correlative analysis difficult. One remarkable characteristic observed from P5, is the absence of the Tiribí Tuff and instead the deposition of pyroclastic fall deposits over the Colima Formation.

3.2.5 Geology of Wells

By means of boreholes (AB-1639, AB-677, AB-764 and AB-1790) it is possible to know more about the geological constitution of the territory of Santo Domingo. In all drill holes, we found a geological composition of volcanic origin. For this group of borehole record, the maximum depth reached was 175.7 meters, which corresponds to an elevation of 994 meters above sea level. The first layer reported in well AB-1639 is 18 meters of a soil and red clay. Under the soil, there is a 9 meter thick brecciated lava (Figure 8). In this well, we found an interlayering of tuffs altered to clay and autoclastic breccia that suggests an alternation in the volcanic activity from effusions of lava and explosive volcanism.

The first layer of the well AB-677 is 26 meters thick and consists of ashes and brown weathered tuffs of and a high content of clay in the upper levels (Figure 8). At lower levels there are fragments of lava with feldspars and ferromagnesian minerals. Then, a layer of scoria and brecciated black and red lava appears. From this layer to 114 meters there is a sequence of lava flows texturally different from scoraceous, vesicular, density and crystal content from aphyric to porphyritic, color variations and autoclastic brecciated sections. At 1075 meters above sea level, there is a transition between the Upper Colima Formation and the Lower Colima Formation as interpreted by correlation with other boreholes from the area. The volcanic succession continues with a tuff, autoclastic breccia, lava flow and closes the sequence with a 6 meter thick tuff from the Lower Colima Formation that appears at 169 meters depth.

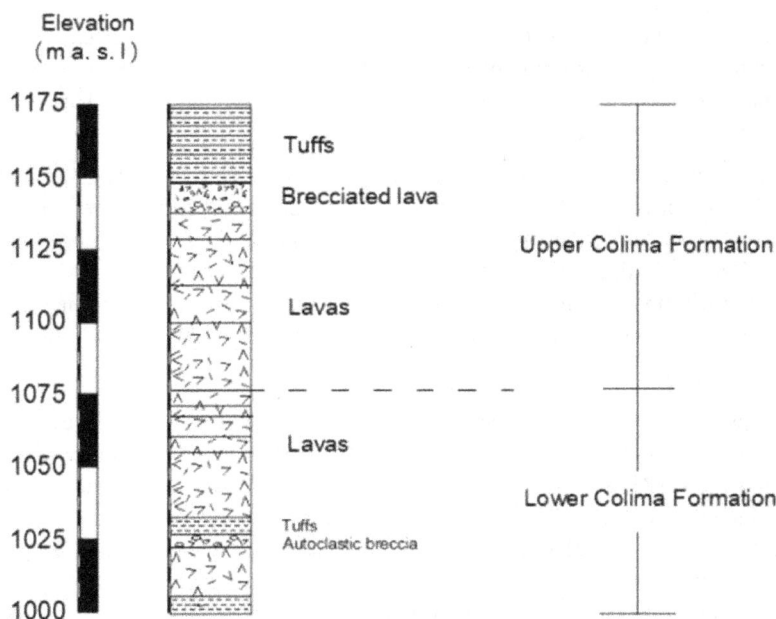

Figure 8. Litho-stratigraphy of the borehole AB-677

The upper layer of the AB-764 well is a vegetal soil and clayey ashes of 15 meters thick. From 15 to 138 meters there is a lava flow sequence occasionally alternated with brecciated lava layers. Characteristics of the sequence are similar of that from borehole AB-677. At the depth of 138.7 meters a tuff layer appears and finally a much fractured lava layer.

The borehole AB-1790 has the maximum relative depth from the group, however not that of the lowest relative elevation as referred from borehole AB-677 (Table 1). It constitutes a similar sequence of interlayering of tuffs, brecciated lava, lavas and autoclastic breccia. The uppermost layer is a tuff altered to soil on top and partially weathered tuff.

The borehole group shows a deposition of tuffs on top of the sequence, suggesting a volcanic ash cover from the latest eruptions from Barba and/or Irazú volcano in the area. Another characteristic from borehole data is the consistency of interlayering of tuffs, lavas, brecciated lavas and autoclastic breccias in the volcanic succession as seen from Pará River canyon geology and according with the descriptions from different authors in the area as explained above.

4. Volcanic Hazard

The International Strategy for Disaster Reduction (ISDR), 2004, defines in the context of disaster risk, hazard as a "potentially damaging physical event, natural phenomenon or human activity that may cause death or injury, property damage, disruption of social and economic activity or environmental degradation". It also indicates that threats or hazards can include latent conditions that maintain or increase the danger. Stratigraphic correlation of

Santo Domingo County shows a boundary between the effusive volcanic deposits and deposits of explosive volcanic activity. The latter are those having the greatest implicit volcanic threat in the geological record of the area, and therefore represent a potential threat of affectation.

A new global assessment of volcanic threat has been proposed in order to assess the hazard and risk at global, regional and country scales to identify significant risk areas, gaps in knowledge and enable prioritize resources (Loughin et al., 2015). The Volcano Hazard Index (VHI) is based on the eruption frequency, registered Volcanic Explosive Index levels (VEI) and occurrence of pyroclastic density currents, lahars and lava flows. Another parameter such the Population Exposure Index (PEI) is derived and weighted from evidence on historical records or distributions of fatalities with distance from volcanoes within 10, 30 and 100 km (Loughin et al., 2015). In order to evaluate these parameters two measures have been proposed to evaluate the threat, understood as a combination of hazard and exposure without considering vulnerability or value, which include or combine the number of volcanoes in a country, the size of the population living within 30 km of active volcanoes and the mean hazard index score (VHI). The measure 1 gives the overall volcanic threat by country, and measure 2 ranks the importance of threat in each country and focuses on the proportion of the population exposed (Loughin et al., 2015). According to the distribution of volcanic threat between countries, Costa Rica ranks number 15 among the top 20 countries with the highest overall volcanic threat, and number 10 for the top 20 countries of territories ranked by proportional threat (Brown et al., 2015). However, these regional rankings are relative to the contexts and different perspectives for each country.

In Central America exist evidence of at least 26 volcanoes that have erupted during the past 10 000 years and probably more than 35 that have had Holocene activity (Alvarado et al., 2007). The volcanic range of Central America from Guatemala to western Panama (with a volcanic gap of 175 – 190 km between Irazú-Turrialba in Costa Rica and Barú-Tisingal in Panamá), has 50 major volcanic centers with more than 400 eruptions occurred historically, and spacing of approximately 26 km on average, which represents one of the highest densities of active volcanic centers along any convergent plate margin (Alvarado et al., 2007).

5. Potential Volcanic Hazard

The most explosive, perhaps, pyroclastic event of the central Costa Rica took place 322 000 years ago. Its products covered an area close to the 900 Km^2 and reached up to 80 km from its source, which was located in the caldera of the Barva volcano (Pérez et al., 2006). In 1952, the deposit was called Nue'e ardente like that of Mt Pelée, in allusion to a similar flow that killed 300 00 people in Martinique by activity of the volcano Mt. Pelée, in 1902. According to Pérez et al. (2006), this eruptive event began with a Plinian eruption that caused the deposition of the layer of Tibás Pumice, with a thickness ranging from 0.5 to 3 meters. The associated pyroclastic flow reached 80 km westward from the source, to the community known as Orotina (not shown).

The minimum mass of magma expelled is estimated at 3.5×10^{13} kg. For this reason, the cataclysmic eruption had a magnitude of 6.6 (VEI) in the Volcanic Explosivity Index. Pérez et al. (2006) established that the Tiribí Tuff represents an Ignimbrite of great size that covered much of the region of the Central Valley of Costa Rica. The eruption began with a Plinian event that originated the layer of Tibás Pumice. According to Reyes, Fernández, Grinesky, & Collings (2014) the territory of Santo Domingo is within the area of the flow.

The geologic record of the uppermost volcanic stratigraphy in the Central Valley, related to explosive eruptions that generate volcanic ash, tephra and pyroclastic density currents (from field work P1, P5, P13 and other authors), and according with the genetic characteristics and emplacement mechanisms of location, volumes and expansion of these pyroclastic flows in the Central Valley and Santo Domingo County, demonstrate the environmental impact that such volcanic eruptions had in the geologic history of the area. The sites subject to analysis clearly show the impact or hazard from volcanic ash and tephra fall from past eruptions of Barba volcano as primary volcaniclastic deposits (P1, P5 and P13). Lahars are present in the lower and upper upstream section of Pará River and Agrá River (P6, P20, P21, P22 and P23), evidence for the spreading capacity of volcanic mudflows in the area from past volcanic eruptions. A preliminary analysis from layer 4 of lithic tuff, a sample from P5, suggests through examination of primary hydrous minerals and pumice fragments, explosive eruptions derived from high pressures in the magmatic chamber that may be related to deeper a magmatic chamber followed by a magmatic eruption, implicit in the geologic record. If this assumption is correct, such determination may be correlated with the possibility of Barba volcano to generate similar volcanic activity in the future.

6. The real hazard

Soto & Paniagua (1986) made a map of volcanic hazard for the Cordillera Volcánica Central based on data from this fall material during eruptions prior to 1986. According to them, the territory of Santo Domingo receives ashes only from Irazú volcano eruptive events. Nevertheless, the recent eruptions of the Turrialba volcano (Figure 9) emitted ash that fell in Santo Domingo and beyond to the west. This confirms that the entire area of the County is exposed to ash fall. Among the expected effects of the ash fall is the coverage and burial of the exposed area, including obstruction of river ways, interruption of pathways of different order and dimming of the sky, which would also affect air transport. The items most affected by the ash would be pastures, crops, and water, all of which would severely affect the economy and the well-being of the population.

The fall of ash in the territory of Santo Domingo can have effects on humans, equipment and infrastructure, agriculture, water, and domestic animals. People with chronic lung problems are particularly vulnerable to the adverse impact of volcanic ash. A common effect in humans is respiratory difficulties by fine suspended particles in the air and an increase of asthma attacks. Other effects include runny nose, sore throat, bronchitis, eye problems, and skin irritation. The volcanic ash can be poisonous to animals and cause diseases, intestinal injury and metabolic disorders. Chronic Fluorosis causes lesions in the body, and death. When fluorine is identified in the pastures, it is recommended to evacuate the livestock whose main threat is the ingestion of ash both inhalation by their respiratory system and eating vegetation covered with ashes.

Figure 9. Exhalative activity of Turrialba volcano on November 13th 2014

The equipment and infrastructure can be seriously affected by the ash fall. Particles that would affect the operation of the cooling system could infiltrate in refrigeration systems. All computer systems could receive ash and stop working, causing power outages and data loss. The supply of drinking water could also have problems. The supply of drinking water in the County comes from wells and the extraction of water from surface springs. Therefore, the fall of large volumes of ash could pollute surface water sources and cause disruption of service by the decrease in the quality of the water and by breakdowns of pumping equipment. In addition, the weight of the ash could cause the collapse of roofs of houses of low resistance and economic condition.

Both traditional and modern agriculture, which operates with computer systems, would be severely affected by ash fall. Traditional agriculture includes basically the cultivation of coffee and some tomato plantations. In the eastern part of the County a small but modern planting of vegetables and herbs and spices (basil, celery, green onion, oregano, parsley, rosemary, thyme) whose irrigation and nutrition is carried out by means of a computer system has been identified. If the ash reaches such a system, the production would be lost completely. Finally, the weight of the ash can affect the survival of plants, increase the costs of the harvest and reduce production.

Alvarado (2008) indicated that in the very recent past our ancestors witnessed large eruptions that affected dozens and even hundreds of square kilometers and that such eruptions will occur in the future so it is likely that this century will witness these eruptions. Today, Turrialba volcano is presenting mainly phreatomagmatic activity with intermitting explosions, expulsing volcanic ash fall-out over one part of the Central Valley. This activity is affecting mainly agriculture (pastures and crops), transportation, water supply, domestic animals, electric lines, equipment and health. Economic losses are now being evaluated for people leaving around surrounding areas of the volcano.

7. Discussion

The environmental impact of volcanism in the area has been prove through examination of local geology from ancient to present volcanic deposits. The volcanic hazard in the area is evaluated on the basis and recognition of recent volcanic deposits. No detailed information about the geology has been made yet, mainly in the river canyons of the Canton where most of the exposing of geology is present. The study provide new data that is in good agreement with previous works at regional scale, and report new local data useful to estimate volcanic hazard in the region, and continuing future work in the field. Imprecisions arise from the lack of continuous correlative analysis in a quite amount of outcrops and sites. Regarding the differences from previews works on geology and volcanic hazard, the present work has a much more detailed information of outcrops and petrographic analysis which strongly suggests explosive volcanic activity in the past.

Based on the compiled data of boreholes, open pit mines and field work geology, the studied area has been affected by an active volcanism since Pleistocene time. The lower contact with the Aguacate Volcanic Group from Pliocene – Miocene was not observed in the field and neither in the drilling core samples. It is clear that this volcanism was very active and dynamic as inferred from the stratigraphic correlation of interlayering of lava flows, autoclastic breccias and tuffs (plateau), pyroclastic flows deposits, pyroclastic fall deposits, debris flows and paleosols. The exposed volcanic sequence along the Pará, Virilla and Agrá rivers, is a clear evidence of this volcanic dynamic process. Different events of flow-laminated to blocky-jointed and massive lava flows interlayering with lithic-crystal tuffs and ancient debris flows in between reveals the intensity of the volcanic activity in the past. On top, there is deposition of tuffs and probably weathering products of volcanic deposits from the most recent volcanic activity. Petrographic analysis confirms explosive volcanism registered in the volcaniclastic deposits of ancient tuffs in the geologic record. The presence of orthopyroxene as a phenocryst fase in sample P3 (Figure 6A), is in good agreement with the presence of this mineral in basaltic lavas in central Costa Rica since the Late Miocene, as a product of change in the source by metasomatic addition of silicate-rich melts to the mantle source (Gazel et al., 2009).

Three main problems that made difficult the interpretation of volcanic deposits in the area are: (1) the rock alteration that mask the real interpretation of volcaniclastic deposits (pyroclastic density currents, tephra-stratigraphy); (2) the considerable thicknesses of soils that covers the well exposing of the uppermost geology and mask in the same way the real interpretation of volcanic deposits, particularly important as deduce from site P5, where a good conserved outcrop allow proper interpretation of volcanic activity; (3) vegetation that play an important role in restricting the access to some places of the area. Despite those limitations, it was possible to recognize a sequence of volcanic events that caused an important environmental impact present in the studied area. This issue should be a concern because the possibility that some kind of similar event could happen in the near or far future cannot be ruled out. In the eastern part of the County, there is evidence of deposition of volcaniclastic deposits (ancient and recent debris flows and tuffs) that can be considered as real threat due to the spreading capacity of this deposits and the development of human settlements. According to the geologic record, there was a transition in the magmatic activity from lava flows (Colima Formation) to explosive activity (Tiribí Formation). The culmination of this activity was the pyroclastic fall of ash and lapilli from the last eruptions of the CVF. The recent volcanic activity of Turrialba and Irazú volcanoes are in the actual context of this threat for the County area.

Further field work is expected to be done on the north, south, central and western part of the County in order to better understand the geology and volcanic hazard for the County. In this sense, this work is a preliminary research of the area that make a step for further investigation in this subject and obtain the geological and hazard maps for the County at the scale 1: 5000.

8. Conclusions

This study reveals that the surface of the territory of Santo Domingo is covered by recent clayey soils from volcanic ashes and partially altered tuffs. According to the borehole data (water-supply wells), open pits, and field observations, the top layer of the geological sequence is a soil composed of clays, volcanic ash, and lahars in small areas. The clayey soil deposits have been probably derived from the ash cover. The observed soil minimum and maximum thicknesses are 5 and 26 meters. Stratigraphic location of the ashes matches the recent eruptions of the volcanoes in the Cordillera Central. The last eruptions of the Turrialba volcano ensure that ashes generated in that volcano can fall in the entire studied area. Beneath the ashes and recent lahars are the deposits of pyroclastic fall and pyroclastic flows from the Tiribí Formation. They were observed in the open pits Vargas Solera, Arizona and Dent. They also were confirmed at the point outcrops P1, P5 and P13. In both the Dent pit and P13 there is evidence of the pumice layer, exploited in the past from digging galleries, which are part of the

Tiribí Formation. Under these pyroclastic deposits are lava flows, autoclastic breccias and tuffs of the Colima Formation.

According to the identified and registered events, the real and current volcanic threat to the population of the County is the fall of ash emitted from the volcanoes of the Cordillera Volcánica Central, in particular the Turrialba and Irazú volcanoes. At least from the local observations along the Virilla and Pará river sections, there is no evidence of younger pyroclastic flows overlying the volcanic sequence.

Acknowledgments

The authors would like to thank the citizens of the Santo Domingo County, Armando Vasquez, Oscar Sojo, Rafael Bolaños, Ileana Murillo and Gonzalo González for their support during the field work planning, historical and archeological information about the County. SENARA for the borehole information supply. The Direction of Geology and Mines of Costa Rica for its support and make available for consulting open pit mines records. Both were useful for stratigraphic correlation during the field work. We gratefully acknowledge Edward Hakanson for the English text revision of this paper.

References

Alvarado, E. G., Soto, G. J., Pullinger, C. R., Escobar, R., Bonis, S., Escobar, D., & Navarro, M. (2007). Volcanic activity, hazards, and monitoring. In: Bundschuh, J., & G. E. Alvarado (Eds.). *Central América: Geology, Resources and Hazards* (pp 1154 – 1187). Taylor & Francis Group plc, London, UK.

Alvarado, G. E. (2008). Los Volcanes de Costa Rica: geología, historia, riqueza natural y su gente. San José: UNED.

Alvarado, G. E., & Gans, P. B. (2012). Síntesis geocronológica del magmatismo, metamorfismo y metalogenia de Costa Rica, América Central. *Revista Geológica de América Central, 46*, 7-122.

Brown, S. K., Loughlin, S. C., Sparks, R. S. J., Vye-Brown, C., Barclay, J., Calder, E., Cottrell, E., Jolly, G., Komorowsky, J. C., Mandeville, C., Newhall, C., Palma, J., Potter, S., & Valentine, G. (2015). Global volcanic hazards and risk. In S. C. Loughlin, S. Sparks, S. K. Brown, S. F. Jenkins & C. Vye-Brown (Eds), *Global volcanic hazards and risk* (pp. 81 – 172). http://dx.doi.org/10.1017/CBO9781316276273

Calvo, V. G. (1998). *Proyecto de explotación en cantera del Tajo Vargas Araya, Santo Domingo de Heredia.* Inmobiliaria Vargas Araya S.A, estudio técnico elaborado por geodesarrollos ambientales (GEODESA). Dirección de Geología y Minas, San José, Costa Rica.

Denyer, P., & Arias, O. (1991). Estratigrafía de la región central de Costa Rica. *Revista Geológica de   América Central, 12*, 1-59.

Echandi, E. (1981). *Unidades volcánicas de la vertiente norte de la cuenca del río Virilla.* Universidad de Costa Rica, San José. Tesis Licenciatura, 123.

Fernández, C. M. (1989). *Proyecto de explotación minera del Tajo y Quebrador Arizona, Santo Tomás de Santo Domingo, Heredia.* Expediente Minero N° 2161. Dirección de Geología y Minas, San José, Costa Rica.

Fernández, M. (1968). *Las unidades hidrogeológicas y los manantiales de la vertiente norte de la cuenca del Río Virilla.* Informe técnico Ministerio de Agricultura y Ganadería 27, 1 – 44, San José, Costa Rica.

Gazel, E., Carr, M., Hoernle, K., Feigenson, M., Szymanski, D., Hauff, F., & Bogaard, P. (2009). Galapagos – OIB signature in southern Central America: Mantle refertilization by arc – hot spot interation. *Geochem. Geophys. Geosyst., 10*(2), 1-32. Q02S11. http://dx.doi/10.1029/2008GC002246

International Strategy for Disaster Reduction (ISDR), 2004. Living with Risk. Vol. I. United Nations, New York and Geneva, 431 p.

Loughlin, S. C., Vye-Brown, C., Sparks, R. S. J., Brown, S. K., Barclay, E., Calder, J., Cottrell, E., Jolly, G., Komorowski, J. C., Mandeville, C., Newhall, C., Palma, J., Potter, S., & Valentine, G. (2015). An introduction to global volcanic hazard. In: S.C, Loughlin, S. Sparks, S.K. Brown, S.F. Jenkins & C. Vye-Brown (Eds). *Global Volcanic Hazards and Risk* (pp 1 – 40). http://dx.doi.org/10.1017/CBO9781316276273

Marin, G. F., & Goic, C. T. (1999). *Estudio geológico geofísico en la zona norte del Tajo Arizona, Santo Tomas de Santo Domingo, Heredia.* Expediente minero N° 2161. Dirección de Geología y Minas, San José, Costa Rica.

Marshall, J. S. (2007). Geomorphology and physiographic provinces. In: Bundschuh, J. & G.E. Alvarado (Eds).

Central América: Geology, Resources and Hazards (pp 131 – 178). Taylor & Francis Group plc, London, UK.

Obando, S. J. (2008). *Informe anual de labores del período diciembre 2007 - noviembre 2008 del Tajo Vargas Solera.* Expediente minero N° 2602. Inmobiliaria Vargas Solera S.A, San José, Costa Rica.

Obando, V. J. (1990). *Plazo recomendado, comprobación de campo y revisión del adendum al programa de explotación expediente minero N° 2161.* Dirección de Geología y Minas, San José, Costa Rica.

Pérez, W., Alvarado, G. E., & Gans, P. B. (2006). The 322 ka Tiribí Tuff: stratigraphy, geochrology and mechanisms of deposition of the largest and most recent ignimbrite in the Valle Central, Costa Rica. *Bull. Volcanol 69*, 25 – 40. http://dx.doi.org/10.1007/s00445-006-0053-x

Prosser, J., & Carr, M. (1987). Poás Volcano, Costa Rica: Geology of the summit region and spatial and temporal variations among the most recent lavas. *Journal of Volcanology and Geothermal Research, 33,* 131 – 146. Elsevier Science Publishers B.V., Amsterdam, The Netherlands.

Reyes, J., Fernández, M., Grinesky, S., & Collings, T. (2014). Natural Hazards in Santo Domingo de Heredia, Costa Rica, Centrral América. *Natural Science. 6*(3), 121-129.

Salazar, G. (1993). *Cálculo de reservas y estudio técnico financiero del Tajo Dent, San Miguel Sur de Santo Domingo, Heredia.* Expediente minero N° 2322. Dirección de Geología y Minas, San José, Costa Rica.

Servicio Nacional de Aguas Subterráneas, Riego y Avenamiento [SENARA-BGS] (1985). *Mapa hidrogeológico del valle central de Costa Rica.* Escala 1: 50 000. British Geological Survey – SENARA, San José, Costa Rica.

Soto, G., & Paniagua, S. (1992). La Cordillera Volcánica Central (Costa Rica): sus Peligros y Prevenciones. *Revista Geográfica de América Central, 25*(26), 291 – 304.

Valverde, M. (1994). *Informe anual de labores del Tajo Dent periodo 1993 - 1994, San Miguel Sur de Santo Domingo de Heredia.* Expediente minero N° 2322. Dirección de Geología y Minas, San José, Costa Rica.

Valverde, M. (2003). *Informe anual de labores del Tajo Dent periodo 2002 - 2004, San Miguel Sur de Santo Domingo de Heredia.* Expediente minero N° 2322. Dirección de Geología y Minas, San José, Costa Rica.

Van Wyk de Vries, B., Grosse, P. & Alvarado, G.E. (2007). Volcanism and volcanic landforms. In: Bundschuh, J. & G. E. Alvarado (Eds). *Central América: Geology, Resources and Hazards* (pp 179 – 210). Taylor & Francis Group plc, London, UK.

Williams, H. (1952). *Volcanic History of the Meseta Central Occidental de Costa Rica.* Univ. California. Public. Geol. Sci., 29(4), 145 – 180. Berkeley and Los Angeles.

Permissions

All chapters in this book were first published in JGG, by Canadian Center of Science and Education; hereby published with permission under the Creative Commons Attribution License or equivalent. Every chapter published in this book has been scrutinized by our experts. Their significance has been extensively debated. The topics covered herein carry significant findings which will fuel the growth of the discipline. They may even be implemented as practical applications or may be referred to as a beginning point for another development.

The contributors of this book come from diverse backgrounds, making this book a truly international effort. This book will bring forth new frontiers with its revolutionizing research information and detailed analysis of the nascent developments around the world.

We would like to thank all the contributing authors for lending their expertise to make the book truly unique. They have played a crucial role in the development of this book. Without their invaluable contributions this book wouldn't have been possible. They have made vital efforts to compile up to date information on the varied aspects of this subject to make this book a valuable addition to the collection of many professionals and students.

This book was conceptualized with the vision of imparting up-to-date information and advanced data in this field. To ensure the same, a matchless editorial board was set up. Every individual on the board went through rigorous rounds of assessment to prove their worth. After which they invested a large part of their time researching and compiling the most relevant data for our readers.

The editorial board has been involved in producing this book since its inception. They have spent rigorous hours researching and exploring the diverse topics which have resulted in the successful publishing of this book. They have passed on their knowledge of decades through this book. To expedite this challenging task, the publisher supported the team at every step. A small team of assistant editors was also appointed to further simplify the editing procedure and attain best results for the readers.

Apart from the editorial board, the designing team has also invested a significant amount of their time in understanding the subject and creating the most relevant covers. They scrutinized every image to scout for the most suitable representation of the subject and create an appropriate cover for the book.

The publishing team has been an ardent support to the editorial, designing and production team. Their endless efforts to recruit the best for this project, has resulted in the accomplishment of this book. They are a veteran in the field of academics and their pool of knowledge is as vast as their experience in printing. Their expertise and guidance has proved useful at every step. Their uncompromising quality standards have made this book an exceptional effort. Their encouragement from time to time has been an inspiration for everyone.

The publisher and the editorial board hope that this book will prove to be a valuable piece of knowledge for researchers, students, practitioners and scholars across the globe.

List of Contributors

Mohammad Hassan Sadeghravesh
Department of Environment, College of Agriculture, Takestan Branch, Islamic Azad University, Takestan, Iran

Hassan Khosravi and Sahar Shekoohizadeghan
Faculty of Natural Resources, University of Tehran, Tehran, Iran

Azam Abolhasani
Natural Resources Engineering - Living with the Desert-University of Tehran, Tehran, Iran

Akaha C. Tse and Adunola O. Ogunyemi
Department of Geology, University of Port Harcourt, Nigeria

Steve Ampofo and Boateng Ampadu
Department of Earth and Environmental Sciences, University for Development Studies (UDS), Ghana

Isaac Sackey
Department of Applied Biology, University for Development Studies (UDS), Ghana

Joel Efiong and Opaminola Nicholas Digha
Department of Geography and Environmental Science, University of Calabar, Calabar, Nigeria

Obianuju Emmanuella Asouzu
Department of Geography and Environmental Management, University of Port Harcourt, Port Harcourt, Nigeria

Akinwumiju, A. S.
Department of Remote Sensing and GIS, Federal University of Technology, Akure, Nigeria

Olorunfemi, M. O.
Department of Geology, Obafemi Awolowo University, Ile-Ife, Nigeria

Adewole J. Adeola and Abisola M. Oyebola
Department of Geology and Mineral Sciences, Crawford University, Igbesa, Nigeria

Eric Clausen
Jenkintown, PA. USA

Wenjing Xu, Sergio Bernardes, Sydney T. Bacchus and Marguerite Madden
Center for Geospatial Research, Department of Geography, University of Georgia, Athens, Georgia 30602-2502, USA

Abdul Hamid Mar Iman
Sustainable Environment and Conservation Cluster, Faculty of Agro-Based Industry, Universiti Malaysia Kelantan, Malaysia

Edlic Sathiamurthy
School of Marine Science and Environment, Universiti Malaysia Terengganu, Kuala Terengganu,Terengganu, Malaysia

Micah J. Hewer
Department of Geography, University of Toronto, Toronto, Ontario, Canada

William A. Gough
Department of Physical and Environmental Sciences, University of Toronto Scarborough, Scarborough, Ontario, Canada

John V. Smith
School of Engineering, Royal Melbourne Institute of Technology University, Victoria, Australia

Douglas B. Sims
Department of Physical Sciences, College of Southern Nevada, North Las Vegas, USA

W. Geoffrey Spaulding
Terra Antiqua Research, Henderson, USA

Martín Rojas-Barrantes
Dirección de Geología y Minas, MINAE, Costa Rica

Mario Fernández-Arce
Escuela de Geografía - PREVENTEC, Universidad de Costa Rica, Costa Rica

Index